Temperature and Humidity Independent Control (THIC) of Air-conditioning System

Xiaohua Liu • Yi Jiang • Tao Zhang

Temperature and Humidity Independent Control (THIC) of Air-conditioning System

Springer

Xiaohua Liu
Yi Jiang
Tao Zhang
Department of Building Science
Tsinghua University
Beijing, People's Republic of China

ISBN 978-3-642-42221-8 ISBN 978-3-642-42222-5 (eBook)
DOI 10.1007/978-3-642-42222-5
Springer Heidelberg New York Dordrecht London

Library of Congress Control Number: 2013957558

Printed on acid-free paper

Springer is part of Springer Science+Business Media (www.springer.com)

Preface

In 1995, Dr. James Freihaut from UTRC (United Technologies Research Center, the R&D base for Carrier, OTIS, and other industries) visited Tsinghua University for a meeting with me. The purpose of the meeting was to establish a cooperation project with our group in Tsinghua University for next-generation HVAC (heating, ventilation, and air-conditioning) system. "Do you have any idea on humidity independent control?" I still remember the question from James, and he tried to persuade me to do something for dehumidifying air without change of the air temperature. However, at that time, I had little idea on independent air dehumidification handling process. This was the time when I started to think about the concept of "temperature and humidity independent control" (THIC). Thanks to the Yung Wing funding from UTRC in 1996, a cooperation project was set up to develop a new approach for air dehumidification by a special membrane with moisture permeable feature. This was the first step for us toward the temperature and humidity independent control of HVAC system.

During the first 10 years since 1996, our major effort was put into new approaches of air dehumidification. We had tried out various ways to achieve this goal, such as using moisture permeable membrane to dehumidify air by heating the air at the sink side, using two rotary desiccant wheels to recover energy from exhaust air to deeply dehumidify the inlet air, using multistage liquid desiccant air handling modules, etc. Fundamental studies were carried out at the same time to understand the real features of humidification/dehumidification process and the ideal minimum energy required for air dehumidification process and where the loss is in each stage of an air dehumidification process.

Accompanied by these studies, a feeling has been getting stronger and stronger in our mind: air temperature and moisture in air (i.e., humidity ratio) are really two independent physical concepts, and unless a phase change process occurs such as evaporation or condensation, temperature and water vapor (or moisture) cannot influence each other. We should not take these two processes into account together in the indoor thermal environment control process as what engineers do in conventional HVAC design and operation procedures. Although both processes are affected by "heat," the sensible heat for temperature regulation (heating or cooling) and the latent heat for moisture regulation (humidification or dehumidification), these two types of heat are not the same. They are not freely convertible with each other, and the latent heat only occurs with phase change. However, the misunderstanding between these two kinds of heat, i.e., the temperature regulation and humidity regulation process, may lead to insufficient or unsatisfactory indoor thermal environment in designs as well as in operations of HVAC system.

For instance, a fan coil unit (FCU) with a total cooling capacity of 1 kW in conventional air-conditioning system can only provide about 50–70 % cooling if it only deals with sensible heat working under a dry condition with no air dehumidification. The cooling capacity for latent heat does not always accompany with that for sensible heat. Another simple example is the set point for supply air state in the all-air system: by setting the enthalpy of the supply air, the total heat removed from the conditioned space can be satisfied. However, either temperature or humidity level may be out of demand, which may result in a hot but dry condition or a cold but humid condition.

If there is no condensation or evaporation in the conditioned space, the temperature regulation and humidity regulation are two independent processes. Why do we still put these two processes together both in design and operation of an HVAC system? It may be better to deal with these two processes separately in analysis, design, and operation. This was how we started to get into the THIC idea for HVAC system, when it was around year 2000.

From then on, with the help of testing and investigating a number of air-conditioning systems in various buildings, it has been found that following the THIC concept, we could get a much clearer understanding and construct a better HVAC system during the design stage.

The sensible load consisted of solar radiation, indoor devices, and so on varies completely different from the moisture load generated from occupants or other sources. If all the loads are undertaken by the air circulation, the supply air state (temperature and humidity) has to be changed all the time to accommodate to the variances of the sensible and moisture loads in order to maintain both the temperature and humidity for the conditioned space. This is quite a hard task. However, if the air circulation system is only used for humidity control and a radiant cooling system (or other cooling system) is adopted for removing the sensible load, things would be much easier. A radiant cooling system can remove sensible load only with little influence on humidity ratio if temperature of the cold water flowing through the radiant cooling device is higher than the indoor air dew-point temperature. In this way, both indoor temperature and humidity can be maintained well regardless of how the sensible and moisture loads change. Furthermore, as the sensible load could be removed by the cold water with a temperature higher than the indoor dew point, e.g., 16–18 °C, which is also higher than the chilled water temperature in most conventional air-conditioning systems (e.g., 7 °C), chiller's efficiency will be much higher than that of conventional chiller if the chiller is specially designed to produce high-temperature chilled water for removing only sensible cooling load and regulating indoor temperature.

This is a new concept for HVAC system, which can provide more comfort thermal environment with less operation energy. This should be the principle and basic design guideline for the air-conditioning system in future. Although the key devices for the THIC system seem to be similar to those for the conventional air-conditioning system, the operating conditions and performances of devices we need in the THIC system are quite different: the air handling processor should be able to dehumidify air to a drier state for removing indoor moisture load but without a too low temperature to be supplied to the conditioned space; the chiller should provide chilled water with a higher temperature (16–18 °C) and higher coefficient of performance (COP); if FCU is adopted for indoor temperature control, the FCU should work efficiently with a smaller temperature difference between circulating water and air but without the worry of condensing water; etc. These indicate that the THIC systems do need a complete set of new types of devices, a new generation of HVAC devices! We need different design approaches, we need different system components, we need different control devices with different logics, and we need different ways to operate and manage the THIC systems.

To push this THIC system to the real construction market, an association for THIC air-conditioning system has been founded since 2007, with members including researchers, manufacturers, building developers, as well as operators. Thanks to the huge construction market in China, different types of THIC systems for different building functions have been adopted in many places of China since 2005. So far, there has been more than 20 million m^2 of buildings with all sorts of functions designed according to the THIC concept in China. Most of them perform well in both thermal comfort level and energy consumption. Manufacturers who provide devices specially developed for THIC system are also increasing year by year. And THIC system has become a new industry both in terms of HVAC construction and device manufacture in China.

Along with the development of the THIC air-conditioning system in China, we wrote the first book to introduce this concept and the design method of THIC system in 2006 in Chinese (published by *China Architecture & Building Press*). Thanks to the continuous study in the last 7 years and experiences accumulated by a large number of practical projects, a clearer understanding of the THIC system has been obtained. We have got a very strong feeling that we should provide a complete description of the THIC system to the HVAC industry in the world as well as China. This is the reason why this book has been written.

Most of the contents in the book are written by Dr. Xiaohua Liu and PhD Candidate Tao Zhang, based on the research work carried out by the THIC group in Tsinghua University. This group has been carrying on relevant research and development for more than 10 years in this field. This book could be regarded as a good summary of part of the research results from this group. There is also a name list for the researchers making great contributions in this THIC group during the last 10 years: Dr. Zhen Li, Mr. Xiaoyang Chen, Dr. Shuanqiang Liu, Dr. Xiaoyun Xie, Mr. Haixiang Li, Mrs. Xiaoqin Yi, Mr. Xiaomin Chang, Mr. Haiqiang Zhang, Mrs. Yidan Tang, Mrs. Zhihong Gao, Mr. Lun Zhang, Miss Jingjing Jiang, Miss Rang Tu, Miss Kang Zhao, etc.

Many thanks to all the members in the THIC group. They have made great contributions to the development of THIC system as well as this book. Thanks should also be given to the designers who have carried out the THIC projects in different types of buildings. Without their great efforts, THIC cannot grow up so fast in China, and this book would not come true.

I hope what have happened in the HVAC field of China today may happen in the world tomorrow since a healthier and more comfortable indoor environment is required and less power consumption and carbon emission from buildings' operation are desired for our earth. THIC system should be one of the possible approaches for this goal, which has made some change to China's HVAC field. Looking forward to see that the THIC system could bring difference to the global HVAC industry and this book may make a bit contribution for this change.

Director and Professor
Building Energy Research Center
Tsinghua University, Beijing,
People's Republic of China

Yi Jiang

Contents

Chapter 1
Characteristics of Conventional Air-Conditioning Systems

Abstract Air-conditioning systems play an important role in maintaining the indoor built environment. Coupled heat and mass handling is usually applied for the current state-of-the-art air-conditioning systems. With the advance of society, conventional air-conditioning methods have been challenged by the demand for a more comfortable indoor environment and a higher system energy efficiency. Continuing to improve the energy efficiency and reducing the energy consumption of air-conditioning systems in order to provide a suitable and comfortable environment are foundations to the development of new strategies for the indoor built environment. Taking these requirements into account, the THIC (temperature and humidity independent control) air-conditioning system is generally considered to be a possible and effective solution.

1.1 Tasks of Indoor Environmental Control Systems

Indoor environmental control systems are responsible for providing a comfortable and healthy indoor environment by regulating indoor temperature, humidity, air velocity, and indoor air quality within appropriate ranges (ASHRAE 2009; Jiang et al. 2011). The indoor thermal environmental condition, which consists of meteorological parameters; heat sources such as equipment, lighting, and occupants; and indoor air flow, is the key factor that determines thermal comfort. As the development of air-conditioning systems in commercial buildings continues to progress, the indoor parameters for comfort in air-conditioned spaces have been established and implemented all over the world (GB 50189 in China, ASHRAE standards 55 and 62 in the United States, etc.). Table 1.1 lists the recommended indoor parameters for winter and summer, and Table 1.2 lists the design outdoor air flow rates for the main functional zones of commercial buildings in China (MOHURD, AQSIQ 2005). Figure 1.1 illustrates the comfort zones according to ASHRAE standard 55 (ASHRAE 2004), and Table 1.3 presents the design outdoor air flow rates from ASHRAE standard 62 (ASHRAE 2010) for different occupied spaces. As indicated

X. Liu et al., *Temperature and Humidity Independent Control (THIC) of Air-conditioning System*, DOI 10.1007/978-3-642-42222-5_1,

Table 1.1 Indoor design parameters for air-conditioning in China (GB 50189)

Parameters		Winter	Summer
Temperature (°C)	Ordinary room	20	25
	Lobby, corridor	18	Indoor and outdoor difference ≤10
Air speed (v) (m/s)		$0.10 \leq v \leq 0.20$	$0.15 \leq v \leq 0.30$
Relative humidity (%)		30–60	40–65

Table 1.2 Design outdoor air flow rates for typical commercial buildings in China

Building type	Hotel room (5 star)	Hotel room (4 star)	Hotel room (3 star)	Theater, music hall	Entertainment hall, ballroom
Outdoor air flow rate (m^3/h/person)	50	40	30	20	30
Building type	**Office**	**Shopping mall, bookstore**	**Restaurant, dining hall**	**Gym**	**Barbershop, bar**
Outdoor air flow rate (m^3/h/person)	30	20	20	20	30

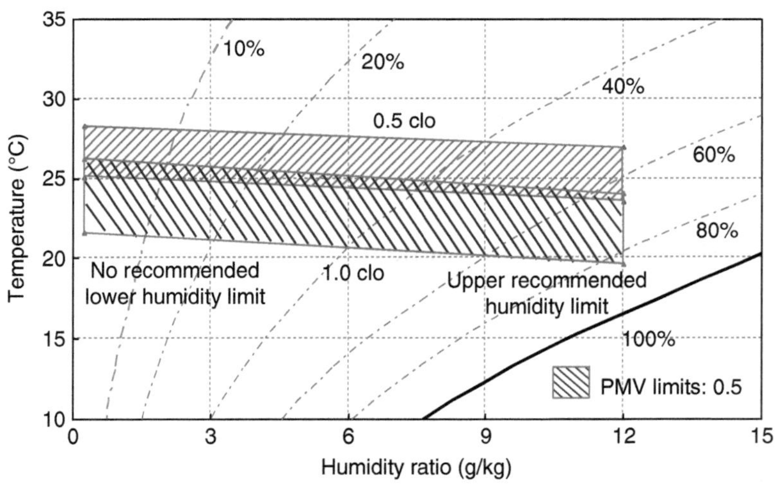

Fig. 1.1 ASHRAE summer and winter comfort zones (ASHRAE standard 55) (*Acceptable ranges of operative temperature and humidity with air speed ≤0.20 m/s for people wearing 1.0 and 0.5 clothing during primarily sedentary activity (≤1.1 met)*)

by these tables and the figure, the indoor design parameters and outdoor air flow rates are similar for different standards, although there are some slight discrepancies.

With recent increases in living standards and a general strengthening of the desire for self-protection, high air quality is being demanded by citizens everywhere. High-quality indoor air is beneficial to our lives – our jobs, our school life, and our daily activities. A series of hygienic standards (GB 9663–GB 9673, CSBTS

Table 1.3 Design outdoor air flow rates for typical commercial buildings (ASHRAE standard 62)

Building type	Office	Library	Supermarket	Museum, art gallery	Auditorium, seating area
Outdoor air flow rate (m^3/h/person)	30.6	30.6	27.4	16.6	9.7
Building type	Stage, studio	Mall	Disco, dance floor	Barbershop	Video arcade
Outdoor air flow rate (m^3/h/person)	19.4	16.6	37.1	18.0	29.9

1996a, b, c, d, e, 2005a, b, c) for public places such as entertainment halls, gymnasiums, shopping malls, and bookstores have been unveiled in China to regulate indoor concentrations of carbon dioxide, carbon monoxide, formaldehyde, inhalable particles, and bacteria. For example, the concentration of carbon dioxide in libraries, museums, art galleries, hotels, and hospital waiting rooms is supposed to be lower than 1,000 ppm (0.10 %), while the standard in cinemas, music halls, video arcades, ballrooms, shopping malls, and bookstores is supposed to be lower than 1,500 ppm (0.15 %). To maintain adequate indoor air quality, ventilation is usually the most efficient way to remove pollutants.

As a whole, the major tasks of regulating the indoor environment are to remove extra heat, moisture, CO_2, odor, and other pollutants (e.g., volatile organic compounds) and to maintain the indoor environmental parameters within appropriate ranges according to the standards mentioned above. However, there exist many different approaches for regulating the indoor environment:

- The sensible load can be removed in many ways, as long as the temperature of the medium used is lower than the room temperature. For example, both indirect contact methods (e.g., radiant panels) and cold air supply methods are applicable for removing the indoor sensible load.
- Extra indoor moisture can be removed only by dry air supply methods; indirect contact methods are not applicable.
- Carbon dioxide, odor, and other pollutants should be removed by ventilation, i.e., lowering the concentration of pollutants by bringing outdoor air into the indoor space.

1.2 Current Air-Conditioning Methods

1.2.1 Air-Conditioning System Categories

Air, water, and refrigerants are common heat transfer fluids in air-conditioning systems for the heat exchange between cooling/heating sources and indoor terminals. The heat or mass transfer processes, which proceed between terminal devices and indoor spaces through convection or radiation, control the indoor thermal environment. According to the selected heat transfer fluid, air-conditioning systems

Table 1.4 Categories of current air-conditioning systems

Building type	Refrigerant system	Air-water system	All-air system
Residential buildings	Household air-conditioner, VRF system		
Small-scale commercial buildings	VRF system	FCU+OA system	CAV system
Large-scale commercial buildings		FCU+OA system	CAV system VAV system

Notes: *FCU* fan coil unit, *OA* outdoor air, *CAV* constant air volume, *VAV* variable air volume

are classified into four distinct systems: all-air systems, air-water systems, all-water systems, and refrigerant-based direct evaporative cooling systems (listed in Table 1.4). Among these four systems, all-water systems indicate that both the indoor sensible load and moisture load are removed by water. However, this kind of system cannot be applied independently from the ventilation system. Refrigerant systems refer to systems where heating and cooling are directly achieved through refrigerant evaporation and condensation. Examples of this kind of system include household split air conditioners and VRF (variable refrigerant flow) systems, which have become widespread in recent years. Air-water systems and all-air systems are the most common central air-conditioning systems in buildings.

Air is chosen as the heat transfer fluid to remove the sensible and moisture loads in all-air systems. The processed air is supplied into the conditioned space to regulate the indoor thermal environment. Due to the low specific heat capacity of air, a high air volume is required, which results in a larger cross-sectional area of (and occupied space for) the air duct. For the air-water systems, both air and water are chosen as the heat transfer fluids, e.g., in systems consisting of a fan coil unit (FCU) and an outdoor air (OA) processing unit. For the FCU+OA system, chilled water and cooled outdoor air are adopted in summer to cool and dehumidify the conditioned space, while hot water and heated outdoor air are used in winter to heat the conditioned space. Meanwhile, the outdoor air sent to the conditioned space can remove pollutants such as CO_2 and meet the occupants' requirement for outdoor air. The supply air flow rate and the occupied space of the FCU+OA system are much smaller than those of all-air systems. Thus, air-water systems are widely used as central air-conditioning systems in China.

1.2.2 Typical Air Handling Processes in Central Air-Conditioning Systems

In current state-of-the-art air-conditioning systems, coupled heat and mass handling is usually applied. Moisture is removed by condensation dehumidification in summer, and the indoor sensible load and moisture load are extracted simultaneously (Zhao et al. 2000). However, although the humidity ratio of the handled air

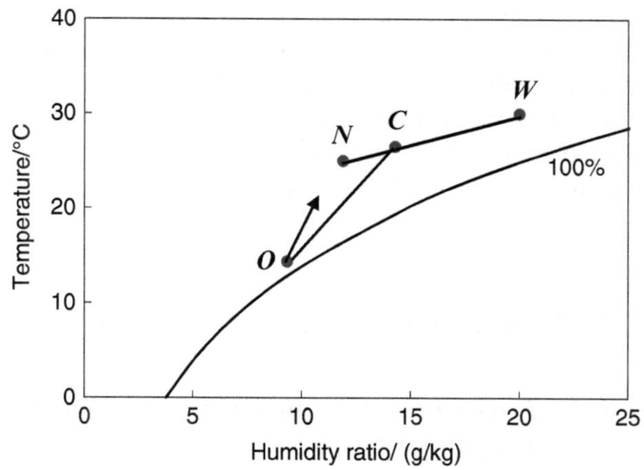

Fig. 1.2 Air handling process in the all-air system with primary air return

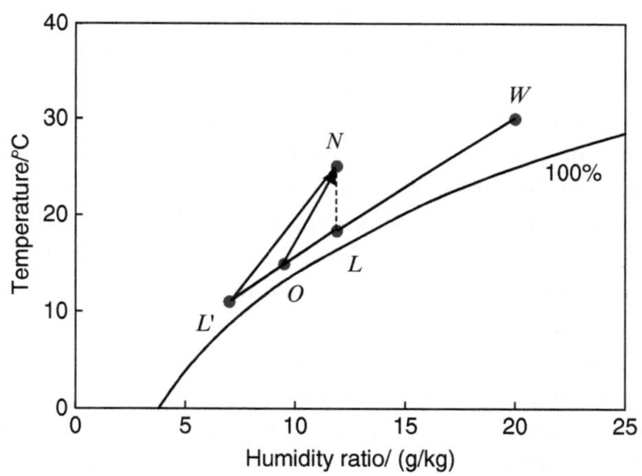

Fig. 1.3 Air handling process in the FCU+OA system

is satisfactory, its temperature is usually too low after dehumidification. Consequently, reheating is sometimes needed to attain an appropriate temperature according to the supplied air temperature requirement. Due to energy conservation considerations, reheating devices that use steam or electricity are prohibited in commercial buildings in China (except for buildings with special requirements). The dehumidified air is supplied directly into the conditioned space. As a result, the indoor temperature and humidity usually cannot be satisfied at the same time, and in most cases, temperature control is given priority. The typical air handling processes are discussed in the remainder of this subsection.

The typical primary return air handling process in the all-air system is illustrated in Fig. 1.2. Outdoor air (state W) and return air (state N) are mixed to state C, and the mixture reaches state O after the condensation dehumidification process. State O is the apparatus dew point, with a relative humidity of 90–95 %. If there is no reheating device, air at state O is sent directly to the conditioned space.

The air handling process in the FCU+OA system (air-water system) is shown in Fig. 1.3. Both the outdoor air and indoor return air are processed using the same cooling source (such as 7 °C chilled water). After condensation dehumidification, the outdoor air is usually processed from state *W* to state *L*, which has the same humidity ratio as the indoor air (state *N*). Meanwhile, the indoor return air is processed from state *N* to state *L'* when flowing through the FCU. The processed outdoor air (state *L*) and the processed return air (state *L'*) are then mixed to state *O*, which is supplied to the conditioned space.

1.3 Problems with Current Air Handling Methods

1.3.1 Loss in the Coupled Heat and Moisture Handling Process

From the perspectives of both thermal comfort and health, indoor temperature control and humidity control are both necessary. In summer, the common indoor design state has a temperature around 25 °C and a relative humidity of 60 %, and the corresponding dew point temperature is 16.6 °C. Therefore, the main tasks of the air-conditioning system are to extract the sensible load from 25 °C and to remove the moisture load from the dew point temperature of 16.6 °C. In current state-of-the-art air-conditioning systems, air is usually cooled and dehumidified simultaneously by cooling coils, and air that is both dry and cool is sent into the room to remove the sensible and moisture loads at the same time.

Taking the FCU+OA system as an example, Fig. 1.4 plots the temperature levels of each heat transfer process in the air-conditioning system. Due to the utilization of a cooling tower, the heat sink temperature is equivalent to the outdoor wet-bulb temperature. The typical operating parameters are as follows: an outdoor wet-bulb temperature of 27 °C, an indoor temperature of 25 °C (with a corresponding dew point temperature of 16.6 °C), a condensing temperature of 38 °C, and an evaporating temperature of 5 °C. For condensation dehumidification, the cooling source temperature must be lower than the indoor dew point temperature (16.6 °C). Considering a 5 °C temperature difference for the heat transfer process between the chiller and the heat transfer fluid and a temperature difference of 5 °C for the heat transfer fluid transportation process, a cooling source with a temperature of about 7 °C is necessary for condensation dehumidification. This is why current air-conditioning systems adopt chilled water with a temperature of around 7 °C (Appendix B), and why the evaporating temperature of the chiller is usually about 5 °C. However, if there is no requirement for dehumidification, the cooling source temperature is only required to be lower than the indoor temperature (25 °C) in theory (Bejan 2006). Considering the temperature differences for heat transfer and heat transfer fluid transportation, a cooling source with a temperature of 15–18 °C is sufficient for removing the sensible cooling load.

As has been stated already, in conventional air-conditioning systems, air is cooled and dehumidified simultaneously by the same cooling source to remove

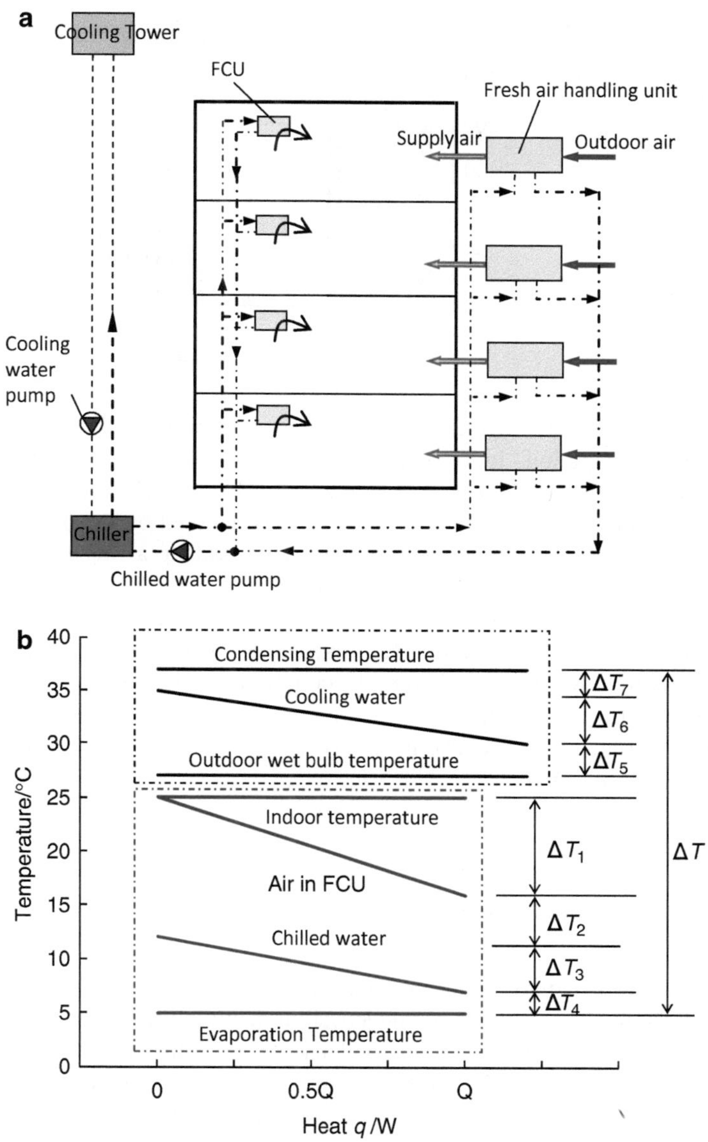

Fig. 1.4 Handling process of a typical central air-conditioning system: (**a**) schematic diagram and (**b**) temperature level (from chilled water to FCU in a typical operating condition)

both sensible and latent (moisture) cooling loads at the same time. Due to coupled heat and moisture handling, the cooling source temperature (usually 5–7 °C) is limited by the indoor dew point temperature. Nevertheless, a cooling source with a temperature of 15–18 °C is sufficient if there is no dehumidification demand. Hence, many natural cooling sources (including underground water, river water, and lake water) can be used. However, most of the time, the 5–7 °C cooling source

Fig. 1.5 Monthly heat consumption of two office buildings in Philadelphia (Zhang 2008)

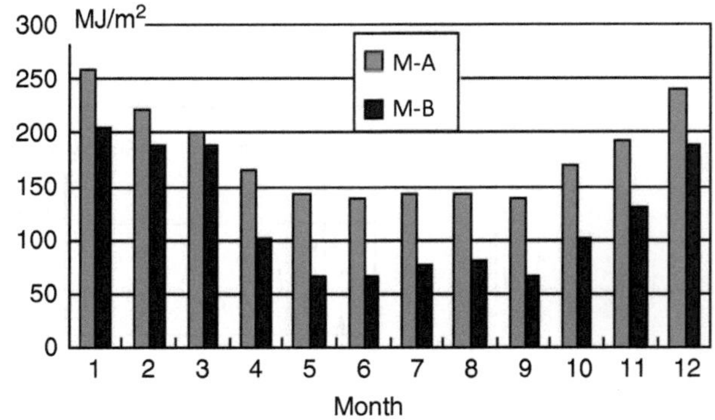

used in the coupled heat and moisture handling process can only be produced by mechanical refrigeration. In typical air-conditioning systems, the sensible load usually accounts for 50–70 % of the entire load. The sensible load, which can be removed by a high-temperature cooling source, is handled by the 5–7 °C cooling source at the same time as the moisture load. This leads to a waste in the cooling source energy grade, which limits the application of natural cooling sources and restricts performance improvement of the chiller. In addition, the air temperature after condensation dehumidification (corresponding to about 90 % relative humidity) is usually too low to be supplied to the conditioned space directly, even though its humidity ratio is satisfactory; this sometimes results in unnecessary energy dissipation caused by the reheating device if the dehumidified air needs to be reheated (Liu et al. 2010).

1.3.2 Energy Dissipation Caused by Offset

In conventional air-conditioning systems, moisture is removed by condensation dehumidification. The principle of condensation dehumidification is to cool the air down to its saturation state and then condense the moisture out. Two typical case studies are discussed in this subsection to illustrate how offset causes energy dissipation in air handling processes. *Case 1*: two office buildings in Philadelphia (in the United States) with reheating devices. Before being supplied to the indoor space, the air is dehumidified by the cooling coil and then reheated by steam, which leads to heating-cooling offset. *Case 2*: an office building in Urumchi (a city in northwest China). 7 °C chilled water is supplied into the FCU to cool and dehumidify the indoor air, and the outdoor air is sent into the room after evaporative cooling, although the outdoor air is dry enough to remove the indoor moisture load.

Case 1: Figure 1.5 shows the monthly heat consumption of the two office buildings in Philadelphia (Zhang 2008). M-A is a large-scale building with an area of 30,000 m^2, and M-B is an ordinary building with an area of 5,500 m^2. The heat consumption shown in the figure refers to the heat consumed by the air-conditioning systems. It can be seen that there is significant heat consumption

Fig. 1.6 Schematic diagram of AHU reheating in the Philadelphia office building M-A

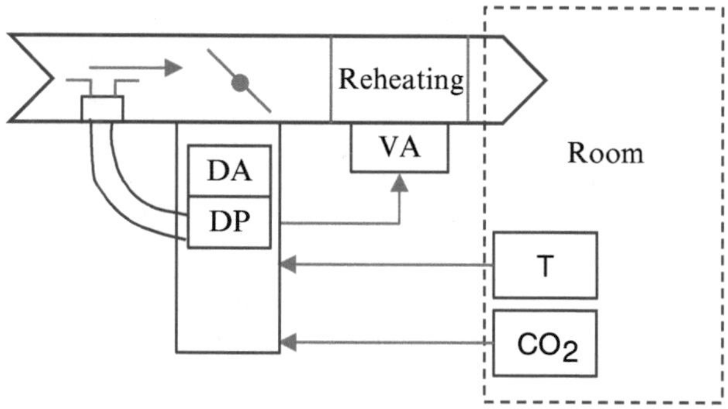

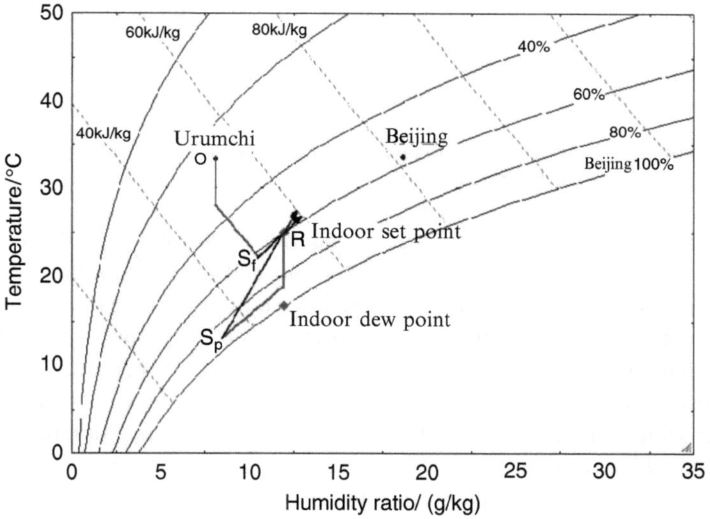

Fig. 1.7 Air handling process of the air-conditioning system in an office building in Urumchi

in these office buildings even in summer. The heat consumption of M-A in the hottest month is 60 % of that in the coldest month; for M-B, the proportion is about 30 %. This is because the air temperature is too low after condensation dehumidification, causing the air to be reheated before being supplied to the room. A schematic of one of the reheating processes is shown in Fig. 1.6. Reheating the dehumidified air leads to unnecessary heating-cooling offset, increasing the cooling and heating loads of the air-conditioning systems and causing tremendous energy dissipation.

Case 2: Figure 1.7 depicts the air handling process of a FCU+OA system in an office building in Urumchi, the capital of China's Xinjiang province. Outdoor air (state O) is sent into the room at state S_f after being cooled and humidified by an indirect evaporative cooling unit and a direct evaporative cooling unit. Chilled water with a temperature of 7 °C is supplied into the FCU to remove the indoor moisture load, which corresponds to the cooling process from state R to state S_p. In fact, the outdoor air in Urumchi is already sufficiently dry, with a humidity ratio

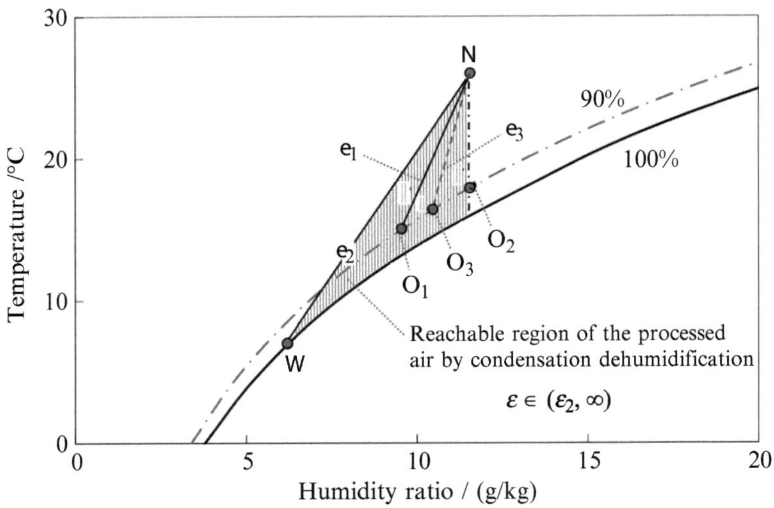

Fig. 1.8 Ratio of enthalpy difference to humidity ratio difference (ε) of the typical air handling process with condensation dehumidification

lower than 8 g/kg. The indoor design air humidity ratio is about 12 g/kg, so the humidity ratio difference between the indoor air and the outdoor air is as high as 4 g/kg. The main indoor moisture source in office buildings is the occupants. Taking an outdoor air flow rate of 30 m^3/h per person and a moisture production of 100 g/h per person, a humidity ratio difference of 2.8 g/kg between the outdoor air state and the indoor design state is sufficient to remove the indoor moisture, which corresponds to supply air with a humidity ratio of 9.2 g/kg. Furthermore, the indoor temperature can be satisfied with a relatively high-temperature cooling source (15–20 °C). However, both humidification of the outdoor air and dehumidification of the indoor air are adopted in this building. This results in offset between humidification and dehumidification and a lower required temperature of the cooling source.

1.3.3 Difficulty Adapting to the Variances of Indoor Sensible and Moisture Loads

The sensible load is composed of heat transfer through the building envelope, solar radiation, and heat gain from indoor occupants, lights, appliances, etc., and the indoor moisture load is derived from occupants, open water surfaces, etc. ε represents the ratio of enthalpy difference (kJ/kg) to humidity ratio difference (g/kg). In actual buildings, ε varies considerably, but the coupled treatment of indoor sensible and moisture loads by the condensation dehumidification method can only satisfy the indoor requirement within a limited range.

Figure 1.8 shows the ε range if the air is processed by chilled water. The figure shows that the coupled heat and moisture handling process cannot meet the requirement if ε is below a certain limit. Without reheating devices, the temperature

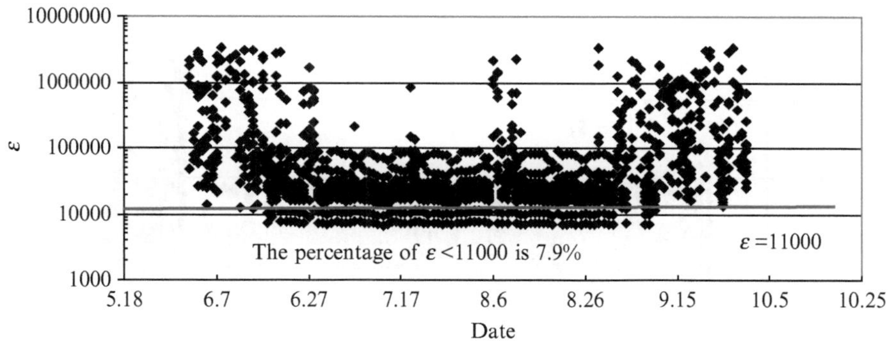

Fig. 1.9 Ratio of enthalpy difference to humidity ratio difference (ε) of shopping malls in summer (6.1–9.30) in Beijing

and humidity requirements of the indoor environment cannot be satisfied simultaneously. In most cases, the indoor temperature is well controlled. However, the humidity level varies, and it may vary beyond the comfortable range. With reheating devices, the indoor temperature and the humidity level can be adjusted simultaneously within the comfortable range, but this leads to substantial (and unnecessary) heat-cold energy dissipation.

The typical ε values of the condensation dehumidification method are shown in Fig. 1.8. To examine the indoor air handling process of conventional air-conditioning systems, the FCU+OA system can be taken as an example. The supply chilled water temperature is usually 7 °C (state W), and the supply air temperature is usually 15 °C (90 % RH, state O_1). When the indoor thermal condition is 26 °C and 55 % RH (state N), the ε_1 of the line connecting O_1 and N is 8,177, and the ε_2 of the line connecting N and W is 6,156. Outdoor air is usually processed to point O_2, which has the same humidity ratio as indoor state N. This means that the outdoor air can remove part of the indoor sensible load, and the ε of the line connecting O_2 and N is infinite. Both processed indoor air O_1 and processed outdoor air O_2 remove the indoor sensible and moisture loads. Air at state O_1 and state O_2 are mixed to state O_3. ε_3 is the ratio of enthalpy difference to humidity ratio difference of the line connecting O_3 and N. If the indoor actual ε is higher than ε_3, the FCU+OA system can meet the demand for dehumidification, but if the ε is lower, the system cannot meet the demands for cooling and dehumidification at the same time.

The actual required ε in buildings varies immensely. Indoor moisture generation generally comes from occupants, so if the number of residents remains unchanged, the indoor moisture load could be regarded as constant (if not taking into account other moisture sources). However, the sensible load varies significantly with both the climate and the condition of indoor equipment. In the opposite situation, the number of indoor occupants varies considerably, but it is not proportional to the variation of the indoor sensible load. In order to analyze the variance of thermal and moisture loads in shopping malls in China, the hour-by-hour building load simulation software DeST was used to investigate the characteristics of the ε in shopping malls in different Chinese cities, with $\varepsilon = 11{,}000$ as the upper limit (Figs. 1.9 and 1.10). The DeST model is discussed in detail in Appendix C. The value of ε is generally

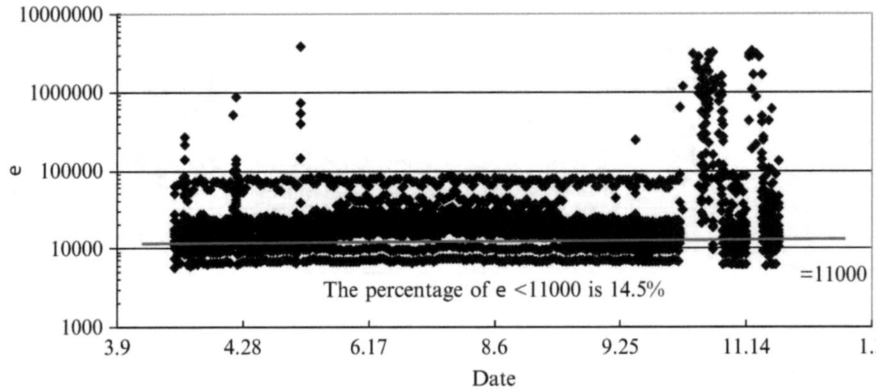

Fig. 1.10 Ratio of enthalpy difference to humidity ratio difference (ε) of shopping malls in summer (4.1–11.30) in Guangzhou

between 10,000 and 100,000 for shopping malls. For about 7.9 % (142 h) of the air-conditioning period in Beijing, the ε is lower than 11,000, as indicated in Fig. 1.9; the proportion is 14.5 % (576 h) in Guangzhou, as shown in Fig. 1.10.

The variance of the ratio of enthalpy difference to humidity ratio difference does not match that of the corresponding ratio of the air handling process with condensation dehumidification, which could be regarded as basically fixed. In most cases, compromises are made to sacrifice humidity control and prioritize temperature control, which might result in overly high or overly low indoor humidity levels. Overly high indoor air humidity leads to occupants' discomfort, and then the indoor temperature set value must decrease in order to improve thermal comfort. This leads to additional and unnecessary energy consumption; an increasing temperature difference between the indoor and outdoor air exaggerates the heat transfer through the building envelope and also increases unnecessary energy consumption for handling the outdoor air. Overly low relative humidity will also increase the energy consumption of the outdoor air handling process due to the increased enthalpy difference between the indoor and outdoor air. Under some circumstances, reheating the cooled and dehumidified air is necessary in order to address the contradiction between temperature control and humidity control, which further intensifies unnecessary energy consumption.

In summary, in conventional air-conditioning systems, temperature and humidity are regulated simultaneously by condensation dehumidification, but this is not the case for the indoor widely varied ratio of enthalpy difference to humidity ratio difference. Therefore, it is necessary to find a new method to match the ratio of enthalpy difference to humidity ratio difference in the air handling process with that required for creating a comfortable indoor environment.

1.3.4 Indoor Terminals

To avoid a supply air temperature that is lower than necessary, a relatively large circulating air flow rate is required to remove the indoor sensible heat and moisture loads. For example, if a sensible load of 80 W/m^2 needs to be extracted and the room temperature is 25 °C, the required circulating air flow rate is 24 m^3/h per unit area (m^2) when the supply air temperature is 15 °C. However, this often leads to a high indoor air flow rate and an uncomfortable sensation of draftiness on the part of occupants. In order to lessen this unwanted sensation, both the position and form of the air vents have to be changed to improve the indoor air flow pattern. However, these adjustments require the construction of air ducts inside the conditioned room, which either reduces the room's net height or enlarges the space between neighboring floors. Moreover, a large circulating air flow rate generates noise that is difficult to eliminate effectively. In winter, such air conditioners are rarely used in China in order to avoid this sensation of draftiness. Instead, radiators are used to heat buildings in winter. Therefore, two different sets of indoor terminal devices are installed for summer and winter. Is there any way to reduce the indoor circulating air flow rate to avoid the sensation of draftiness? And could the same terminals used in winter for air-conditioning through radiation and natural convection be used to supply cooling in summer, so that one set of indoor terminal devices could be used in both seasons? With the increasing prevalence of air conditioners, these questions are now being put forward.

As for radiant cooling terminals, one of the premises for their application is to ensure that the surface temperature of the radiant panels is higher than the dew point of the indoor air, i.e., there must not be any condensed water on the surfaces of the panels. In current air handling systems that utilize condensation dehumidification, the chilled water temperature is around 7 °C, which is far lower than the dew point of the indoor air. If chilled water with a temperature of around 7 °C is supplied directly to the radiant panels, moisture will generate on their surfaces due to condensation. To make sure that this does not occur, the surface temperature of the radiant panels must be kept higher than the dew point temperature of the indoor air. As the indoor air temperature is 25 °C and relative humidity is 60 %, the corresponding dew point temperature is 16.6 °C. Factoring in a margin for error, the temperature difference between the surface of the radiant panels and the indoor air is only about 6–7 °C. If there is no shortwave radiation (e.g., solar radiation) irradiating the panels directly, the cooling capacity of the radiant panels will not be much higher than 60 W/m^2. If the sensible load of the conditioned space is higher than that, the area of radiant panels may not be large enough, and extra terminal devices like fan coil units will then be needed. Therefore, as radiant cooling terminals continue to be adopted more frequently, improving the cooling capacity per unit area as much as possible and avoiding water vapor condensation remain key issues to be addressed.

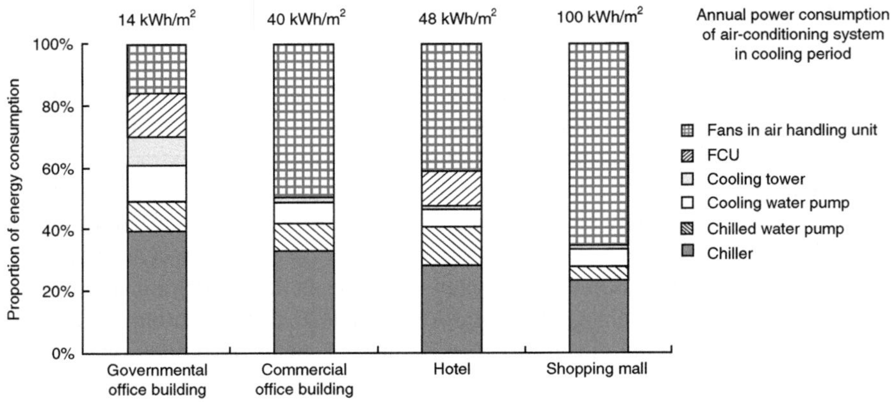

Fig. 1.11 Annual operation electricity consumption of air-conditioning systems in typical buildings in Beijing

1.3.5 Energy Consumption of Transportation

In order to implement the task of controlling the indoor environment and ensuring comfort for occupants, transportation and distribution systems are needed to extract the indoor sensible heat, moisture, CO_2, etc. In central air-conditioning systems, the energy consumption of fans and pumps represents a considerable proportion of their total energy consumption (50–70 % in some cases, which exceeds the chiller's energy consumption). Figure 1.11 shows the annual operating power consumption of the air-conditioning systems in some typical buildings in Beijing (Building Energy Research Center of Tsinghua University 2009), including a government office building, a commercial office building, a hotel, and a shopping mall. As it accounts for a large proportion of the system's total energy consumption, the energy consumption of fans and pumps is of particular importance, and it should be reduced as much as possible.

When different heat transfer fluids are adopted, there are significant differences among transportation systems in terms of their performance. In all-air systems, all the cooling output is transported by air, which leads to a low efficiency of the transportation and distribution system. In the following subsection, air and water are compared as intermediate heat transfer fluids with regard to their energy consumption.

When air and water adopt the same temperature difference and transport the same quantity of cooling capacity, the mass flow rate ratio between air and water is calculated by Eq. (1.1):

$$\frac{\dot{m}_{\mathrm{a}}}{\dot{m}_{\mathrm{w}}} = \frac{c_{p,\mathrm{w}}}{c_{p,\mathrm{a}}} \tag{1.1}$$

where $\dot{m}$ is the mass flow rate, kg/s; c_p is the specific heat at atmospheric pressure, kJ/(kg·°C); and subscripts a and w stand for air and water, respectively. The specific heat of air and water at atmospheric pressure is 1.01 kJ/(kg·°C) and 4.18 kJ/(kg·°C), respectively. Thus, the mass flow rate of the circulating air is about four times that of the circulating water. Ratios between cross-sectional areas of ducts/pipes transporting air and water are then calculated by Eq. (1.2):

$$\frac{A_a}{A_w} = \frac{c_{p,w} \cdot \rho_w \cdot v_w}{c_{p,a} \cdot \rho_a \cdot v_a} \tag{1.2}$$

where A is the cross-sectional area of the duct/pipe, m^2; ρ is the density, kg/m^3; and v is the velocity, m/s. The ratio between energy consumption for the intermediate heat transfer fluids of air and water is then calculated by Eq. (1.3):

$$\frac{W_a}{W_w} = \frac{(\dot{m}_a/\rho_a) \cdot \Delta p_a}{(\dot{m}_w/\rho_w) \cdot \Delta p_w} = \frac{\rho_w \cdot c_{p,w} \cdot \Delta p_a}{\rho_a \cdot c_{p,a} \cdot \Delta p_w} \tag{1.3}$$

where W is the power consumption, W, and Δp is the pressure drop, Pa. In conventional air-conditioning systems, air is adopted to cool and remove the indoor moisture, i.e., extract the total cooling load. The ratio γ between the total cooling load and the sensible cooling load is calculated by Eq. (1.4):

$$\gamma = \frac{\Delta h_a}{c_{p,a} \cdot \Delta t_a} \tag{1.4}$$

where Δh is the enthalpy difference, kJ/kg, and Δt is the temperature difference, °C. Therefore, considering both cooling and extracting moisture by air, Eqs. (1.2) and (1.3) can be revised as

$$\frac{A_a}{A_w} = \frac{1}{\gamma} \cdot \frac{c_{p,w} \cdot \rho_w \cdot v_w}{c_{p,a} \cdot \rho_a \cdot v_a} \tag{1.5}$$

$$\frac{W_a}{W_w} = \frac{1}{\gamma} \cdot \frac{\rho_w \cdot c_{p,w} \cdot \Delta p_a}{\rho_a \cdot c_{p,a} \cdot \Delta p_w} \tag{1.6}$$

As for the transportation system, the air velocity is around 3 m/s and the water velocity is approximately 1 m/s. If $\gamma = 2$, according to Eq. (1.5), the cross-sectional area of a transporting duct/pipe with air as the intermediate medium is nearly 600 times larger than when water is the intermediate medium for the same cooling capacity and temperature difference. If the transport cooling capacity is 5 kW with a temperature difference of 5 °C, the diameter of the air duct is about 420 mm, whereas the water pipe diameter is less than 20 mm. If the temperature difference is 8 °C, the diameter of the air duct is about 330 mm. Even when the removal of both the sensible load and the moisture load is considered, the air duct diameter is still about 280 mm, much more than the water pipe diameter if water is selected as the heat transfer fluid.

If the pressure drop of air and water are 25 mmH_2O and 10 mH_2O, respectively, the energy consumed when air is the intermediate fluid is 4.3 times greater than when water is the intermediate fluid with a temperature difference of 5 °C (without considering the influence of the air's ability to remove the moisture load). Therefore, water is recommended to be the transporting heat transfer fluid instead of air for cooling/heating, as the transportation systems for the former take up less space and are more energy efficient than those for the latter.

1.3.6 *Influence on Indoor Air Quality*

As the use of air-conditioning systems has become more widespread, increasing attention has been devoted to indoor health issues, especially after the SARS crisis in 2003. A common concern is whether air conditioners can result in air pollution and create health problems. It is believed that indoor health problems are mainly caused by mildew, dust, and VOCs (volatile organic compounds) from the indoor environment. In conventional air-conditioning systems, air flows through the cooling coil to be cooled and dehumidified, which is where the condensate can accumulate; such surfaces are the best places for mildew to grow. Research shows that environments with a relative humidity of 70–95 % are suitable for the growth of bacteria, where it is possible to detect 10,000 germs/cm^2 (Yanagi and Ikeda 2005; Yanagi et al. 2008). Mildew reproduction and diffusion has been a cause of many emerging health problems. As one of the main tasks of air-conditioning is to dehumidify the air, dehumidification by condensation dehumidification inevitably generates wet surfaces. Many recent studies have focused on the sterilization of air-conditioning systems. However, these methods not only make the systems more complicated; they also increase initial investment and operating expenses. Choosing an alternative method of dehumidification that does not result in wet surfaces has become the primary issue in the development of healthy air-conditioning systems.

The most effective method for removing VOCs produced by furniture and other indoor ornaments, as well as reducing the concentration of indoor CO_2, is ventilation, which involves introducing outdoor air and removing the indoor air to realize effective air exchange. Nevertheless, importing a large amount of outdoor air consumes a great deal of cooling capacity for cooling and dehumidification (or heating in winter). The proportion of the cooling capacity consumed by handling outdoor air relative to the entire cooling capacity can be higher than 50 %, when the properties of the building envelope are relatively good and the quantity of the indoor heat source is not very large. A further increase of the outdoor air flow rate often results in increasing the energy consumption of the air-conditioning system. In the past 40 years, both Chinese and international air-conditioning standards for average outdoor supply air flow rate have been changing constantly. For example, the US standard ranged from 25 to 10 m^3/h per person after energy crisis and finally to the current level of 30 m^3/h per person. In contrast, Denmark set their standard to 90 m^3/h per person due to its climate being neither too hot nor too humid. The question of how to increase the outdoor air flow rate without increasing the energy consumption of the air handling process is another important issue to consider.

1.4 Requirements for a New Air-Conditioning Solution

Air-conditioning systems play an important role in the indoor built environment. As society has continued to develop, conventional air-conditioning methods have been challenged by the demand for a more comfortable indoor environment, and more emphasis has been placed on efficient energy use. Continuing to improve the energy efficiency and reducing the energy consumption of air-conditioning systems in order to provide a suitable and comfortable environment are fundamental to the development of new strategies for the indoor built environment. Considering the handling methods and problems of current air-conditioning systems, the requirements for a new type of air-conditioning system include the following:

- Being able to adapt to the continuously changing indoor sensible load and moisture load, achieving regulation of indoor temperature and humidity at the same time
- Avoiding energy loss due to offset by cooling and reheating or dehumidifying and humidifying at the same time
- Providing opportunities for the utilization of natural cooling sources and low-grade energy
- Choosing suitable heat transfer fluids and reducing the energy consumption of transportation and distribution systems as much as possible
- Reducing the indoor supply air flow rate, partly sharing the terminal devices for both cooling and heating
- With a sufficient flow rate of outdoor air, reducing the additional energy consumption due to the increased outdoor air flow rate with the help of effective heat recovery methods
- Avoiding indoor wet surfaces by utilizing new methods of dehumidification

Taking these requirements into account, the THIC (temperature and humidity independent control) air-conditioning system is generally considered to be a possible effective solution. The second chapter of this book will reexamine the task of the indoor environmental control system, i.e., removing the indoor sensible load, moisture, CO_2, odor, and other pollutants based on certain comfort levels (environmental parameters). Certain approaches for satisfying the air-conditioning requirements will be introduced, and quantitative analysis of the requirements for removing the indoor moisture load, CO_2, and odors (which can all be handled by dried outdoor air) is conducted. The following concept is proposed: removing extra moisture, CO_2, and odors with dried outdoor air to meet the requirements of indoor humidity control and indoor air quality, in conjunction with removing extra heat

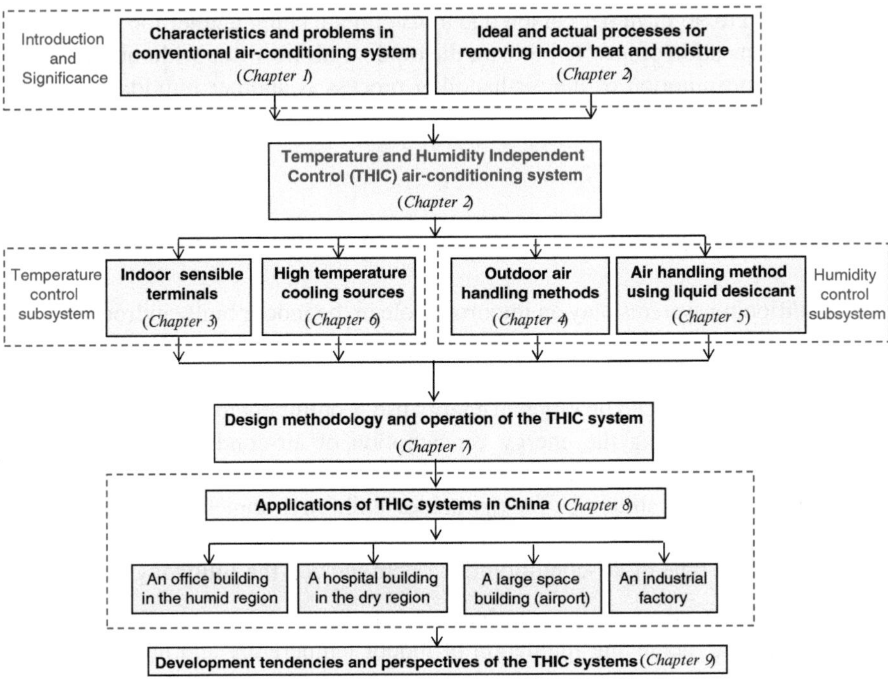

Fig. 1.12 Framework of this book

using another independent system to satisfy the requirement of indoor temperature control. In other words, the THIC system can be utilized to fully control and regulate the indoor thermal environment.

The framework of this book is illustrated in Fig. 1.12.

References

ASHRAE (2004) ASHRAE standard 55 thermal environmental conditions for human occupancy. American Society of Heating, Refrigerating and Air-Conditioning Engineers, Inc., Atlanta

ASHRAE (2009) ASHRAE handbook- fundamentals. American Society of Heating Refrigerating and Air-conditioning Engineer, Inc., Atlanta

ASHRAE (2010) ASHRAE standard 62.1 ventilation for acceptable indoor air quality. American Society of Heating, Refrigerating and Air-Conditioning Engineers, Inc., Atlanta

Bejan A (2006) Advanced engineering thermodynamics, 3rd edn. Wiley, New York

Building Energy Research Center of Tsinghua University (2009) 2009 annual report on China building energy efficiency. China Architecture & Building Press, Beijing (in Chinese)

CSBTS (1996a) GB 9663–1996. Hygienic standard for hotel. China Standards Press, Beijing (in Chinese)

CSBTS (1996b) GB 9664–1996. Hygienic standard for public place of entertainment. China Standards Press, Beijing (in Chinese)

CSBTS (1996c) GB 9668–1996. Hygienic standard for gymnasium. China Standards Press, Beijing (in Chinese)
CSBTS (1996d) GB 9669–1996. Hygienic standard for library, museum, art gallery and exhibition. China Standards Press, Beijing (in Chinese)
CSBTS (1996e) GB 9673–1996. Hygienic standard for public means of transportation. China Standards Press, Beijing (in Chinese)
CSBTS (2005a) GB 9666–1996. Hygienic standard for barber shop and beauty shop. China Standards Press, Beijing (in Chinese)
CSBTS (2005b) GB 9670–1996. Hygienic standard for shopping centre and book store. China Standards Press, Beijing (in Chinese)
CSBTS (2005c) GB 9672–1996. Hygienic standard for waiting room of public transit means. China Standards Press, Beijing (in Chinese)
Jiang Y, Liu XH, Xie XY (2011) Thermological analysis frame in thermal-hygro environment building. Chin HV&AC 41(3):1–12 (in Chinese)
Liu XH, Jiang Y, Liu SQ, Chen XY (2010) Research progress on liquid desiccant air-conditioning devices and systems. Front Energy Power Eng China 4(1):55–65
MOHURD, AQSIQ (2005) GB 50189–2005. Design standard for energy efficiency of public buildings. China Architecture & Building Press, Beijing (in Chinese)
Yanagi U, Ikeda K (2005) A study on the behavior and control of microbial contamination in an air conditioning system: Part 1 Growth environment and contamination status of microbes. J Environ Eng 70(593):49–56 (in Japanese)
Yanagi U, Kagi N, Ikeda K (2008) Research on the behavior and control of microbial contamination in an air conditioning system: Part 3 Fundamental experiments on the sterilization performance of Ozone. Trans Archit Inst Jpn 73(632):1197–1200 (in Japanese)
Zhang YN (2008) Case studies in energy performance investigation and analysis on public buildings of U.S.A. Master thesis, Tsinghua University, Beijing (in Chinese)
Zhao RY, Fan CY, Xue DH, Qian YM (2000) Air conditioning, 3rd edn. China Architecture & Building Press, Beijing (in Chinese)

Chapter 2
The Basic Idea of the THIC Air-Conditioning System

Abstract The basic principle of the THIC (temperature and humidity independent control) air-conditioning system is proposed in this chapter. The indoor requirements in terms of heat and moisture extraction together with air quality are investigated first, including the sources and characteristics of the indoor sensible load and latent load. Then the ideal cooling and dehumidifying process as well as the real process of removing heat and moisture are presented. Besides, the main reasons leading to the performance discrepancies between ideal and real processes are discussed. On the basis of these analyses, the THIC air-conditioning system is proposed, and its operating strategy as well as requirements to the components are introduced.

2.1 Indoor Requirements for Heat and Moisture Extraction Along with Air Quality

2.1.1 Sources and Characteristics of the Indoor Sensible Load

2.1.1.1 Sources of the Indoor Sensible Load

To analyze the performance of air-conditioning systems removing the indoor sensible load and moisture load, the primary tasks involve determining the sources of the loads, the differences between various sources, and the approaches through which they become loads that need to be removed by the air-conditioning system. The factors that affect the indoor thermal environment in buildings in the most common circumstances are illustrated in Fig. 2.1.

The illustration in the above figure – how heat and moisture from various influencing sources become the indoor sensible and moisture loads, respectively, by conduction, convection, and radiation processes through the building envelope and internal sources – is described in detail in *Building Thermal Process* (Yan and Zhao 1986). The various influencing sources in Fig. 2.1 are generally called either heat sources or moisture sources, and they can be divided into external factors and

X. Liu et al., *Temperature and Humidity Independent Control (THIC) of Air-conditioning System*, DOI 10.1007/978-3-642-42222-5_2,

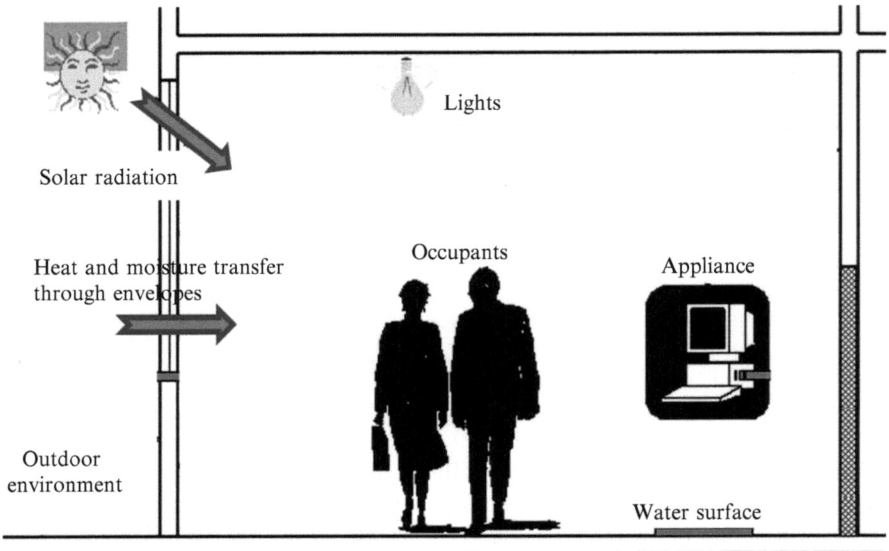

Fig. 2.1 Various sources that influence the indoor thermal environment in buildings

internal factors according to their spatial characteristics. External factors include walls, roofs, windows, doors, ceilings, and floors, and internal factors include lights, occupants, appliances, and equipment.

According to Fig. 2.1, the heat sources responsible for the extra heat in the indoor environment can be divided into external and internal sources based on spatial characteristics and into different temperatures that represent grade differences. The grades of various heat sources are shown in Fig. 2.2. The temperature of the heat source is an important parameter that impacts the heat dissipation efficiency of the air-conditioning system. The higher the temperature of the heat source, the higher the required temperature of the cooling source (to take the indoor heat away to the outdoor environment). As indicated in Fig. 2.2, it can be seen that the temperatures of different heat sources vary widely. Furthermore, the temperatures of the majority of heat sources are higher than the dry-bulb ambient temperature; thus, in theory, they can be removed directly by outdoor natural cooling sources without any energy consumption. Therefore, the ways in which heat sources influence the indoor built environment should be studied first, and how to effectively remove the extra heat from conditioned spaces can then be determined.

2.1.1.2 Ways to Remove the Indoor Sensible Load

According to Fig. 2.2, only the internal surfaces of external walls, occupants' exposed surfaces, clothing surfaces, and surfaces of internal walls, floors, and roofs are sources whose temperatures are indeed lower than the outdoor ambient environment.

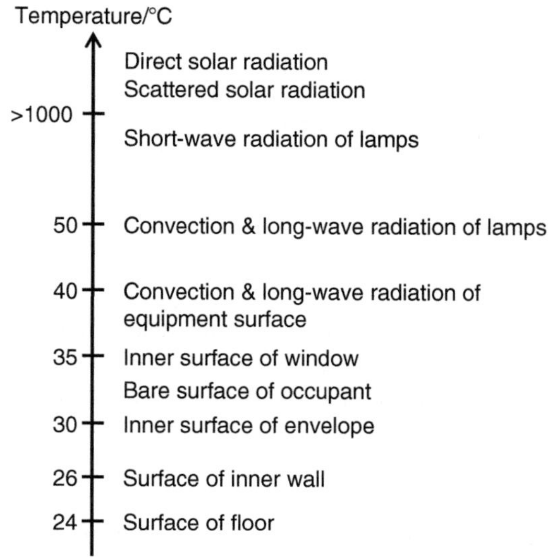

Fig. 2.2 Temperature levels of various heat sources

The heat produced from these different sources needs to be taken away by the cooling sources. The surface temperatures of electrical devices are close to those of human surfaces due to their shell-like construction, which affects their heat production. Therefore, their heat production should also be absorbed by the cooling sources. However, the internal surfaces of the building envelope and human surfaces are quite different. The former has a large superficial area and a fixed spatial location and is thus a secondary object of indoor environmental control. The latter has a smaller volume and superficial area in comparison to the entire indoor space, but it is the primary object of the indoor environmental control system. As a result, it is recommended that these two kinds of heat sources be removed through different approaches.

Based on their temperature level and characteristics, various heat sources can be divided into different categories. To prevent the mixing of the heat sources, which would affect their temperature, different approaches should be used to remove them. The main task of air-conditioning systems in terms of temperature control is to discharge indoor heat, and this can be accomplished in the following ways:

- For heat sources whose temperature is higher than the outdoor ambient temperature, in conjunction with building envelope construction and air-conditioning system integration, outdoor natural cooling sources should be utilized to extract the heat directly as much as possible.
- For the internal surfaces of the building envelope with a temperature higher than the indoor environment due to the effects of solar radiation and heat transfer, cooling sources with a temperature close to that of the indoor environment should be utilized. In this way, the temperatures of all surfaces can be maintained close to the indoor temperature, thereby satisfying the control requirements for the indoor thermal environment.

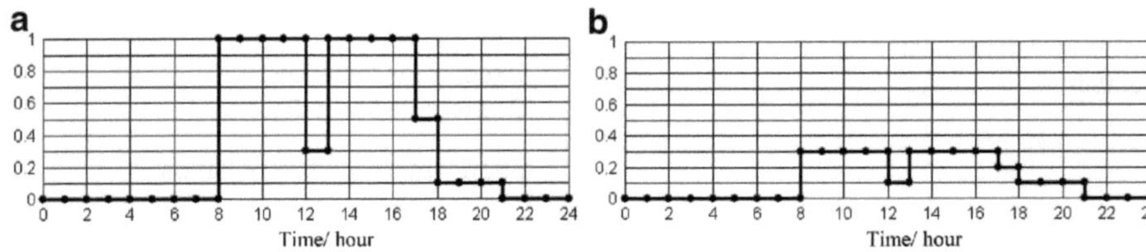

Fig. 2.3 Schedules of occupant density in typical office buildings: (**a**) workdays and (**b**) weekends

- For the task of removing the heat generated by human beings, cooling sources with a temperature lower than the human surface temperature should be utilized to discharge this kind of heat locally and directly.

2.1.2 Sources and Characteristics of the Indoor Moisture Load

Indoor moisture mainly comes from indoor occupants, evaporation from open water surfaces and plant transpiration, infiltration moisture from the building envelope, etc. (The 10th Design and Research Institute of Ministry of Electronics Industry 1995). The calculations for moisture production for different moisture sources are shown in detail in Appendix A. The most common air-conditioned rooms contain few or no indoor plants. For these rooms, since there are either no open water surfaces or only a negligible amount of scattered moisture from the few existing open water surfaces, it is clear that the waste moisture in the indoor environment is mainly from the indoor occupants (the infiltration moisture from the building envelope can be ignored). Thus, the amount of waste moisture is proportional to the number of occupants. However, the number of indoor occupants changes dramatically throughout a typical day. The daily schedules of occupants in several typical offices and shopping malls are shown in Figs. 2.3 and 2.4, respectively. The schedule of the offices generally follows a similar pattern every week: differences between workdays and weekends but generally remaining the same throughout the different seasons. For malls, the occupants' density is generally low in the summer due to a lack of sales and relatively high during Spring Festival in China due to a peak in sales. In other words, the occupants' density schedule in malls is a year-round cycle, and seasonal effects should be taken into account.

Removing the extra moisture in the indoor environment is a mass transfer process that has to be realized by supplying dry air. The fundamental problem to be solved in the process of removing the waste moisture is to determine an appropriate supply air flow rate, according to the sources and characteristics of the indoor moisture load. In order to regulate the removal of extra moisture in the indoor environment, the characteristics of the extra moisture should be considered. Determining the ways in which the dry air flow rate should be adjusted in proportion to the number of occupants in offices is an example of this essential consideration.

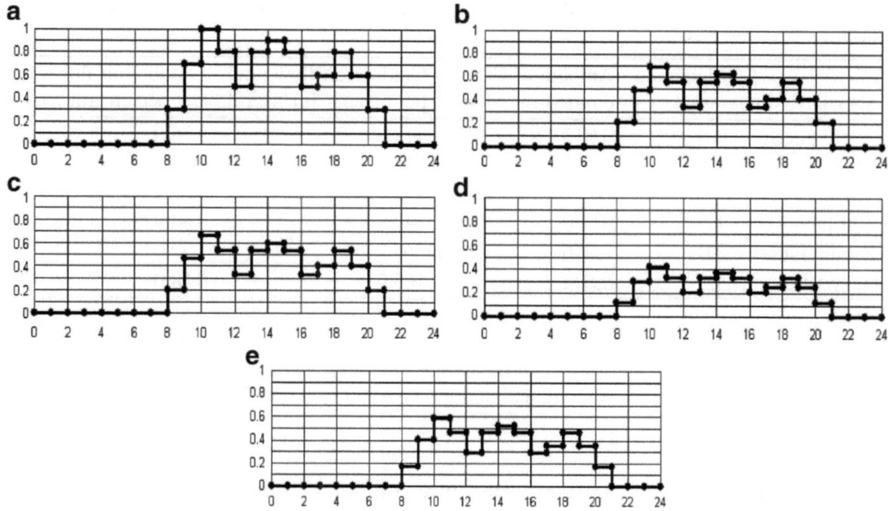

Fig. 2.4 Schedules of occupant density in typical shopping malls in China: (**a**) national holidays (International Labor Day, National Day, Spring Festival, etc.), (**b**) from early Dec. to late Feb., (**c**) from early Mar. to late May, (**d**) from early Jun. to late Aug., and (**e**) from early Sep. to late Nov.

2.1.3 Requirements for Indoor Air Quality (IAQ), Including Extracting CO_2

2.1.3.1 Sources of CO_2 and Odors

One of the major tasks for indoor environmental control is to remove the indoor CO_2 and the odors produced there so that health demands can be satisfied. Supplying outdoor air is an effective method for extracting CO_2. Indoor contaminants are mainly from occupants and indoor air pollution sources (e.g., finishing materials and furniture in buildings). According to ASHRAE standards, the minimum design outdoor air flow rate is as follows:

$$\mathrm{DVR} = R_p P_D D + R_b A_b \tag{2.1}$$

where DVR (design ventilation rate) is the design outdoor air flow rate, l/s; R_p is the minimum outdoor air flow rate per person, l/s; P_D is the number of persons; D is the variation coefficient; R_b is the required minimum outdoor air per unit floor area, l/(s·m^2); and A_b is the floor area, m^2. As indicated by this equation, the required amount of outdoor air can be divided into two parts: one for people and the other for the building. The former is proportional to the number of indoor occupants. In other words, diluting contaminants produced by people and their activities can be regulated based on the varying number of indoor occupants. The latter is proportional to the building area, and this part of the outdoor air is used to dilute the contaminants

Table 2.1 Amount of CO_2 exhaled by occupants in different states and required outdoor air flow rate for extracting CO_2

Labor intensity	Metabolic rate	Amount of CO_2 emitted [$m^3/(h \cdot p)$]	Outdoor air flow rate per person[a] [$m^3/(h \cdot p)$]	Outdoor air flow rate per person[b] [$m^3/(h \cdot p)$]
Seated quietly	0.0	0.013	18.6	26.0
Extremely light labor	0.8	0.022	31.4	44.0
Light labor	1.5	0.030	42.9	60.0
Medium labor	3.0	0.046	65.7	92.0
Heavy labor	5.5	0.074	105.7	148.0

Notes:
[a]CO_2 concentrations in the outdoor and indoor environment are 300 ppm and 1,000 ppm, respectively
[b]CO_2 concentrations in the outdoor and indoor environment are 500 ppm and 1,000 ppm, respectively

emitted by the building materials, furniture, indoor activities, etc. Moreover, this second part of the outdoor air flow rate is relatively constant and accounts for 15–30 % of the total design ventilation rate. Therefore, the outdoor air flow rate discussed below refers especially to the first part. The amount of CO_2 produced by occupants is related to the human body state, and Table 2.1 provides the amount of CO_2 exhaled by occupants in different states (Zhao et al. 2000).

2.1.3.2 Outdoor Air Required for Extracting CO_2

The main approach for ensuring adequate indoor air quality is ventilation, which displaces the indoor contaminant-laden air using outdoor air with a low concentration of contaminants. The required outdoor air flow rate should be based on diluting the indoor contaminants to the standard concentration. To extract indoor CO_2 and other contaminants, the required outdoor air flow rate can be described mathematically as

$$G_{\mathrm{w}} = \frac{g}{C_{\mathrm{in}} - C_{\mathrm{out}}} \tag{2.2}$$

where G_w is the outdoor air flow rate per person, $m^3/(h \cdot p)$; g is the CO_2 emission load per person, $m^3/(h \cdot p)$; C_{in} is the indoor CO_2 concentration; and C_{out} is the outdoor CO_2 concentration.

The World Health Organization recommends that CO_2 concentration be maintained below 2,500 ppm. Table 2.2 summarizes the regulations for indoor CO_2 concentration according to Chinese national standards. Natural CO_2 concentration is generally 300–500 ppm. Table 2.1 lists the outdoor air flow rate required to extract the CO_2 generated by occupants in a conditioned space where the indoor CO_2 concentration is 1,000 ppm and the corresponding outdoor air CO_2 concentration is either 300 or 500 ppm. Personal CO_2 emission is 0.022 $m^3/(h \cdot p)$ for

Table 2.2 CO_2 concentration standards for certain typical buildings (GB9663–GB9673)

Limit of CO_2 concentration (ppm)	Type of building
700	3–5-star restaurants and hotels
1,000	1–2-star restaurants and hotels, guesthouses, libraries, museums, art galleries, beauty salons, hospital waiting rooms
1,500	Theaters; concert halls; entertainment halls; dance halls; bars; cafes; coffee shops; shopping malls; bookstores; exhibition halls; stadiums; waiting rooms for trains, ships, and airplanes; passenger train cars; ship cabins; aircraft cabins; swimwear shops

common offices. Based on this amount, when the outdoor CO_2 concentration is 300 ppm, the outdoor air flow rate required per person is 31.4 $m^3/(h{\cdot}p)$, and the required ventilation rate for the outdoor air is 1.2 if the area is 8 m^2 per person. When the outdoor air CO_2 concentration is 500 ppm, the outdoor air flow rate required per person is 44.0 $m^3/(h{\cdot}p)$, and the required ventilation rate is 1.7.

During the design stage of air-conditioning systems, although the design of air distribution systems is important apart from the outdoor air and indoor contaminants, it is also related to IAQ. If the air distribution system is designed well, the outdoor air can be supplied to the workplace properly, the contaminants can be quickly extracted to the outdoor environment, and the indoor air quality will be improved.

2.1.3.3 Consistency in the Outdoor Air Flow Rate for Extracting Waste Moisture and CO_2

Occupants' metabolic processes can produce CO_2 and water vapor, so in buildings for which occupants' activities are the main uses, CO_2 and water vapor mainly come from occupants. Figure 2.5 illustrates how the scattered moisture amount and the CO_2 emission amount vary with labor intensity when the indoor air temperature is 25 °C. As shown in this figure, the variation trend of the amount of scattered moisture is consistent with that of the amount of CO_2. Therefore, the number and distribution of occupants in the conditioned space can be predicted based on the moisture content or CO_2 concentration, and the outdoor air flow rate can be adjusted accordingly.

Table 2.3 shows the variation trend of the outdoor air flow rate required to extract the waste moisture of the indoor environment as a function of labor intensity. The humidity ratio difference between the indoor air and supply air is 2.5 g/kg (the indoor humidity ratio is 10.8 g/kg, and the required humidity ratio of the supply air is 8.3 g/kg). Figure 2.6 shows the comparison of CO_2 emission amount per person with the CO_2 extracted by the outdoor air whose flow rate is determined by controlling the indoor moisture content. When the CO_2 concentration of the outdoor ambient environment is 300 ppm, the outdoor air flow rate determined by extracting the waste moisture can keep the indoor CO_2 concentration

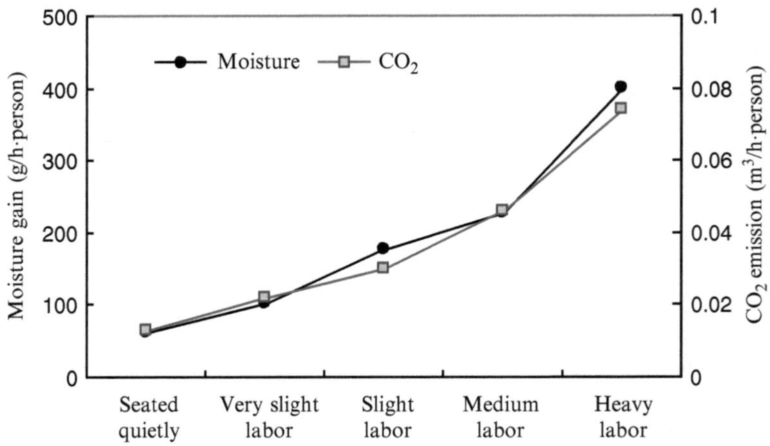

Fig. 2.5 Variation trend of scattered water amount and CO_2 emission amount with labor intensity

Table 2.3 Outdoor air required to extract the waste moisture of the indoor environment

Labor intensity	Scattered moisture amount per person [g/(h·p)]	Outdoor air flow rate per person [m^3/(h·p)]	CO_2 amount taken away by outdoor air[a] [m^3/(h·p)]	CO_2 amount taken away by outdoor air[b] [m^3/(h·p)]
Seated quietly	61	20.3	0.014	0.010
Extremely light labor	102	34.0	0.024	0.017
Light labor	175	58.3	0.041	0.029
Medium labor	227	75.7	0.053	0.038
Heavy labor	400	133.3	0.093	0.067

Notes:
[a]CO_2 concentrations in the outdoor and indoor environment are 300 ppm and 1,000 ppm, respectively
[b]CO_2 concentrations in the outdoor and indoor environment are 500 ppm and 1,000 ppm, respectively

within 850–950 ppm. When the CO_2 concentration of the outdoor ambient environment is 500 ppm, the outdoor air flow rate determined by extracting the waste moisture can keep the indoor CO_2 concentration within 1,000–1,150 ppm, which can probably meet the indoor air quality demand.

Table 2.4 describes the waste moisture in the indoor environment removed by the outdoor air, whose flow rate is determined by the demand for extracting CO_2. In this scenario, the indoor relative humidity can be maintained in a range of 52–59 %, which can meet the indoor humidity demand. That means the supply air flow rate of the outdoor air can be determined according to the measured CO_2 concentration, and it can control the indoor air quality and humidity to meet both demands at the same time. The supply outdoor air flow rate can be also determined based on the air moisture content, and it can control the indoor humidity and CO_2 concentration simultaneously.

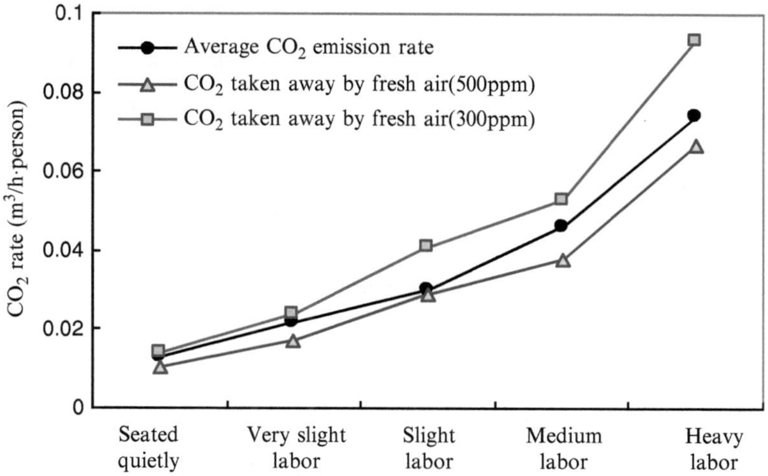

Fig. 2.6 Comparison of CO_2 emission amount per person with the CO_2 discharged by the outdoor air

Table 2.4 Dehumidification capacity of the outdoor air flow rate determined by CO_2 concentration

Labor intensity	Scattered moisture amount per person [g/(h·p)]	Waste moisture amount discharged by outdoor air[a] [g/(h·p)]	Waste moisture amount discharged by outdoor air[b] [g/(h·p)]
Seated quietly	61	56	78
Extremely light labor	102	94	132
Light labor	175	129	180
Medium labor	227	197	276
Heavy labor	400	317	444

Notes:
[a]CO_2 concentrations in the outdoor and indoor environment are 300 ppm and 1,000 ppm, respectively
[b]CO_2 concentrations in the outdoor and indoor environment are 500 ppm and 1,000 ppm, respectively

As indicated by the analysis above, in the thermal built environment, the processed outdoor air can be supplied to the indoor space for humidity control and CO_2 removal, and effective indoor temperature control approaches are not limited to air supply. Thus, we introduce a new type of system designed for the indoor built environment; this approach is different from conventional systems that regulate indoor temperature, humidity, and IAQ. The basic concept of this new system is illustrated in Fig. 2.7.

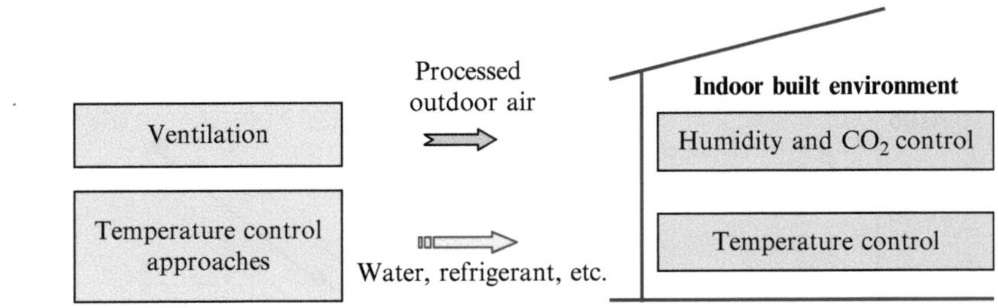

Fig. 2.7 A new system designed for the indoor built environment

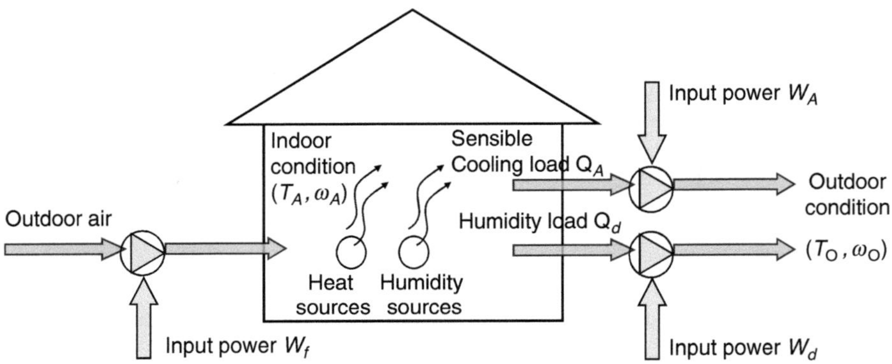

Fig. 2.8 Schematic diagram of the ideal cooling and dehumidifying process

2.2 The Ideal Cooling and Dehumidifying Process

In the ideal cooling and dehumidifying process, sensible cooling load Q_A and moisture load (latent load) Q_d are treated by two separate systems, as shown in Fig. 2.8 (Zhang et al. 2012a, b). Moreover, all the heat sources are treated independently. The ideal process of dehumidification and outdoor air handling is used. Since the focus of this section is to consider the concept of the ideal process, some influences in the actual system are ignored: (1) the energy consumptions of transportation and distribution of the heat transfer fluids are not considered in the ideal process, (2) the chiller is an ideal Carnot chiller, and (3) the heat and mass transfer area and coefficient are not limited for heat and mass exchangers.

2.2.1 *Ideal Cooling Process*

The ideal cooling process refers to removing the sensible cooling load from the indoor environment to the outdoor environment using the minimum amount of energy. For different heat sources with different temperature levels, when

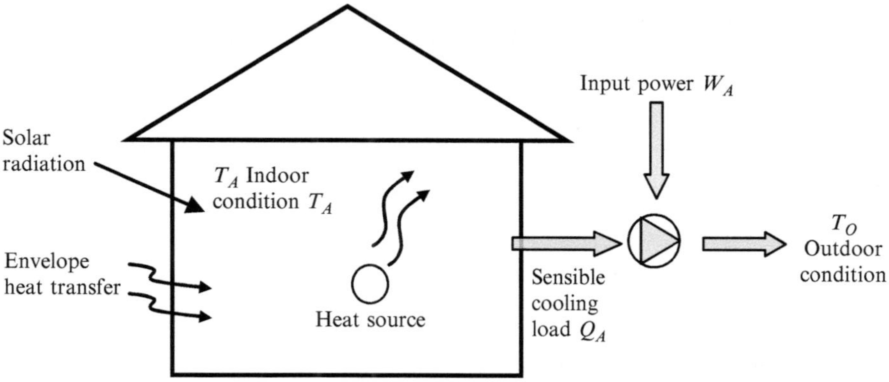

Fig. 2.9 Ideal cycle of the exhaust heat removal process for buildings

Fig. 2.10 Ideal cooling process from the indoor environment to the outdoor environment

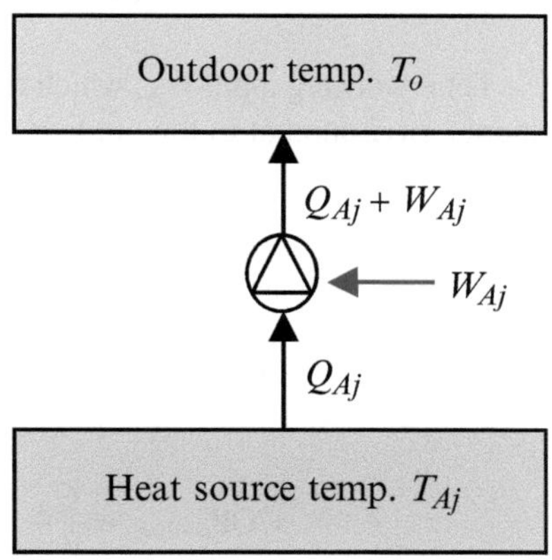

temperature of the heat source T_{Aj} is higher than ambient temperature T_O, as described by Eqs. (2.3) and (2.4), the outdoor environment can be used as the cooling source to cool the heat source, so the chiller (and its energy consumption) is not required. When temperature of the heat source T_{Aj} is lower than ambient temperature T_O, as described by Eq. (2.5), a Carnot engine needs to be set up between T_{Aj} and T_O, absorbing heat Q_{Aj} at temperature T_{Aj} while releasing heat Q_{Aj} and consuming energy W_{Aj} at temperature T_O (Figs. 2.9 and 2.10).

Therefore, the sensible heat to be removed from the indoor environment to the outdoor environment is equal to

$$Q_A^{'} = \sum_j Q_{Aj} \cdot \mathrm{sign}(T_O - T_{Aj}) \tag{2.3}$$

where Q_{Aj} is the heat load of each heat source, and the definition of sign($T_O - T_{Aj}$) is

Table 2.5 Efficiency of the ideal cooling process

Heat Q_{Aj}	Temp. T_{Aj}	sign($T_O - T_{Aj}$)	COP_j	W_{Aj}
q	40	0	–	0
q	38	0	–	0
q	32	1	101.7	0.010 q
q	30	1	60.6	0.017 q
q	25	1	29.8	0.034 q

$$\text{sign}(T_O - T_{Aj}) = \begin{cases} 1 & \text{if } T_{Aj} < T_O \\ 0 & \text{if } T_{Aj} \geq T_O \end{cases} \tag{2.4}$$

Therefore, $Q_A^{'} \leq Q_A$. For the Carnot engine working between T_{Aj} and T_O, the working input and the coefficient of performance (COP_j) are

$$W_{Aj} = \frac{Q_{Aj}}{COP_j}, \quad \text{where} \quad COP_j = \frac{T_{Aj}}{T_O - T_{Aj}} \tag{2.5}$$

Total working input W_A, which removes total sensible cooling load Q_A from the indoor environment to the outdoor environment, can be expressed as

$$W_A = \sum_j W_{Aj} = \sum_j \frac{Q_{Aj}}{COP_j} \cdot \text{sign}(T_O - T_{Aj}) \tag{2.6}$$

Therefore, the ideal energy efficiency for cooling is

$$COP_{cooling} = \frac{Q_A}{W_A} = \frac{\sum_j Q_{Aj}}{\sum_j \frac{Q_{Aj}}{COP_j} \cdot \text{sign}(T_O - T_{Aj})} \tag{2.7}$$

If the entire heat of various heat sources is transferred to the indoor air first, and then the cooling load is removed from the indoor environment to the outdoor environment by the chiller, the efficiency of the cooling process will be

$$COP_{cooling}^{'} = \frac{T_A}{T_O - T_A} \tag{2.8}$$

Table 2.5 shows some examples to compare Eqs. (2.7) and (2.8). Outdoor temperature T_O is 35 °C. The temperature of the indoor heat sources, COP_j of the Carnot cycle, and input work W_{Aj} are all listed in Table 2.5.

When the heat source temperature is higher than the outdoor ambient temperature, the indoor heat can be removed by outdoor free cooling. Only when the heat source temperature is lower than the outdoor temperature, the heat should be removed by the Carnot chiller. When the total heat in the room is 5 q, the required

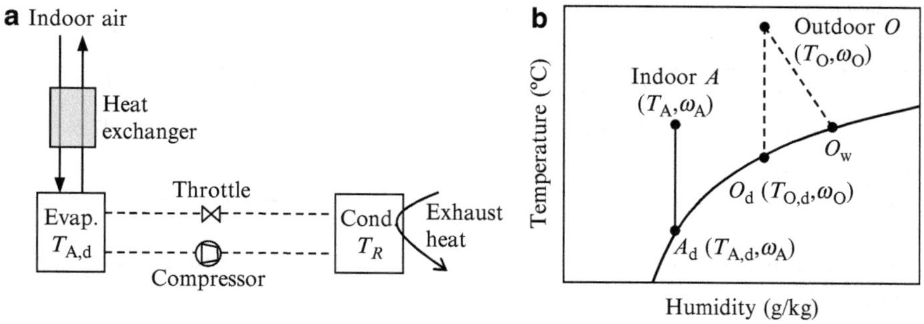

Fig. 2.11 The ideal condensation dehumidification process: (**a**) principle of the air handling process and (**b**) air handling process in psychrometric chart

input power is only 0.06 q, so the ideal cooling efficiency of the system ($COP_{cooling}$) is 83.5. When all the heat is first sent to the air and then extracted, the ideal cooling efficiency of the system ($COP'_{cooling}$) is 29.8, which is only 35.7 % of $COP_{cooling}$. Since the difference between the two cooling efficiencies is caused by the mixing losses of indoor terminals, the terminal efficiency can be defined as

$$\eta_{terminal} = \frac{COP'_{cooling}}{COP_{cooling}} \tag{2.9}$$

For the example given in Table 2.5, terminal efficiency $\eta_{terminal}$ is 35.7 %, which means the ideal cooling efficiency decreases by about two-thirds when all the heat is mixed with the indoor air before cooling the conditioned space. Thus, avoiding mixing heat of different temperature levels and enhancing terminal efficiency are the key points for improving the energy efficiency of air-conditioning systems. For heat sources above room temperature, the heat could be extracted by jacketed exhaust, glass interlayer ventilation, hierarchical ventilation (in large space buildings), etc., to utilize the natural cooling source to extract the indoor heat.

2.2.2 *Ideal Dehumidification Process*

2.2.2.1 The Condensation Dehumidification Method

Figure 2.11 depicts the principle of the ideal condensation dehumidification method. Indoor air $A(T_A, \omega_A)$ goes into an ideal heat exchanger to be cooled to its dew point temperature A_d $(T_{A,d}, \omega_A)$. The air is then dehumidified in the evaporator. When the efficiency of the heat exchanger is 1, and the changes of water content in the air are ignored, the capacity of the evaporator is $Q_d = r \cdot \Delta\omega$,

Table 2.6 Ideal COP of condensation dehumidification with different heat sinks

Heat sink	Equation of ideal COP_{deh}	Condition 1	Condition 2
Dry-bulb temperature of outdoor air	$COP_{deh} = \frac{T_{A,d}}{T_O - T_{A,d}}$	13.7	13.7
Wet-bulb temperature of outdoor air	$COP_{deh} = \frac{T_{A,d}}{T_{O,w} - T_{A,d}}$	21.6	27.6
Dew point temperature of outdoor air	$COP_{deh} = \frac{T_{A,d}}{T_{O,d} - T_{A,d}}$	26.1	45.6

and the heat is exhausted into the outdoor environment at temperature T_R by a heat pump:

$$COP_{deh} = \frac{T_{A,d}}{T_R - T_{A,d}} \tag{2.10}$$

The temperature of heat extraction (T_R) in Fig. 2.11 can be outdoor dry-bulb temperature T_O, wet-bulb temperature $T_{O,w}$ from the cooling tower, or dew point temperature $T_{O,d}$ with indirect evaporative cooling. Table 2.6 shows the ideal COP_{deh} using different methods of heat extraction. The indoor air condition is 25 °C and 10 g/kg.

Condition 1: The outdoor temperature is 35 °C, humidity ratio is 20 g/kg, wet-bulb temperature is 27.3 °C, and dew point temperature is 25.0 °C.

Condition 2: The outdoor temperature is 35 °C, humidity ratio is 15 g/kg, wet-bulb temperature is 24.4 °C, and dew point temperature is 20.3 °C.

From Table 2.6, the heat sink (or condensing temperature) influences the ideal COP_{deh} directly. When the outdoor air condition is 35 °C, 20 g/kg, the COP_{deh} is 13.7, 21.6, and 26.1 when the condensing temperature is dry-bulb, wet-bulb, and dew point temperature, respectively. Therefore, releasing heat at the dew point temperature can realize the highest efficiency.

2.2.2.2 The Liquid Desiccant Dehumidification Method

The iso-concentration line of the ideal solution coincides with the iso-relative humidity line of the humid air, as shown in Fig. 2.12. An ideal process can be established for the liquid desiccant dehumidification method (Li et al. 2010). Figure 2.13 illustrates the ideal dehumidification process, where points A and B are on the iso-relative humidity line of the humid air (or the iso-concentration line of the ideal solution). In the ideal solution, moisture is absorbed at point A and desorbed at point B. The latent heat of absorption at point A is removed by the evaporator of a Carnot chiller with evaporating temperature T_A, and the latent heat of desorption at point B is supplied by the condenser of the Carnot chiller with condensing temperature T_B. Between point A and point B, there is a heat recovery process for the solution, and between point O and point B, there is a heat recovery process for the air. The input work of the Carnot chiller is exhausted into reference state (or heat sink) T_R. Temperature of the heat sink (T_R) can be the dry-bulb (T_O),

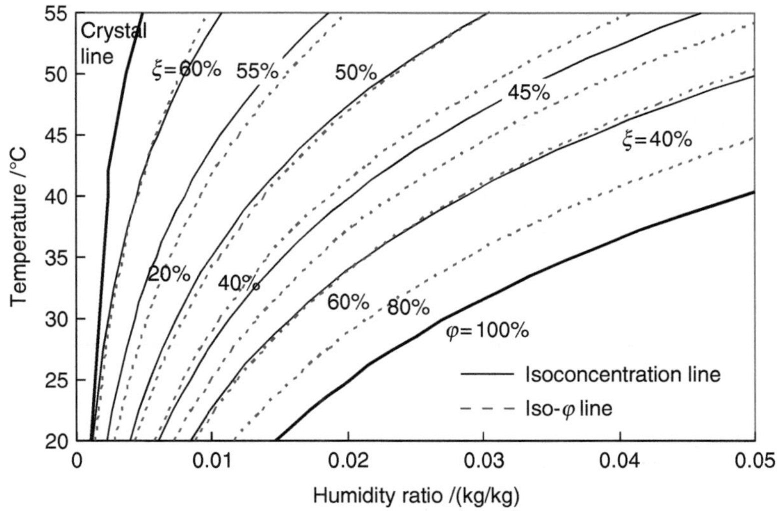

Fig. 2.12 State of LiBr aqueous solution in psychrometric chart

wet-bulb ($T_{O,w}$), or dew point ($T_{O,d}$) temperature of the outdoor air by using an air-cooled heat exchanger, direct evaporative cooling, or indirect evaporative cooling of the outdoor air, respectively. The flow chart of this liquid desiccant dehumidification process is shown in Fig. 2.13b, where $T_R = T_O$.

Input work W_1 required to remove the moisture from the indoor humidity ratio to the outdoor humidity ratio can be expressed as the efficiency of the Carnot chiller. The value of W_2 required to exhaust the heat is determined by the relationship between T_B and T_R ($T_R = T_O$):

$$\mathrm{COP_{deh}} = \frac{Q_d}{W_d} = \frac{Q_d}{W_1 + W_2} \tag{2.11}$$

$$W_1 = Q_d \frac{T_B - T_A}{T_A}, \quad W_2 = W_1 \frac{T_O - T_B}{T_B} \tag{2.12}$$

W_2 is negative if $T_B > T_O$, which means the process of exhausting heat from T_B to T_O can generate work W_2. If $T_B \le T_O$, then W_2 is nonnegative, which means the process of exhausting heat from T_B to T_O will consume work W_2.

The total input work of the ideal process using liquid desiccant with the outdoor dry-bulb temperature as the heat sink is

$$\mathrm{COP_{deh}} = \frac{1}{T_O} \cdot \frac{T_A T_B}{T_B - T_A} \tag{2.13}$$

The ideal COP of the liquid desiccant dehumidification process can also be expressed by the dew point temperature:

$$\mathrm{COP_{deh}} = \frac{1}{T_O} \cdot \frac{T_{A,d} T_{O,d}}{T_{O,d} - T_{A,d}} \tag{2.14}$$

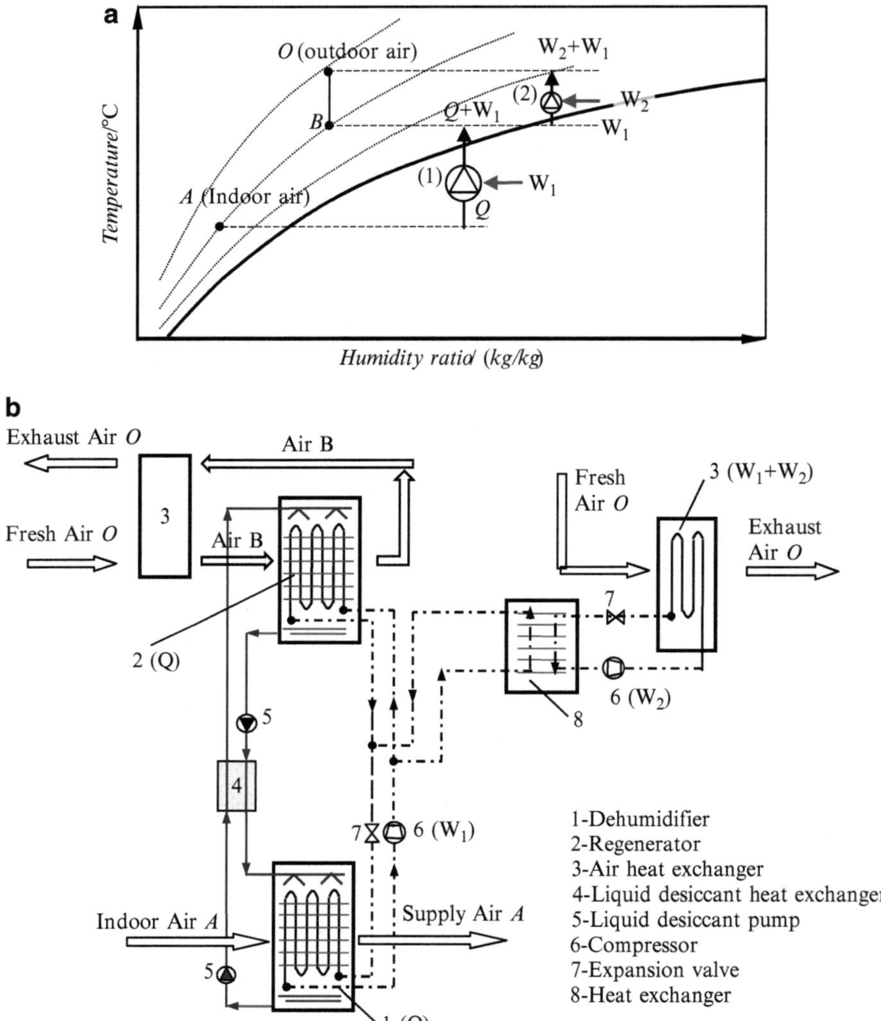

Fig. 2.13 The ideal dehumidification process using liquid desiccant: (**a**) handling process in psychrometric chart and (**b**) schematic diagram of the liquid desiccant dehumidification process

The temperature of the heat sink can also be the wet-bulb or dew point temperature of the outdoor air. Table 2.7 shows the influence of the heat sink on the efficiency of the liquid desiccant dehumidification process; conditions 1 and 2 are the same as in Table 2.6. From the table, it can be seen that the heat sink's influence on the efficiency of liquid desiccant dehumidification is minimal. The reason is that most of the heat in liquid desiccant dehumidification is exhausted by regeneration, which is equal to exhaust heat at the dew point temperature.

Figure 2.14 illustrates the data of two tables to compare the efficiency of condensation dehumidification and liquid desiccant dehumidification. When the

Table 2.7 COP values of liquid desiccant dehumidification processes with different heat sinks

Heat sink	Equation of ideal COP_{deh}	Condition 1	Condition 2
Dry-bulb temperature of outdoor air	$COP_{deh} = \frac{1}{T_O} \cdot \frac{T_{A,d} T_{O,d}}{T_{O,d} - T_{A,d}}$	25.3	43.4
Wet-bulb temperature of outdoor air	$COP_{deh} = \frac{1}{T_{O,w}} \cdot \frac{T_{A,d} T_{O,d}}{T_{O,d} - T_{A,d}}$	25.9	45.0
Dew point temperature of outdoor air	$COP_{deh} = \frac{T_{A,d}}{T_{O,d} - T_{A,d}}$	26.1	45.6

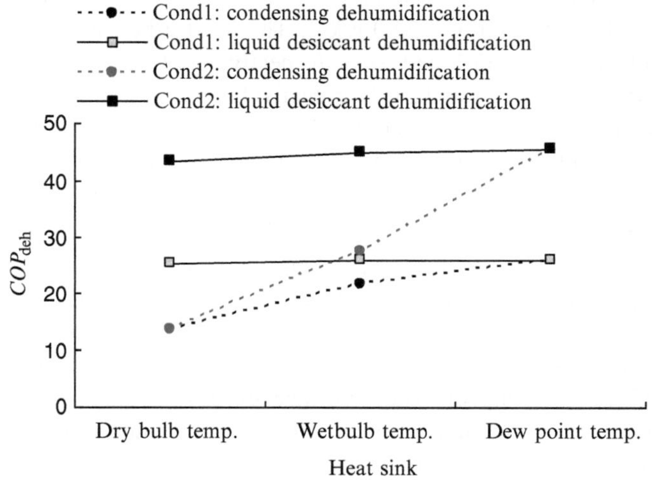

Fig. 2.14 COP_{deh} comparison of condensation dehumidification and liquid desiccant dehumidification

heat sink is the dew point of the outdoor air, condensation dehumidification and liquid desiccant dehumidification have the same efficiency. When the heat sink is the dry-bulb or wet-bulb temperature of the outdoor air, condensation dehumidification has lower efficiency than liquid desiccant dehumidification. Thus, the heat sinks influence condensation dehumidification to a greater extent. Both processes experience a decrease in efficiency when the outdoor humidity ratio increases.

2.2.3 *Total Efficiency of Cooling and Dehumidification*

The total efficiency of cooling and dehumidification (COP_{sys}) is defined as the ratio of the sensible and moisture loads to the input work of the system, as shown in Eq. (2.15); x_1 and x_2 are the ratio of sensible load and moisture load to the entire load, respectively:

$$COP_{sys} = \frac{Q_A + Q_d}{W_A + W_d} = \frac{1}{x_1/COP_{cooling} + x_2/COP_{deh}} \tag{2.15}$$

The system efficiency calculations for several cities are summarized below. The indoor air condition is 25 °C, 60 %, and 11.8 g/kg, and all the heat sources are mixed into the indoor environment and then removed together by the cooling source. $COP_{cooling}$ and COP_{deh} are calculated by Eqs. (2.8) and (2.13) respectively.

- Beijing outdoor design condition (dry-bulb = 33.2 °C, relative humidity = 59.3 %, humidity ratio = 19.1 g/kg): The ideal efficiencies of cooling and dehumidification are 36.4 and 37.6, respectively. When $x_2 = 20$ %, system efficiency is 36.6.
- Shanghai outdoor design condition (dry-bulb = 34.0 °C, relative humidity = 65.1 %, humidity ratio = 22.0 g/kg): The ideal efficiencies of cooling and dehumidification are 33.1 and 28.8, respectively. When $x_2 = 20$ and 30 %, system efficiency is 32.1 and 31.7, respectively.
- Guangzhou outdoor design condition (dry-bulb = 33.5 °C, relative humidity = 64.8 %, humidity ratio = 21.3 g/kg): The ideal efficiencies of cooling and dehumidification are 35.1 and 30.5, respectively. When $x_2 = 20$ and 30 %, system efficiency is 34.1 and 33.6, respectively.

In buildings, there is a requirement of the outdoor air, and an ideal handling process of the outdoor air can be established as well. We can take outdoor air dehumidification as an example, where the water content in the outdoor air needs to be removed and exhausted to the outdoor environment. The efficiency of outdoor air dehumidification is higher than the efficiency of removing moisture from the indoor environment to the outdoor environment, as indicated in Eq. (2.10). When outdoor air is considered, the system efficiencies of the above three cities are higher than 30. Moreover, the nonuniform condition of heat sources is not considered, meaning the efficiency will be higher if the nonuniform condition is considered.

2.3 Actual Process of Removing Heat and Moisture

Section 2.2 established the ideal process of removing sensible and moisture loads in buildings and analyzed the ideal efficiency. The current section analyzes the actual process of removing sensible and moisture loads and compares the efficiencies between ideal and actual processes.

2.3.1 From the Ideal Process to the Actual Process

The previous section showed that the ideal efficiency of removing the heat and moisture in buildings is above 30, but in actual buildings, the efficiency is much lower than the ideal value. Table 2.8 summarizes the required performance values of key components in HVAC (heating, ventilation, and air-conditioning) systems according to Chinese standards. As can be seen from the table, the efficiency of the central air-conditioning plants is around 4.0 without considering the energy

Table 2.8 Coefficients of performance (COPs) of actual air-conditioning systems

Efficiency	Equation	Reference value	Source
COP_c of chiller	Cooling capacity of chiller/ energy consumption of chiller (compressor)	5.1–5.6	GB 19577–2004 (level II and level III of chiller)
Transportation coefficient of chilled water (TC_{chw})	Cooling capacity of chilled water/energy consumption of chilled water pump	41.5	GB 50189–2005 (transportation efficiency of chilled water 0.0241)
Transportation coefficient of cooling water (TC_{cdp})	Cooling capacity of cooling water/energy consumption of cooling water pump	41.5	GB 50189–2005 (transportation efficiency of cooling water 0.0241)
Transportation coefficient of cooling tower (TC_{ct})	Heat exhausted by cooling tower/energy consumption of cooling tower	150–200	Sample of cooling tower
Transportation coefficient of FCU (TC_{fc})	Cooling capacity of FCU/energy consumption of FCU	50–60	GB/T 19232–2003 for FCU
Transportation coefficient of fans (TC_{fan})	Cooling capacity of AHU/energy consumption of AHU fans	20	GB/T 17987–2007
Efficiency of cooling plant	Cooling capacity of cooling plant/energy consumption of cooling plant	4.0–4.3	Energy consumption of the cooling plant includes chillers, chilled water pumps, cooling water pumps, and cooling towers
Efficiency of HVAC system (COP_{sys})	Cooling capacity of HVAC system/energy consumption of HVAC system	3.6–3.9	Energy consumption of the HVAC system includes cooling plants and fans

consumption of fans (outdoor air fans, exhaust air fans, fan coils, etc.). The efficiency drops to 3.6–3.9 when the energy consumption of fans is considered, which means the actual efficiency is only 10 % of the ideal efficiency.

Why is there such a significant difference between the ideal efficiency and the actual efficiency? If ideal HVAC systems are examined and compared to actual HVAC systems, four main explanations emerge, summarized as follows:

(A) The actual system uses the same chilled water for cooling and dehumidifying, while the ideal HVAC system is a temperature and humidity independent control (THIC) system. For example, if the indoor air state is 25 °C and 60 % relative humidity, the ideal efficiency of cooling and dehumidifying is over 30 for Beijing's design outdoor condition. In the actual system, if the condensation dehumidification method is used, the efficiency is 17.5, even with the ideal Carnot cycle (between outdoor temperature 33.2 °C and indoor dew point temperature 16.6 °C).

(B) COP of the actual chiller is lower than the efficiency of the ideal Carnot cycle. The thermodynamic perfectness of the actual chiller is around 65 %.

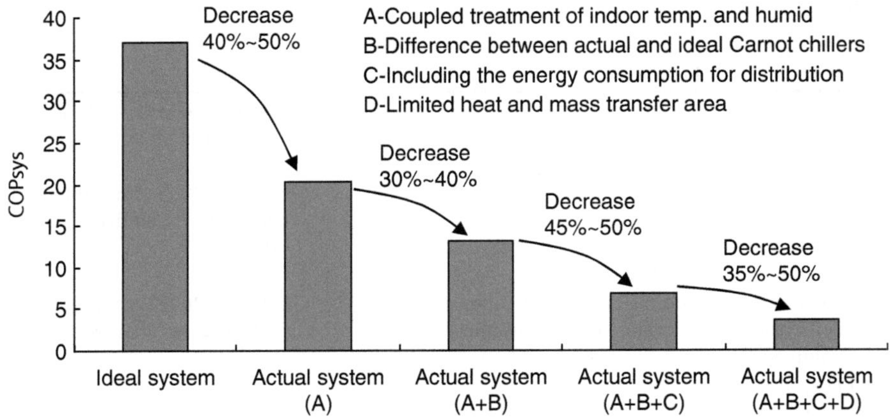

Fig. 2.15 Variance of COP_{sys} in ideal and actual air-conditioning systems

(C) The energy consumption of pumps and fans must be taken into account in actual systems. Energy consumption of transportation represents a large proportion of the total energy consumption of HVAC systems; in some buildings, it can be greater than 50 % (even greater than the chillers' power consumption). Reducing the flow rates of air and water can reduce the energy consumption of fans and pumps, but this will increase the temperature difference in the system such as temperature difference from indoor temperature to the evaporating temperature of the chiller and lower the efficiency of the chillers as a result.

(D) The heat transfer ability of the actual system is lower than that of the ideal system. In actual systems, the heat transfer ability, e.g., the area of the heat exchanger, is proportional to the initial investment of the building. The influence of the heat transfer area on heat transfer efficiency is significant at first, but it decreases when the heat transfer area becomes sufficiently large. The flow rates also influence the performance of the heat exchangers. When the flow rates of both sides of the heat exchanger are unmatched, there always exists heat transfer loss (Zhang et al. 2012e), even if the heat transfer area is infinite. In actual systems, the temperatures of the condenser and evaporator are constant, but the flow rates of the heat transfer fluids are always finite. As a result, heat transfer loss always exists in actual systems.

The above four explanations influence each other. For instance, the COP of the actual chiller is lower than that of the ideal Carnot cycle, which means the condenser capacity of the actual chiller increases compared to the ideal case for the same cooling capacity. If the condenser capacity increases, the heat transfer area of the condenser and the flow rate of the cooling water should be adjusted accordingly. In condensation dehumidification systems, the evaporating temperature is lower than the dew point temperature of the indoor air (which is around 8 °C lower than the dry-bulb temperature of the indoor air). This temperature difference of 8 °C can cover the losses caused by the limited heat transfer area and finite flow rate of the heat transfer fluid to some extent. Figure 2.15 demonstrates the influence of these factors on the efficiency of HVAC systems to some extent. In the ideal system, the chiller is an ideal Carnot cycle, temperature and humidity are controlled separately, the flow rates of air and water are infinite, and the heat transfer ability is infinite, too. However, the efficiency of the actual system is only 10 % of the ideal system.

Based on these four influencing factors, the efficiency of HVAC systems can be improved as follows:

Factor A: Separate the temperature control and humidity control in the HVAC system. This method can avoid the loss associated with the mixing treatment of heat and moisture.

Factor B: Improve the thermodynamic perfectness of the chiller. Currently, the thermodynamic perfectness of chillers can reach 65 %, which is already very high. Therefore, the potential for improvement is limited.

Factor C: Reduce the energy consumption of fans and water pumps. Section 1.3.5 indicates that water is more energy efficient than air for the transportation of the cooling capacity. Utilizing air as the transportation fluid should be avoided, except for necessary outdoor air.

Factor D: Increase the heat transfer ability. Heat transfer ability is limited by initial investment. When the heat transfer efficiency is above 70 %, it will result in greater initial investment (heat transfer area), which will further improve the efficiency.

2.3.2 The Temperature Levels of Actual HVAC Systems

In Fig. 1.4 from Chap. 1, the FCU+OA system depicts the temperature levels of a typical HVAC system. Figure 2.16 shows the temperature change under the summer condition when the indoor temperature and outdoor wet-bulb temperature are 25 °C and 27 °C, respectively. If all the heat sources are mixed into the indoor environment and removed together by the same cooling source, the ideal efficiency of the sensible cooling process (Carnot cycle) is

$$\mathrm{COP}'_{\mathrm{cooling}} = \frac{T_{\mathrm{A}}}{T_{\mathrm{O,w}} - T_{\mathrm{A}}} = \frac{273 + 25}{27 - 25} = 149 \tag{2.16}$$

In the actual HVAC system, the evaporating temperature and the condensing temperature of the actual chiller are 5 °C and 37 °C, respectively. Under this condition, the ideal efficiency of the Carnot cycle is

$$\mathrm{COP}_{\mathrm{Carnot}} = \frac{T_{\mathrm{evap}}}{T_{\mathrm{cond}} - T_{\mathrm{evap}}} = \frac{273 + 5}{37 - 5} = 8.7 \tag{2.17}$$

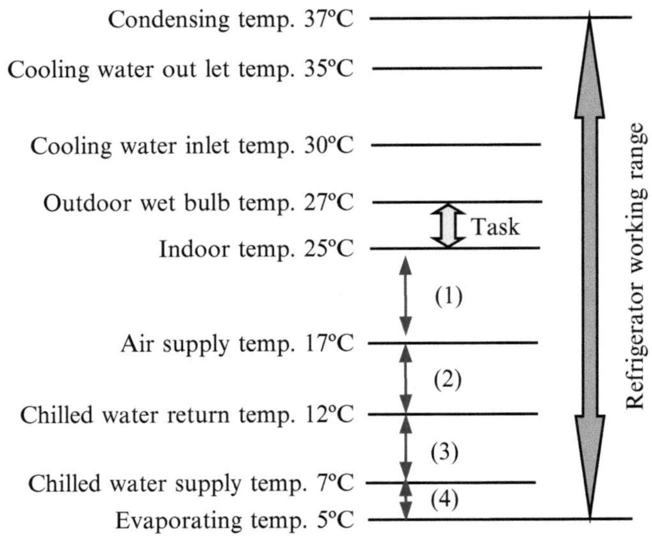

Fig. 2.16 Temperature levels of the central HVAC system under the cooling and dehumidification condition in summer

In order to show the difference between the actual efficiency and the ideal efficiency, a coefficient γ is defined as the ratio of the actual efficiency to the ideal Carnot efficiency:

$$\gamma_{\text{system}} = \frac{\text{COP}_{\text{actual}}}{\text{COP}'_{\text{cooling}}} \times 100\ \% \tag{2.18}$$

$$\gamma_{\text{chiller}} = \frac{\text{COP}_{\text{actual}}}{\text{COP}_{\text{Carnot}}} \times 100\ \% \tag{2.19}$$

For an evaporating temperature of 5 °C and a condensing temperature of 37 °C, the COP of the actual chiller is 5.5–6.0. The γ of the chiller is already 63–69 %, but the γ of the HVAC system is only 4 %, which indicates that the chiller performance is approaching the Carnot limit, even though the performance of the HVAC system is quite low. Considering the temperature differences in Fig. 2.16, segment (4) results in 2 °C of temperature difference loss, which is limited by the heat transfer area of the evaporator; segment (3) results in 5 °C of temperature difference loss, which is limited by the flow rate of the chilled water. If this temperature difference loss is reduced, the energy consumption of the chilled water pump will increase. Segment (2) corresponds to the temperature difference between the chilled water and the supply air, which is limited by the heat transfer area of AHU; segment (1) is limited by the flow rate of the supply air. If the temperature difference loss is reduced, the energy consumption of the supply air fan will increase.

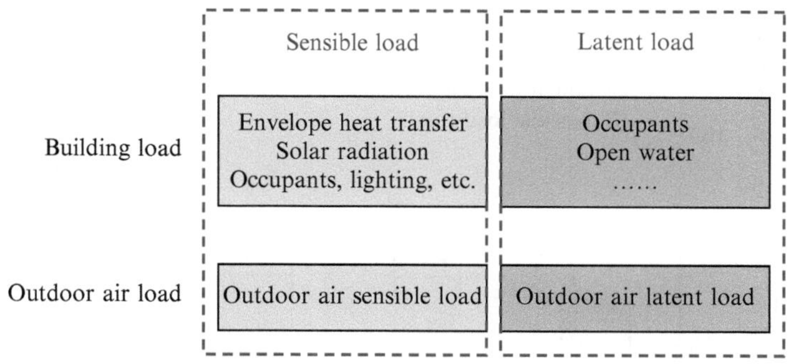

Fig. 2.17 The cooling load composition in buildings

In current HVAC systems, the heat is transferred from the refrigerant through the chilled water and the supply air to the conditioned space, where each heat transfer process causes a temperature difference. When all the temperature differences add up, this results in a low evaporating temperature and a high condensing temperature. Moreover, if the sensible and moisture loads are removed simultaneously, the temperature of the cooling source needs to be lower than the dew point of the indoor air. Together, these reasons account for the low efficiency of current HVAC systems.

2.3.3 The Ratios of Sensible and Moisture Loads in Practical Buildings

Figure 2.17 demonstrates the proportion of cooling loads in typical buildings. The sensible cooling load consists of the heat transferred through the building envelope, solar radiation, occupants, equipment, lights, the sensible cooling load of the outdoor air, etc. The moisture load consists of indoor humidity sources and the outdoor air moisture load. The cooling coil is the most commonly used piece of equipment for cooling and dehumidifying the air in HVAC systems. Cold and dry air is supplied to the indoor environment in order to maintain comfortable indoor temperature and humidity. For the purpose of cooling, any cooling source with a temperature lower than the indoor temperature (25 °C) can be used to cool the indoor air. If the temperature difference of transportation is considered, cooling sources around 15–18 °C can satisfy this requirement. For the purpose of dehumidification, the temperature of the cooling source should be lower than the dew point of the indoor air (16 °C) with condensation dehumidification. Taking the temperature difference of transportation into account, the temperature of the chilled water has to be as low as about 7 °C.

In most cases, the sensible cooling load accounts for the majority of the total cooling load. Figure 2.18a shows the cooling load of typical office buildings located in different cities (the model is described in detail in Appendix C). The latent

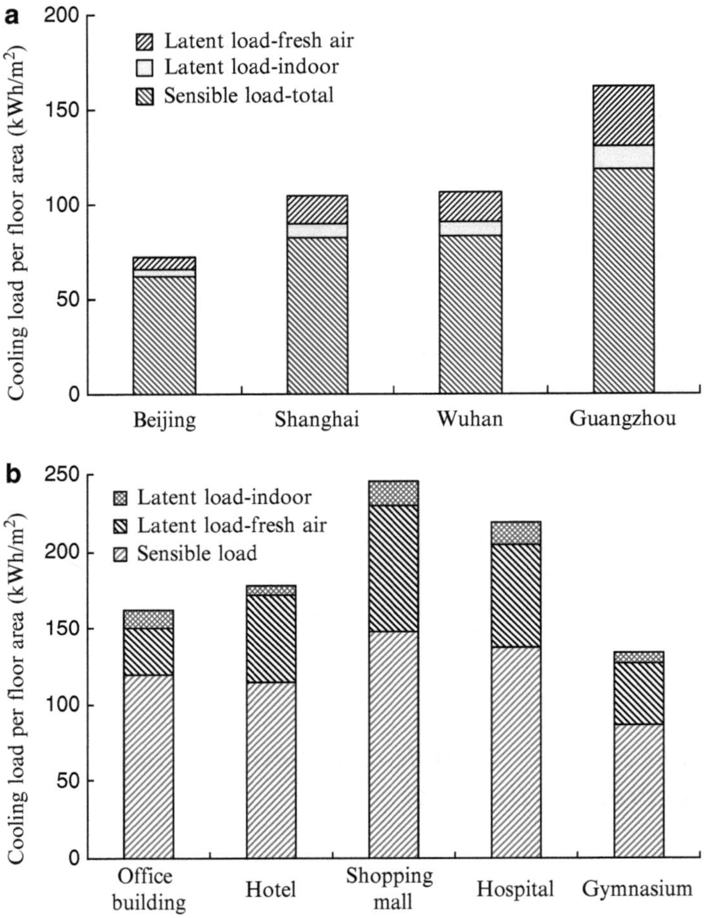

Fig. 2.18 Cooling loads of typical buildings: (**a**) typical office buildings in different locations and (**b**) different building types in Guangzhou

cooling load is within 30 % of the total cooling load. Figure 2.18b shows the cooling load of different building types in Guangzhou. The moisture load is 30–40 % of the total cooling load, and the sensible cooling load is more than 50 % of the total cooling load in the buildings, which can be treated by high-temperature cooling sources. In current systems, the combined handling process of sensible and moisture loads will result in unnecessary waste of the cooling source, such as 7 °C chilled water.

If sensible cooling were the only purpose of HVAC systems, the temperature of the chilled water would not be restricted by the dew point temperature of the indoor air. Chilled water with a temperature lower than the dry-bulb temperature of the air can be used, which can be much higher than 7 °C considering the temperature difference of the heat transfer and transportation process. High-temperature chilled

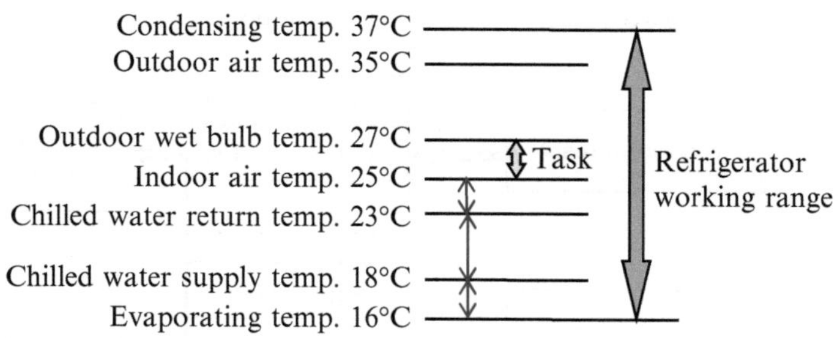

Fig. 2.19 Temperature difference in the temperature control subsystem in HVAC systems

Table 2.9 Energy consumption comparison of the THIC system and the conventional system

System	$COP'_{cooling}$	COP_{Carnot}	COP_{actual}	$\gamma_{chiller}$	γ_{system}
Figure 2.16	149	8.7	5.5–6.0	63–69 %	4 %
Figure 2.19	149	13.8	8.5–9.0	62–65 %	6 %

water increases the evaporating temperature of the chiller and the chiller's COP. For example, if the entire cooling load is generated from a 28 °C indoor surface, 18/23 °C chilled water can be used to absorb the heat from the heat source. The corresponding evaporating temperature can be as high as 16 °C, as indicated in Fig. 2.19.

The COP of the Carnot chiller can be 13.8 with a condensing temperature of 37 °C and an evaporating temperature of 16 °C, while the COP of the actual chiller ranges from 8.5 to 9.0. Table 2.9 shows a comparison between the THIC system and the conventional HVAC system. The energy consumption of the chiller can be 40 % lower due to a higher COP.

2.4 The Core Idea of Temperature and Humidity Independent Control

2.4.1 Operating Principle of the THIC Air-Conditioning System

The basic tasks of air-conditioning systems are to extract indoor sensible heat, moisture, CO_2, etc., effectively. For these tasks, different solutions are proposed as follows:

- The extraction of sensible heat can be achieved by adopting various approaches; it is not limited to the direct contact method. Radiant heat transfer is applicable for heat extraction, as well as convective heat transfer.

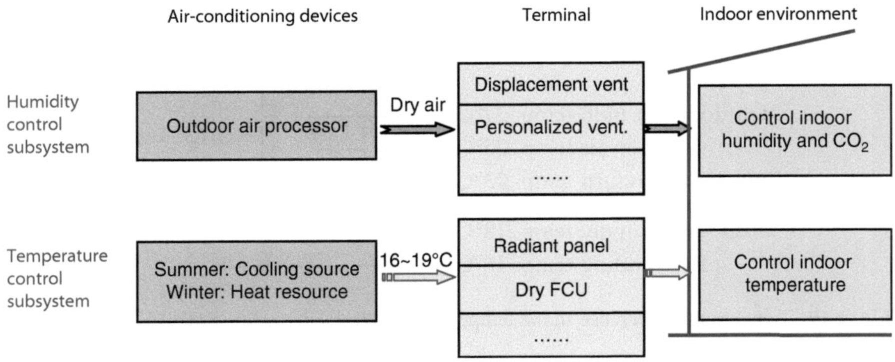

Fig. 2.20 Operating principle of the THIC air-conditioning system

- The extraction of indoor moisture, CO_2, or other gases has to be achieved by supplying air with a lower humidity ratio or concentration. This is a mass transfer process between the indoor air and the supply air.

As indicated in Sect. 2.1.3, the required outdoor air flow rate and variation tendency of the task of indoor moisture extraction are consistent with those of the task of indoor CO_2 extraction. The indoor temperature can therefore be regulated by an independent system, and the required temperature of the cooling source increases compared to that in the conventional system.

Based on the analysis of the conventional air-conditioning system in Sect. 1.3, the temperature and humidity independent control (THIC) air-conditioning system is proposed as an effective solution. Figure 2.20 illustrates the operating principle of the THIC air-conditioning system, with an outdoor air handling subsystem and a relatively high-temperature cooling source subsystem that can separately regulate the indoor temperature and humidity, respectively. As indoor temperature and humidity are regulated by independent subsystems, the THIC system can satisfy the variance of the indoor heat moisture ratio. Therefore, the THIC system can avoid the imbalance of indoor parameters, and it addresses the conventional system's inability to meet temperature and humidity requirements simultaneously. Table 2.10 shows a detailed comparison of THIC systems and conventional systems.

In the humidity control subsystem, the handled outdoor air assumes the two main tasks of the air-conditioning system, i.e., supplying enough outdoor air to meet the requirements of indoor air quality and supplying sufficiently dry outdoor air to remove the total moisture load in the building to control the indoor humidity. The humidity control subsystem consists of the outdoor air handling processors and the indoor air supply terminals. As the task can only be completed by supplying air, outdoor air is usually chosen as the medium; however, the required air flow rate will be much lower than in the CAV or VAV systems, where air is used for extracting the entire load.

As for the indoor sensible load, another subsystem with a relatively higher-temperature cooling source can be utilized to control the indoor temperature. To reduce the energy consumption of transportation systems, the minimum air

Table 2.10 Basic characteristics of the THIC air-conditioning system

	Problem in conventional systems	THIC air-conditioning system
1	Loss in coupled temperature and humidity handling process	Avoids this kind of loss; regulates indoor temperature and humidity separately
2	Offset in cooling and heating, dehumidification, and humidification	Avoids offset
3	Insufficiently accommodates variance of heat moisture ratio	Separate subsystems are responsible for temperature and humidity; accurate regulation
4	Indoor terminals	The same terminal is adopted both for cooling and heating; radiant terminals are a feasible solution
5	High energy consumption of transportation	Supplying air is only for moisture extraction; water or refrigerant is recommended as the temperature control medium
6	Influence on indoor air quality	Indoor sensible terminals operate under dry conditions without condensation

flow rate is recommended because water pumps consume only 10–20 % of the energy consumed by fans when transporting the same cooling or heating capacity. Therefore, either water or refrigerant, but not air, is the recommended fluid in the temperature control subsystem. The cooling source temperature needed for temperature control could be increased to about 16–18 °C, which is much higher than that of conventional systems. This makes it more convenient to adopt a natural cooling source with the appropriate temperature; if this is not permitted, the mechanical refrigeration method can satisfy the requirement. The energy efficiency of the mechanical cooling source will be improved considerably due to the increased evaporating temperature. Both radiant terminals and common FCUs are feasible for extracting the sensible load. Moreover, there will be no risk of condensation, as a high-temperature cooling source is adopted and the indoor humidity is regulated by the other subsystem.

Figure 2.21a illustrates the operating schematic of a THIC system. Taking an outdoor air processor and a dry FCU as examples, the air handling processes in this THIC system are illustrated in Fig. 2.21b. The outdoor air is dehumidified to a sufficiently dry state, and then supplied to the conditioned space to extract the indoor moisture, which is quite different from the conventional system. For the outdoor air handling processor, the high-temperature cooling source, and the indoor terminals, there are different requirements in the THIC system compared to conventional systems.

2.4.2 *Annual Handling Requirements of Outdoor Air*

2.4.2.1 Determining the Outdoor Air Flow Rate

The original intention of supplying outdoor air is to meet the occupants' health requirement. Furthermore, the supplied outdoor air in the THIC system is also

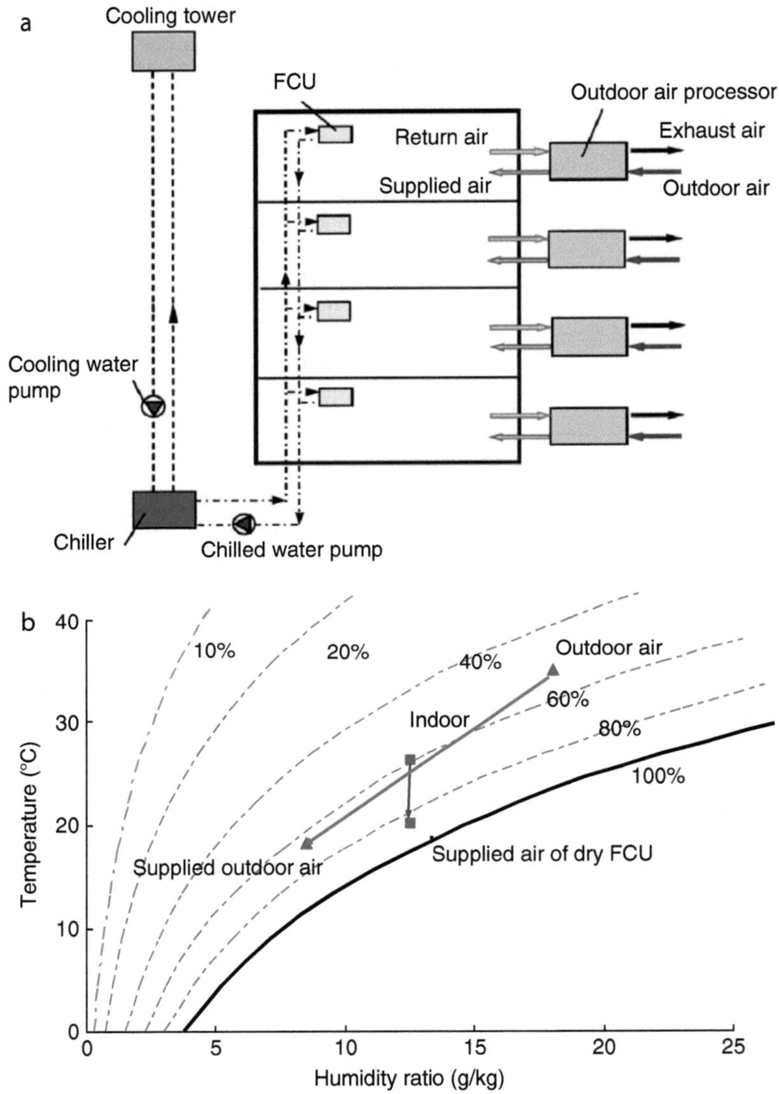

Fig. 2.21 Typical air handling process of the THIC air-conditioning system: (**a**) schematic diagram of the system and (**b**) air handling process in psychrometric chart

responsible for removing indoor moisture. The selection of outdoor air flow rate should follow the principles listed below.

1. The selected outdoor air flow rate should meet satisfy the minimum requirement according to relevant codes and standards.
2. The selected outdoor air flow rate should be sufficient for extracting the indoor moisture. As the outdoor air flow rate is determined, the humidity ratio of the supplied air (ω_s) should be low enough to extract all the indoor moisture

produced from the moisture sources. The relation between the required supply air humidity ratio (ω_s) and the indoor design humidity ratio ($\omega_{a,in}$) can be expressed by Eq. (2.20):

$$\omega_s = \omega_{a,in} - \frac{W}{\rho G} \tag{2.20}$$

where W is the indoor total moisture generation, g/h; G is the chosen outdoor air flow rate, m^3/h; and ρ is the outdoor air density, kg/m^3.

3. The chosen outdoor air flow rate should take into account the handling ability of the outdoor air processor. The required supply air humidity ratio based on the flow rate must not exceed the handling ability. In addition, the relation between the air flow rate and the energy efficiency of the handling device needs to be checked during the design process to maximize energy performance.

In order for the minimum outdoor air flow rate to meet the requirement for occupants' health with predetermined indoor parameters, the required supply air humidity ratio will be lower when a lower flow rate is selected, which makes it more challenging for the outdoor air handling device. Otherwise, if a higher flow rate is chosen, the required supply air humidity ratio will be higher, leading to a lower handling ability demand for the outdoor air handling device. Consequently, the loads of the outdoor air and fan power will increase as a result of the higher air flow rate. Thus, the determination of the outdoor air flow rate should have an overall consideration of these factors during system design. Furthermore, it is better to regulate the outdoor air flow rate according to the variance of moisture sources, just like the number of occupants. This can also avoid supplying too much dry air, leading to energy dissipation as the flow rate is beyond the requirements for occupants' health and extracting moisture.

2.4.2.2 Parameter Requirements in Summer

Required Supply Air Humidity Ratio

In the THIC system, the outdoor air is chosen as the medium for removing the indoor moisture, and the supplied air humidity ratio should be sufficiently dry. There are numerous indoor moisture sources, and their variances may be quite different. If occupants are the main moisture sources, the moisture generation is influenced by their activity intensity and the indoor design state, while the humidity ratio of the supply air (ω_s) is determined by the outdoor air flow rate, as shown in Eq. (2.20). Figure 2.22a illustrates the indoor design humidity ratio as a function of the indoor temperature and relative humidity ratio (RH_{in}). Taking an office as an example, the required ω_s during the cooling season as a function of different indoor states is shown in Fig. 2.22b, with an outdoor air flow rate of 30 m^3/h per person when only occupants are taken into account. If the outdoor air flow rate varies, the

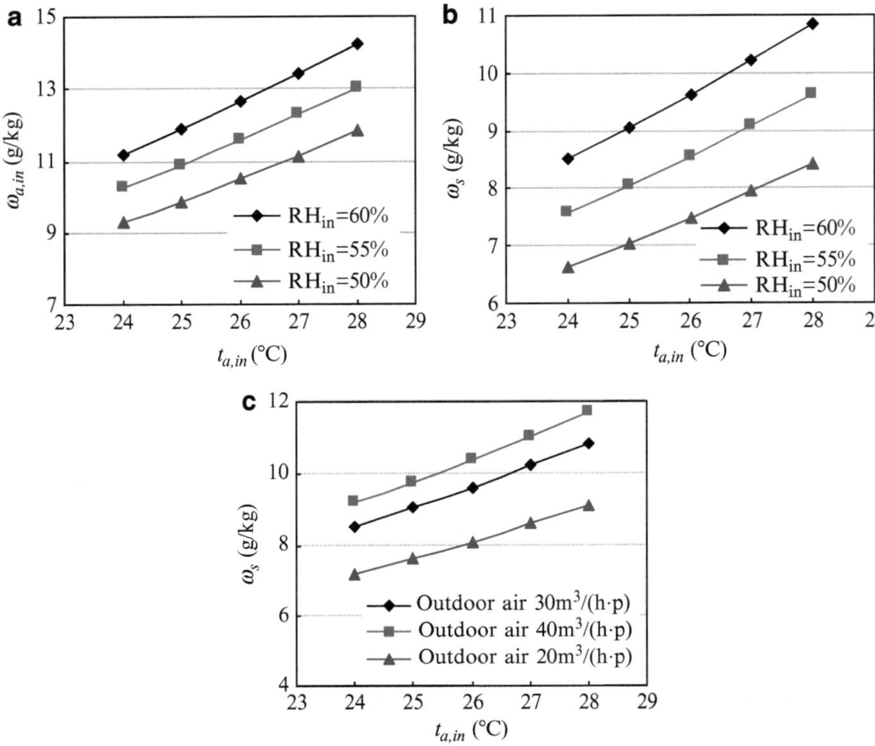

Fig. 2.22 $\omega_{a,in}$ and required ω_s for different indoor states (cooling season): (**a**) indoor design $\omega_{a,in}$, (**b**) ω_s with 30 m^3/h outdoor air per person, and (**c**) ω_s with varying outdoor air rates per person

required ω_s for extracting the indoor moisture will be different. Figure 2.22c shows the required ω_s with a varying outdoor air flow rate per person and a RH_{in} of 60 %. As indicated by the figure, the required ω_s values are 8.1, 9.6, and 10.4 g/kg for the outdoor air flow rate per person of 20 m^3/h, 30 m^3/h, and 40 m^3/h, respectively, with an indoor temperature of 26 °C and RH_{in} of 60 %.

Required Supply Air Temperature

The outdoor air has to be handled to a state that is sufficiently dry to extract the indoor moisture in the THIC system, indicating that the building moisture load is extracted by the processed outdoor air. Meanwhile, if the supplied outdoor air temperature is different from the indoor temperature, the indoor sensible load will be affected. When supply air temperature t_{sa} is lower than indoor temperature $t_{a,in}$, part of the indoor sensible load is then extracted by the humidity control subsystem, and the temperature control subsystem is responsible for the remainder of the total indoor sensible load. Figure 2.23 shows the load composition of the THIC system when t_{sa} is lower than $t_{a,in}$.

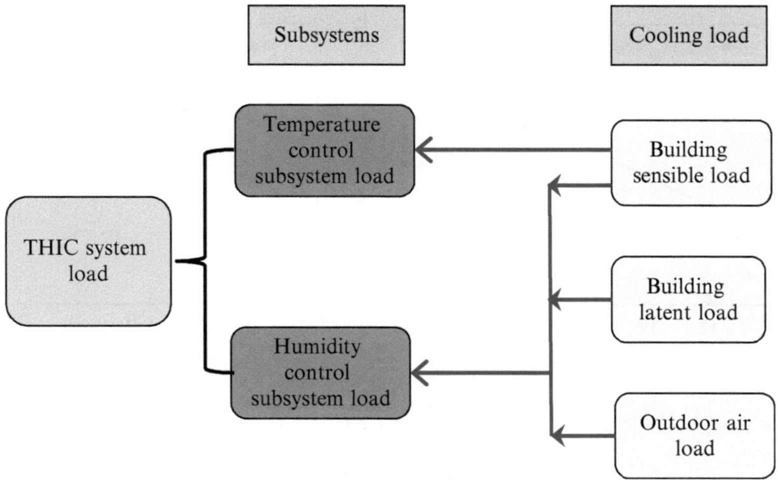

Fig. 2.23 Load composition of the THIC system ($t_{sa} < t_{a,in}$)

Table 2.11 Sensible load assumed by the supplied outdoor air

	$t_{a,in}$ = 26 °C		$t_{a,in}$ = 25 °C		$t_{a,in}$ = 24 °C	
t_{sa} (°C)	Q_{person} (W)	Q_{sa} (W)	Q_{person} (W)	Q_{sa} (W)	Q_{person} (W)	Q_{sa} (W)
17	61.0	90.5	66.0	80.4	70.0	70.4
18		80.4		70.4		60.3
19		70.4		60.3		50.3
20		60.3		50.3		40.2
21		50.3		40.2		30.2
22		40.2		30.2		20.1

For common office buildings, the required outdoor air flow rate can be regarded as proportional to the number of occupants. Therefore, the sensible load extracted by the supplied outdoor air is proportional to the sensible load generated by occupants. Thus, it would be advantageous to extract the sensible load of occupants by the supplied outdoor air, and its feasibility is investigated here. Table 2.11 lists the sensible load extracted by supplied outdoor air in a typical office building with a varying indoor temperature of 24–26 °C and an outdoor air flow rate of 30 m^3/h per person. As indicated by this table, for indoor temperatures of 25 and 26 °C, the sensible load extracted by the supplied outdoor air (Q_{sa}) corresponds to the occupants' sensible load (Q_{person}) if t_{sa} is 18 °C and 20 °C, respectively. In other words, the required t_{sa} of the supplied outdoor air is 18 and 20 °C for extracting occupants' sensible load at an indoor temperature of 25 °C and 26 °C, respectively.

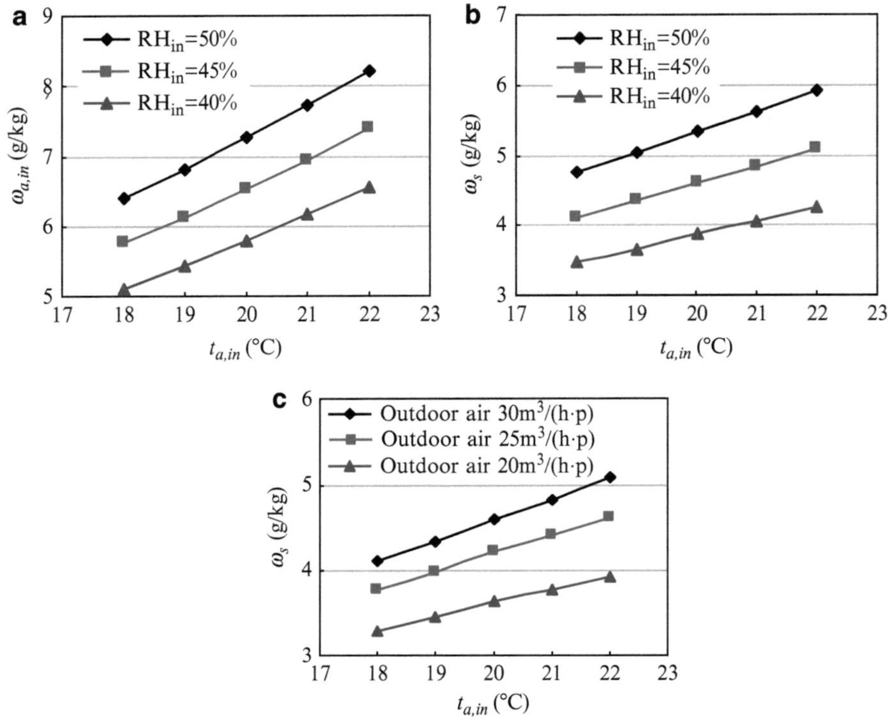

Fig. 2.24 $\omega_{a,in}$ and the required ω_s for different indoor states (heating season): (**a**) indoor design $\omega_{a,in}$, (**b**) ω_s with 30 m^3/h outdoor air per person, and (**c**) ω_s with varying outdoor air rates per person

2.4.2.3 Parameter Requirements in Winter

The temperature and humidity ratio of the outdoor air are both relatively low in winter, and the outdoor air is usually heated and humidified before being supplied to the conditioned space. In the THIC system, the required ω_s is calculated by Eq. (2.20) based on the indoor design state and the outdoor air flow rate per person. The indoor design humidity ratio varying with the indoor temperature and RH_{in} is shown in Fig. 2.24a. Taking an office as an example, Fig. 2.24b shows the required ω_s during the heating season as a function of different indoor states, with an outdoor air flow rate of 30 m^3/h per person when only occupants are taken into account. The required ω_s for extracting the indoor moisture varies with the supplied outdoor air flow rate. The required ω_s is shown in Fig. 2.24c, with a varying outdoor air flow rate per person and a RH_{in} of 45 %.

As indicated by the figure, the required ω_s values are 3.6, 4.2, and 4.6 g/kg as the outdoor air flow rate per person is 20 m^3/h, 25 m^3/h, and 30 m^3/h, respectively, with $t_{a,in}$ of 26 °C, RH_{in} of 45 %, and $\omega_{a,in}$ of 6.5 g/kg. Taking the outdoor design air humidity ratio (0.7 g/kg) of Beijing in winter as an example, the supplied outdoor air is supposed to extract moisture; however, for the outdoor air handling device, it works as a humidification process.

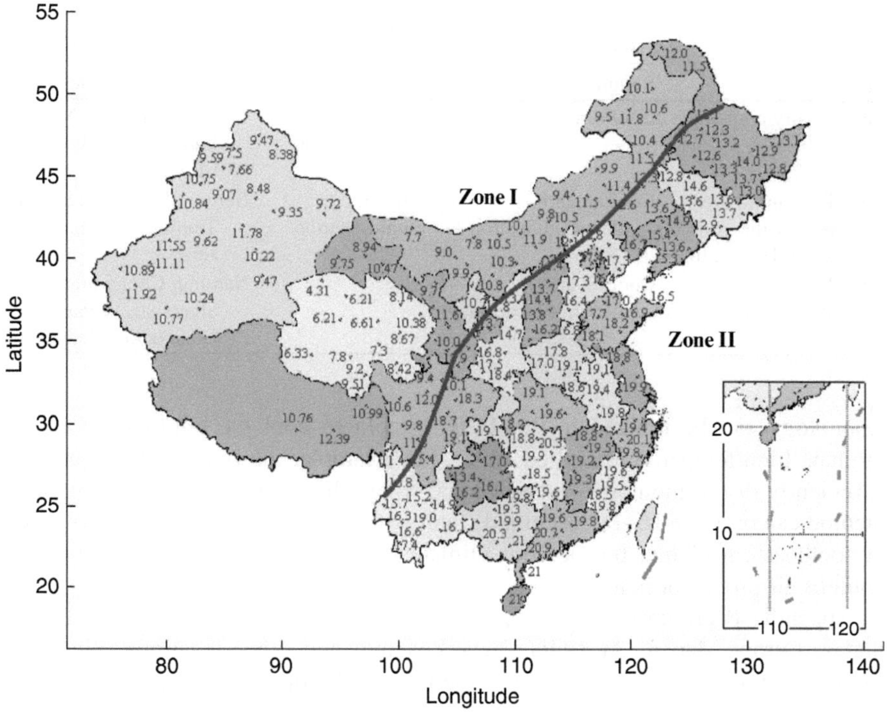

Fig. 2.25 Monthly average outdoor air humidity ratio during the most humid month in China (g/kg)

As for the supplied outdoor air temperature in winter, it should be able to accommodate the requirement for thermal comfort. Generally speaking, the supply air temperature can be 2–3 °C higher than that of the indoor state in winter, which should be determined by the actual device and based on other considerations.

2.4.3 Analysis of Global Outdoor Climate Conditions

Various kinds of devices can be combined together to form the THIC system, and different natural resources can be utilized for different climate regions. The outdoor conditions in Chinese cities vary tremendously, and Fig. 2.25 shows the monthly average outdoor air humidity ratio during the most humid month (Meteorological Information Center of China Meteorological Administration and Tsinghua University 2005). According to the average humidity ratio of 12 g/kg, climates can be classified as humid climates or dry climates. Table 2.12 lists the climate regions and representative cities in China. In northwestern China, the outdoor climate is sufficiently dry, and the main task of the outdoor air processor is to cool the outdoor air rather than dehumidify it. In dry areas, direct or indirect evaporative cooling

Table 2.12 Climate regions and representative cities in China

	Handling requirement for outdoor air		
Regions	Summer	Winter	Representative cities
Zone I – dry area	Cooling	Heating and humidification	Urumchi, Hohhot, Xining, Lanzhou, Kashgar, Lhasa, Turpan
Zone II – humid area (north of the Qinling-Huaihe line)	Cooling and dehumidification	Heating and humidification	Beijing, Harbin, Shenyang, Tianjin, Dalian, Xi'an, Jinan
Zone II – humid area (south of the Qinling-Huaihe line)	Cooling and dehumidification	–	Nanjing, Chongqing, Shanghai, Wuhan, Guangzhou, Shenzhen

technology can be utilized in the THIC system to cool the outdoor air and to produce high-temperature chilled water. For example, the outdoor air is already sufficiently dry in the northwest of China, where the indirect evaporative cooling method can produce 16–18 °C chilled water with high energy efficiency. However, in southeastern China, the outdoor climate is humid, and the main task of the outdoor air processor is to dehumidify it efficiently.

Appendix B presents the summer outdoor design humidity ratios of the United States, Europe, Japan, Australia, New Zealand, and other locations. Since the summer ambient air is relatively dry, the major mission of outdoor air processors in the western United States, Northern Europe, southwestern Australia, New Zealand, and other places is to cool the air. When the outdoor air is humid, the main task of the outdoor air processor is to dehumidify the air efficiently.

This paragraph analyzes the outdoor air handling requirements in winter. As part of the national standard for HVAC in China, where district heating is provided only in the north, the humid region can be divided further into two areas. The boundary is the Qinling-Huaihe line, which is an important geographical boundary in China. These two main regions are distinguished in Table 2.12, and the handling requirements for outdoor air in these regions are also listed. Figure 2.26 depicts the monthly average outdoor humidity ratio in the coldest month of China. As indicated by the figure, the outdoor air humidity ratio is relatively low (lower than 1.5 g/kg in most cities), which is lower than the required humidity ratio of the supply air. Thus, the outdoor air has to be heated and humidified before being supplied to the conditioned space.

Based on these analyses, the annual handling requirements for the outdoor air in different climate regions of China can be summarized as follows:

- Zone I (the dry region in the northwest): The humidity ratio of the outdoor air is relatively low in summer, and the major requirement for the outdoor air handling process is to cool the air; however, the outdoor air needs to be heated and humidified in winter.
- Zone II (north of the Qinling-Huaihe line): The outdoor air has to be cooled and dehumidified in summer, and it needs to be heated and humidified in winter.

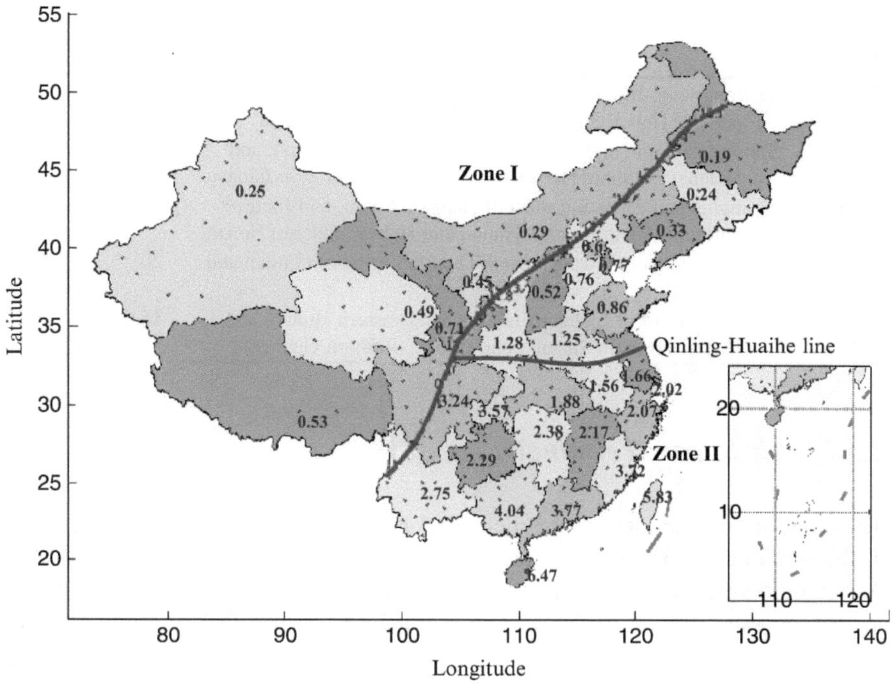

Fig. 2.26 Monthly average outdoor air humidity ratio during the coldest month (g/kg)

- Zone II (south of the Qinling-Huaihe line): The outdoor air has to be cooled and dehumidified in summer, and there is almost no heating requirement for the outdoor air in winter.

2.5 Requirements of Devices Needed for the THIC System

2.5.1 Sensible Heat Terminals

In the THIC system, the required chilled water temperature for indoor temperature control is about 16–18 °C (compared to 7 °C in conventional systems), and extracting the indoor sensible load effectively with a high-temperature cooling source is the new focus of the indoor terminals. For different indoor sensible heat sources, including the inner surfaces of external walls, surfaces of human bodies and clothes, equipment, and lights, there are several key issues to address. These issues include how to select the appropriate approaches for heat extraction according to temperature characteristics and distribution, how to extract the heat from sources directly, how to reduce the links in the heat transfer process, and how to increase the efficiencies of indoor terminals for heat extraction.

Table 2.13 Underground water temperature for different regions (Zhu et al. 1994)

Zone	Regions	Temperature (°C)
Zone A	Heilongjiang, Jilin, and Inner Mongolia; most of Liaoning; parts of Hebei, Shanxi, and Shaanxi; eastern Ningxia	6–10
Zone B	Beijing, Tianjin, and Shandong; most of Hebei, Shanxi, and Shaanxi; southern Henan; eastern Qinghai; northern Jiangsu	10–15
Zone C	Shanghai and Zhejiang; most of Jiangxi, Anhui, and Jiangsu; northern Fujian; eastern Hunan and Hubei; southern Henan	15–20
Zone D	Guangdong and Taiwan; most of Guangxi; southern Fujian and Yunnan	20
Zone E	Guizhou; most of Sichuan and Yunnan, western Hunan, and Hubei; southern Shaanxi and Gansu; northern Guangxi	15–20

2.5.2 *Air Supply Terminals*

The outdoor air is adopted as the medium for extracting the indoor moisture in the THIC system and is responsible for extracting CO_2 as well as pursuing a good IAQ. As the outdoor air is only supplied for the requirement of IAQ and for extracting indoor moisture, the air exchange rate is much lower than that of all-air systems (i.e., CAV or VAV systems). For example, if the outdoor air flow rate is 30 m^3/h and the area is 8 m^2 per person, the air exchange rate is about 1.2 per hour, with a floor height of about 3 m. Designing the air distribution to supply the outdoor air effectively is crucial for the indoor air supply terminals.

Given that the outdoor air is supplied only for the IAQ requirement and for extracting indoor moisture, there are several important concerns for the indoor air supply terminals in THIC systems. These include how to arrange the locations of air supply terminals, how to design the indoor air distribution system, and how to regulate the air flow rate (varying with the number of occupants) so as to extract the indoor moisture and pollutants with high efficiency.

2.5.3 *High-Temperature Cooling Sources*

In conventional systems, chilled water with a temperature of about 7 °C is utilized for both sensible cooling and dehumidification. In contrast, the required temperature of the chilled water could be increased to 16–18 °C for cooling in the temperature subsystem of the THIC system. This temperature of the cooling source makes it feasible to utilize natural cooling sources, such as deep phreatic water, high-temperature chilled water collected through soil heat exchangers, and direct or indirect evaporative cooling methods in some dry regions. The underground water temperatures for different regions in China are listed in Table 2.13, which indicates that there are a number of regions where the temperature of the underground water could satisfy the temperature requirement for cooling in the THIC system.

If the natural cooling source cannot satisfy the cooling temperature requirement, the vapor compression refrigeration cycle will then be adopted. The temperature increase for chilled water helps to increase the evaporating temperature of the refrigeration cycle, resulting in a decrease in the compression ratio under a similar condensing condition. Based on the characteristics of the reverse Carnot cycle, the ideal COP of the refrigeration cycle is then increased by a large margin. For example, if the evaporating temperature increases from 5 °C in the conventional system to about 14–16 °C in the THIC system, the ideal COP of the reverse Carnot cycle will be improved from 8.7 to 13, with a condensing temperature of 37 °C. Furthermore, for the current chillers adopting the vapor compression cycle, improving the structures of key components to achieve relatively satisfactory efficiency under a lower compression ratio is a key issue for both researchers and manufacturers.

2.5.4 Outdoor Air Handling Devices

For the air handling devices of the humidity control subsystem, the major task is to supply the conditioned space with sufficiently dry air to extract the indoor moisture. In the humid region, the outdoor air has to be dehumidified in summer; there are various dehumidification methods for the air handling process, including condensation dehumidification, liquid desiccant dehumidification, and solid desiccant dehumidification. Selecting the proper dehumidification method to fit the handling requirements and improving the performance of the handling process are the main considerations in determining the air handling devices for the indoor humidity control of THIC systems. For the dry region, the main task for handling the outdoor air is to cool it (rather than to dehumidify it) in summer. The remaining task is to choose the most effective solution for cooling.

If indoor exhaust air is available, the heat recovery device operating between the outdoor air and the indoor exhaust air facilitates the energy recovery process and reduces the energy consumption of handling the outdoor air. Then, the main consideration is to choose a heat recovery method with satisfactory efficiency to avoid cross-contamination. In current air handling units, there are usually different sections with various functions that work together to meet the air handling requirement, leading to complex devices and regulating solutions. Are there any outdoor air handling processes that can satisfy the annual handling requirement with greater simplicity?

2.6 Review on Relative Research Concerning THIC Systems

2.6.1 Relevant Research Progress of THIC Air-Conditioning Methods

The idea of temperature and humidity independent control for air-conditioning offers numerous solutions and handling processes, and both feasible and creative

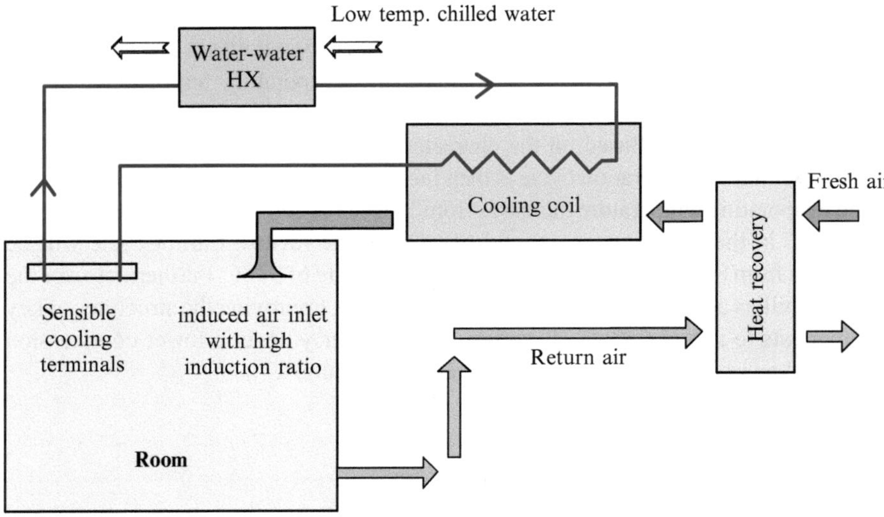

Fig. 2.27 System schematic of DOAS combined with an ice storage method

approaches for the construction of the built environment are being explored all the time. Some researchers have suggested effective methods that adopted the idea of THIC for buildings with different functions and dimensions, in which different devices are utilized to achieve the independent control of indoor temperature and humidity. The following subsection summarizes these existing THIC solutions.

2.6.1.1 The Dedicated Outdoor Air System (DOAS)

The dedicated outdoor air system (DOAS) was first proposed in the United States in the 1990s (Yin 2003). It is an air-conditioning solution that combines a low-temperature air diffuser with other sensible cooling devices. The condensation dehumidification method is adopted to handle the outdoor air. In order to dehumidify the air to a sufficiently dry humidity ratio, the outdoor air handling process can be combined with an ice storage approach, with a supplied chilled water temperature as low as 4 °C. If there is no ice storage, the direct expansion method can be adopted for dehumidification.

Different devices are adopted for humidity control and temperature control. It incorporates the low-temperature air supply devices and sensible cooling terminals where the outdoor air is processed to a state dry enough to satisfy the requirement of extracting the indoor moisture load. As the temperature of the handled air is usually quite low, an induced air inlet with a high induction ratio is recommended to relieve thermal discomfort as much as possible. Then, the sensible terminals are responsible for extracting the remaining sensible load and regulating the indoor temperature. The operating schematic of this kind of DOAS is shown in Fig. 2.27. The typical operating parameters of the DOAS system can be summarized as follows:

the indoor design temperature is 25 °C with a relative humidity of 40 %, and the corresponding dew point is 10.5 °C. The chilled water can be cooled to as low as 2.5 °C and then distributed to the cooling coil to dehumidify the outdoor air. The outlet water temperature can be as high as 11 °C, and the chilled water is supplied into the sensible terminal (e.g., a radiant panel) with an outlet chilled water temperature of about 14 °C. This kind of system configuration utilizes chilled water in series to help increase the circulating water temperature difference and reduce the water flow rate. The inlet water temperature for the sensible terminal is higher than the indoor dew point, ensuring no risk of condensation.

The original intention for proposing DOAS in the United States was to overcome the failure to satisfy the ASHRAE standard on outdoor air (Meckler 1987; "Dedicated outdoor air systems"; Mumma 2001). With the help of DOAS, the outdoor air could be supplied to the indoor environment according to the requirement and the standard, avoiding the insufficient flow rate of the outdoor air common in conventional VAV systems. DOAS conforms to the ASHRAE standard on outdoor air in an energy-efficient manner, and both operating cost savings and significant improvements in indoor air quality and thermal comfort are achieved. DOAS developed quickly, and many applications have been investigated. As indicated by the system configuration and air handling method, DOAS is a kind of THIC air-conditioning system that adopts handled air that is sufficiently dry for indoor humidity control and adopts low-temperature supplied air or radiant terminals for indoor temperature control.

For the kind of DOAS adopting low-temperature air supply terminals, the utilization of chilled water in series both for dehumidification and in sensible terminals leads to a discord between the outdoor air climate and the chilled water flowing into the sensible terminal. Although the chilled water temperature could be regulated if a heat exchanger were adopted, the chilled water temperature is still quite low. Moreover, conventional chillers are adopted in DOAS for both humidity control and temperature control; no special chillers supplying high-temperature chilled water are utilized for temperature control. This limits the further improvement of the energy performance of DOAS.

2.6.1.2 Radiant Method for Air-Conditioning

Radiant cooling terminals, an effective way to extract sensible heat for temperature control, are springing up all over Europe (Bjarne 2008; IEA-ECBCS Annex 37, IEA-ECBCS Annex 49). If chilled water of an appropriate temperature is imported into the radiant terminals, the indoor sensible heat can be extracted through the radiant heat exchanger and natural convection. One significant reason for the popularity of radiant terminals in Europe is that a suite of terminals can be used for both cooling and heating. The outdoor climate in most parts of Europe is dry in summer, and the dehumidification requirement can often be omitted. The major air-conditioning demand is cooling or regulating indoor temperature. Thus, there is little risk of condensation in the application of radiant cooling in summer. Thanks especially to research on low exergy systems in the Annexes (Annex 37 and Annex 49) of the International Energy Agency's (IEA) Implementing Agreement on Energy in Buildings and Communities Programme (EBC, formerly ECBCS), radiant terminals are seen as good choices for low-temperature heating systems and high-temperature cooling systems. Radiant cooling has been adopted in some buildings in Europe, and there have been a number of theoretical and on-site performance calculations and investigations.

The outdoor condition in most of Europe is sufficiently dry that condensation on the surface of radiant terminals in summer is not a major concern. However, if the outdoor air is humid, like it is in the southeast of China, humidity control and temperature control are both required for air-conditioning systems in summer. If radiant cooling is adopted for temperature control, humidity control will become a more challenging task. There is a risk of condensation on the surface of the radiant terminal if the dew point related to the indoor humidity ratio is higher than the surface temperature. Thus, a humidity control device or subsystem is required to regulate the indoor humidity effectively, which can avoid condensation on the radiant terminals for temperature control.

2.6.1.3 Dehumidification Methods Using Liquid or Solid Desiccant

Humidity control is an important task for indoor built environment, and air dehumidification is commonly required for air-conditioning systems in most instances, especially in the humid region. Recently, liquid desiccant or solid desiccant dehumidification becomes more and more popular, which is regarded to be an alternative solution to condensation dehumidification adopted in conventional air-conditioning systems (Yadav 1995; Bergero and Chiari 2011; Xiong et al. 2010; Subramanyam et al. 2004; Aynur et al. 2010; Jeong et al. 2010; Niu et al. 2010; Lee et al. 2012; Ling et al. 2011, 2013; Zhang et al. 2011, 2012c, 2012d; Zhang 2005, 2006, 2009). As to the desiccant dehumidification method, humid air could be dehumidified to the required humidity ratio with no need for cooling the air to a temperature lower than its dew point. Compared with the conventional condensation dehumidification method, superiorities for adopting desiccant dehumidification method are mainly embodied in no reheating requirement after dehumidification and no wet surface accumulating condensate water.

Many researchers focused on advanced desiccant material, enhanced heat and mass transfer performance between air and desiccant, and proposing new system configurations. From 1970 till now, there are over 6,400 patents about "desiccant" according to the ISI Web of Knowledge. As described in Fig. 2.28, more and more patents on desiccant were proposed in recent years, due to a rapid development of the desiccant dehumidification devices and corresponding systems. The desiccant dehumidification system can be regenerated by low-grade heat, such as solar energy, waste heat from cogeneration system, and waste heat from condenser of heat pump. With these desiccant dehumidification methods, indoor humidity control could be achieved independent from indoor temperature control, and solutions for constructing a THIC process for air-conditioning have also been proposed and put into application recently.

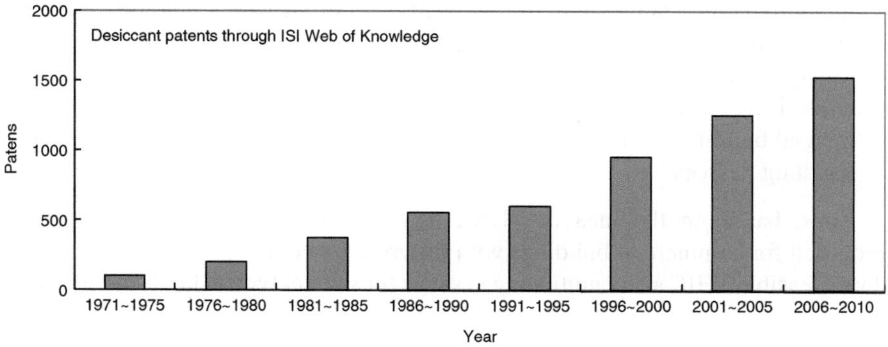

Fig. 2.28 Number of desiccant patents through ISI Web of Knowledge

These air-conditioning systems and solutions are appropriate for buildings with different functions and in different situations. All of these air-conditioning systems are based on the idea of THIC for the indoor built environment. Moving forward, there could be many more air-conditioning solutions based on THIC for buildings with different functions or in different climate regions. It is clear that improving the energy efficiency of THIC air-conditioning systems based on satisfying the requirement for regulating indoor temperature and humidity will be the key issue for advancing the popularization and practical application of THIC systems.

2.6.2 Possible Ways to Construct THIC Systems

As indicated by the analysis of air-conditioning systems in Chap. 1 (Table 1.4), there are significant discrepancies among the air-conditioning systems of buildings of different scales, and the heat transfer fluids for transporting the cooling capacity are also different. Based on the notion of THIC, different solutions can be utilized to construct THIC systems for buildings with different scales or functions:

- In commercial buildings of a relatively small scale, VRF (variable refrigerant flow) systems are being widely adopted because of their flexible and convenient arrangement, as shown in Table 1.4. A VRF unit with a high evaporating temperature (high-temperature type) can satisfy the indoor temperature regulation requirement; an independent outdoor air handling processor can be adopted to handle the outdoor air, and it meets the requirement of indoor humidity control. This solution for small-scale commercial buildings represents one possible application of THIC systems.
- For commercial buildings of a larger scale in which central air-conditioning systems are adopted, indoor temperature and humidity control can be realized independently in a THIC system. In the temperature control subsystem, the heat transfer fluid (water or refrigerant) with a temperature (16–18 °C) higher than that of conventional systems is distributed to the indoor terminals and is responsible for extracting the indoor sensible load and for temperature control. In the

humidity control subsystem, the supplied outdoor air with a humidity ratio dry enough to extract the indoor moisture is supplied to the conditioned space, satisfying the humidity control requirement. Compared with conventional systems, THIC air-conditioning systems could bring about great changes for commercial buildings related to natural cooling sources, system processes, and even handling devices.

Thus, based on the idea of THIC, new air-conditioning systems could be proposed for commercial buildings with different functions. This subsection introduces feasible THIC air-conditioning systems for several typical kinds of buildings:

2.6.2.1 Office Buildings

As Chinese society has continued to develop rapidly, the number of large-scale, high-grade office buildings has increased dramatically. Not surprisingly, the air-conditioning solutions for these buildings are facing unprecedented opportunities for development. Currently, conventional air-conditioning systems, including the FCU and outdoor air type and the all-air type, are usually adopted for these office buildings (Building Energy Research Center of Tsinghua University 2010). From the perspective of the THIC air-conditioning system, the indoor moisture generated inside office buildings is mainly from occupants, and the variance of the required outdoor air flow rate is related to the variance of occupants. The handled outdoor air with a lower humidity ratio is supplied to the occupied area and is responsible for the requirements for outdoor air and indoor humidity regulation. Meanwhile, the other subsystem extracts the remainder of the indoor sensible load to meet the requirement of temperature regulation. In this manner, the handled outdoor air is used to extract the indoor moisture and regulate indoor humidity; sensible terminals such as radiant panels and dry FCUs are then adopted for indoor temperature regulation. According to this method, the indoor requirements for outdoor air, humidity, and temperature are satisfied and regulated separately.

As the required indoor temperature is 25 °C, any cooling source with a temperature lower than 25 °C could (in theory) be adopted for indoor temperature control. Consequently, this makes it possible to utilize certain natural cooling sources or mechanical cooling sources with an outlet chilled water temperature of about 15–20 °C. The higher energy efficiency of a mechanical cooling source used only for temperature control will result in an increased evaporating temperature compared to a conventional mechanical cooling source used for coupled temperature and humidity control. Indeed, the required temperature of the cooling source in conventional systems is usually much lower; for example, 7 °C is a common design supply chilled water temperature, since the cooling source is responsible for temperature control combined with humidity control. To extract the indoor moisture, the cooling source temperature is restricted and must be very low if the condensation dehumidification method is adopted. During the cooling season, there are hardly any natural or low-cost cooling sources with a temperature of 7 °C, so the mechanical compression method is adopted to produce chilled water.

The energy performance of a mechanical cooling source with a lower temperature is inferior to that of a source with a chilled water temperature of 15–20 °C. Besides, although the outdoor air temperature is high (over 35 °C) during summer in the northwest of China, the outdoor air humidity ratio is quite low, with a dew point temperature lower than 15 °C in most cases. Therefore, the indirect evaporative cooling method could be adopted to produce chilled water with a temperature of about 2–3 °C higher than the dew point using outdoor dry air. Thus, in the dry region, high-temperature chilled water could be produced for temperature control without adopting the vapor compression method.

Using high-temperature chilled water with a temperature of 15–20 °C to extract the indoor sensible load, the indoor terminals are required for the regulation of the indoor temperature. As the terminals are only for extracting the indoor sensible load, there will be no moisture condensation (or condensate) when high-temperature chilled water is adopted. The radiation method (e.g., radiant panels) and the forced convection method (e.g., FCUs operating in the dry condition) are both feasible for indoor temperature control.

2.6.2.2 Large Spaces

As for large-scale spaces such as airport terminals, railway stations, and building atriums, indoor occupants are mostly under 2 m in height. The main tasks for the air-conditioning systems in these areas are to meet the temperature and humidity requirements of the occupied regions. Currently, most of these types of locations adopt all-air systems, and nozzles set up in the top or in the middle of these large spaces supply the air. The handled air is responsible for both temperature and humidity control, with the occupants in the return flow zone. In summer, this kind of solution leads to a relatively high cooling load, reaching about 150–200 W/m^2. In winter, the temperature difference in the vertical direction is a key issue, and although the total heat supply is relatively high, the temperature of the area near the floor is still low due to the fact that low-density hot air is difficult to supply to lower spaces. Furthermore, the fan power of this kind of nozzle-based air supply system is usually quite high, which can be about 2–4 times the chiller power in some cases. At present, the annual power consumption of the air-conditioning systems in these kinds of large spaces is usually more than 150 kWh/m^2, which is significantly higher than in other common commercial buildings.

The key issues for constructing the thermal built environment of large spaces are listed as follows:

1. Realizing the temperature gradient in the vertical direction and constructing a nonuniform temperature distribution system. In other words, the temperatures of spaces higher than 2 m could be higher in summer and lower in winter than those of areas at 2 m and below (where occupants are). This kind of nonuniform temperature profile helps to reduce the air-conditioning load compared to the nozzle-based air supply system, which relies on a more uniform temperature distribution.

2. Avoiding the need to heat or cool the entire large space by using terminals to achieve local control, which will reduce the fan power. Local control will also be effective for extracting the load in crowded areas.

According to the concept of THIC air-conditioning systems, local control and the stratified air-conditioning method could be adopted for the construction of the thermal built environment of large spaces, and the power consumption of air-conditioning could be cut significantly.

With the help of outdoor air handling devices, the outdoor air is processed to a state with an appropriate temperature and humidity ratio (e.g., 18–20 °C and 8–10 g/kg in summer). Then, the air supply devices such as displacement ventilation terminals and other floor-based air supply devices are responsible for supplying the handled outdoor air to the occupied areas. It is recommended that the supply of outdoor air be supplied directly to the occupants and avoid mixing with the indoor air. The supplied outdoor air flow rate could be in proportion to the number of occupants in most cases, and then the supplied dry air would be able to meet the requirement for extracting indoor moisture.

In the occupied areas of large spaces, high-temperature chilled water (18–20 °C) and low-temperature hot water (about 35–40 °C) are supplied into the indoor terminals for temperature control in summer and winter, respectively. Thus, the indoor terminals will be responsible for extracting the sensible load in summer and the supply heat in winter. For this kind of requirement (i.e., only for temperature control), radiant floors are a relatively feasible and effective solution. By setting the coils under the floor with a certain density and reducing the heat resistance between the coil and the floor surface as much as possible, the floor surface temperature could be about 20–22 °C in summer, which is low enough to extract the indoor sensible load and satisfy the indoor temperature regulation requirement. The coils under the floor could be divided into certain regions, and on-off control could be suitable for regulating the indoor temperature according to the divided regions. In each region, the time ratio of "on" and "off" could be determined by the indoor temperature controller, achieving effective temperature control.

For large spaces where the floor is mostly covered by objects, where there is not enough room to set up radiant coils, or where too much equipment results in a high sensible load, FCUs could be adopted for the extraction of the indoor sensible load. As the high-temperature chilled water is supplied, the FCUs will operate under the dry condition without any condensate. This kind of dry FCU is used only for temperature control, not for humidity control, and the FCU should be set up in the floor or in columns. Side air supply is adopted, and both side and top returns are acceptable. Figure 2.29 shows a dry FCU combined with displacement ventilation.

For other commercial buildings with different functions, including hotels, shopping malls, and hospitals, there are also new THIC-based solutions for the indoor built environment. In contrast to conventional air-conditioning systems that combine temperature regulation and humidity regulation, THIC provides a feasible approach for regulating the indoor temperature and humidity separately. Moreover, the THIC concept could also be effective for the thermal built environment of buildings with special functions, such as archives and industrial factories. Based on

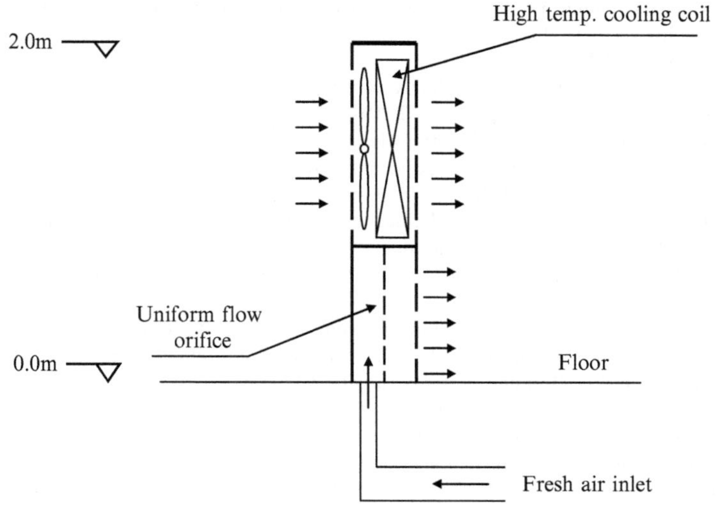

Fig. 2.29 An operating schematic of the indoor terminal for large spaces

an analysis of the operating features and functional characteristics of these buildings, new THIC-based solutions could result in better energy efficiency compared to existing conventional air-conditioning systems. The next chapters focus on THIC systems adopted in commercial buildings.

References

AQSIQ, SAC (2003) GB/T 19232–2003 Fan-coil unit. China Standards Press, Beijing (in Chinese)
AQSIQ, SAC (2004) GB 19577–2004 The minimum allowable values of the energy efficiency and energy efficiency grades for water chillers. China Standards Press, Beijing (in Chinese)
AQSIQ, SAC (2008a) GB/T 17981–2007 Economic operation of air conditioning systems. China Standards Press, Beijing (in Chinese)
AQSIQ, SAC (2008b) GB/T 7190.1–2008 Glass fiber reinforced plastic cooling tower-Part 1: Middle and small glass fiber reinforced plastic cooling tower. China Standards Press, Beijing (in Chinese)
Aynur TN, Hwang Y, Radermacher R (2010) Integration of variable refrigerant flow and heat pump desiccant systems for the cooling season. Appl Therm Eng 30:917–927
Bergero S, Chiari A (2011) On the performances of a hybrid air-conditioning system in different climatic conditions. Energy 36:5261–5273
Bjarne O (2008) Radiant floor cooling systems. ASHRAE J 50:16–22
Building Energy Research Center of Tsinghua University (2010) 2010 Annual report on China building energy efficiency. China Architecture & Building Press, Beijing (in Chinese)
Dedicated outdoor air systems. http://doas-radiant.psu.edu
IEA-ECBCS Annex 37. Low exergy systems for heating and cooling. http://www.ecbcs.org/annexes/annex37.htm
IEA-ECBCS Annex 49. Low exergy systems for high-performance buildings and communities. http://www.ecbcs.org/annexes/annex49.htm
Jeong J, Yamaguchi S, Saito K, Kawai S (2010) Performance analysis of four-partition desiccant wheel and hybrid dehumidification air-conditioning system. Int J Refrig 33:496–509

Lee WL, Chen H, Leung YC, Zhang Y (2012) Decoupling dehumidification and cooling for energy saving and desirable space air conditions in hot and humid Hong Kong. Energy Convers Manage 53:230–239
Li Z, Liu XH, Zhang L, Jiang Y (2010) Analysis on the ideal energy efficiency of dehumidification process from buildings. Energy Build 42:2014–2020
Ling JZ, Kuwabara O, Hwang Y, Radermacher R (2011) Experimental evaluation and performance enhancement prediction of desiccant assisted separate sensible and latent cooling air-conditioning system. Int J Refrig 34:946–957
Ling JZ, Kuwabara O, Hwang Y, Radermacher R (2013) Enhancement options for separate sensible and latent cooling air-conditioning systems. Int J Refrig 36:45–57
Meckler G (1987) Innovative ways to save energy in new buildings. Heat Pip Air Cond Eng 59:59–64
Meteorological Information Center of China Meteorological Administration, Tsinghua University (2005) Dedicated meteorological data sets for thermal built environment analysis in China. China Architecture & Building Press, Beijing (in Chinese)
Mumma SA (2001) Designing dedicated outdoor air systems. ASHRAE J 43:28–31
Niu XF, Xiao F, Ge GM (2010) Performance analysis of liquid desiccant based air-conditioning system under variable fresh air ratios. Energy Build 42:2457–2464
Subramanyam N, Maiya MP, Murthy SS (2004) Application of desiccant wheel to control humidity in air-conditioning systems. Appl Therm Eng 24:2777–2788
The 10th Design and Research Institute of Ministry of Electronics Industry (1995) Design handbook for air-conditioning, 2nd edn. China Architecture & Building Press, Beijing (in Chinese)
Xiong ZQ, Dai YJ, Wang RZ (2010) Development of a novel two-stage liquid desiccant dehumidification system assisted by $CaCl_2$ solution using exergy analysis method. Appl Energy 87:1495–1504
Yadav YK (1995) Vapour-compression and liquid-desiccant hybrid solar space-conditioning system for energy conservation. Renew Energy 6:719–723
Yan QS, Zhao QZ (1986) Building thermal process. China Architecture & Building Press, Beijing (in Chinese)
Yin P (2003) Study of the dedicated outdoor air system (1): a review. Chin HV&AC 33(6):44–49 (in Chinese)
Zhang LZ (2005) Dehumidification technology. Chemical Industry Press, Beijing (in Chinese)
Zhang LZ (2006) Energy performance of independent air dehumidification systems with energy recovery measures. Energy 31:1228–1242
Zhang LZ (2009) Total heat recovery: heat and moisture recovery from ventilation. Nova Press, New York
Zhang XJ, Yu CY, Li S, Zheng YM, Xiao F (2011) A museum storeroom air-conditioning system employing the temperature and humidity independent control device in the cooling coil. Appl Therm Eng 31:3653–3657
Zhang L, Liu XH, Jiang Y (2012a) A new concept for analyzing the energy efficiency of air-conditioning systems. Energy Build 44:45–53
Zhang L, Liu XH, Jiang Y (2012b) Ideal efficiency analysis and comparison of condensing and liquid desiccant dehumidification. Energy Build 49:575–583
Zhang L, Hihara E, Saikawa M (2012c) Combination of air-source heat pumps with liquid desiccant dehumidification of air. Energy Convers Manage 57:107–116
Zhang XJ, Xiao F, Li S (2012d) Performance study of a constant temperature and humidity air-conditioning system with temperature and humidity independent control device. Energy Build 49:640–646
Zhang T, Liu XH, Zhang L, Jiang Y (2012e) Match properties of heat transfer and coupled heat and mass transfer processes in air-conditioning system. Energy Convers Manag 59:103–113
Zhao RY, Fan CY, Xue DH, Qian YM (2000) Air conditioning, 3rd edn. China Architecture & Building Press, Beijing (in Chinese)
Zhu YS, Zhang XF, Shan ZJ (1994) Air-conditioning method using underground water. Aviation Industry Press, Beijing (in Chinese)

Chapter 3
Key Components of the THIC System: Indoor Terminals

Abstract Terminal devices of the temperature control subsystem consist of a heat exchanger and a heat transfer fluid, which could be a high-temperature chilled water or refrigerant. The heat transfer fluid is circulated inside the heat exchanger, where heat is transferred to the conditioned space through convection or radiation. In the humidity control subsystem, handled outdoor air with a lower humidity ratio is supplied into the conditioned space to remove the indoor moisture load. Terminal devices of the temperature control subsystem and humidity control subsystem complement each other, together achieving the goal to construct the indoor thermal built environment. This chapter introduces different terminal devices for handling the indoor sensible load, including radiant terminal devices and dry FCUs.

The indoor heat and moisture control process for the THIC air-conditioning system is illustrated in Fig. 3.1. Terminal devices of the temperature control subsystem include a heat exchanger and high-temperature chilled water or refrigerant that is transported to the end of the heat exchanger and then transfers heat with the indoor air or surfaces through convection or radiation. In the humidity control subsystem, handled outdoor air with a lower humidity ratio is supplied into the indoor space to extract the indoor moisture load. Terminal devices of the temperature control subsystem and humidity control subsystem complement each other, completing the task of constructing the indoor thermal built environment together. This chapter introduces the terminal devices for handling the sensible load, including radiant terminal devices and dry FCUs.

X. Liu et al., *Temperature and Humidity Independent Control (THIC) of Air-conditioning System*, DOI 10.1007/978-3-642-42222-5_3,

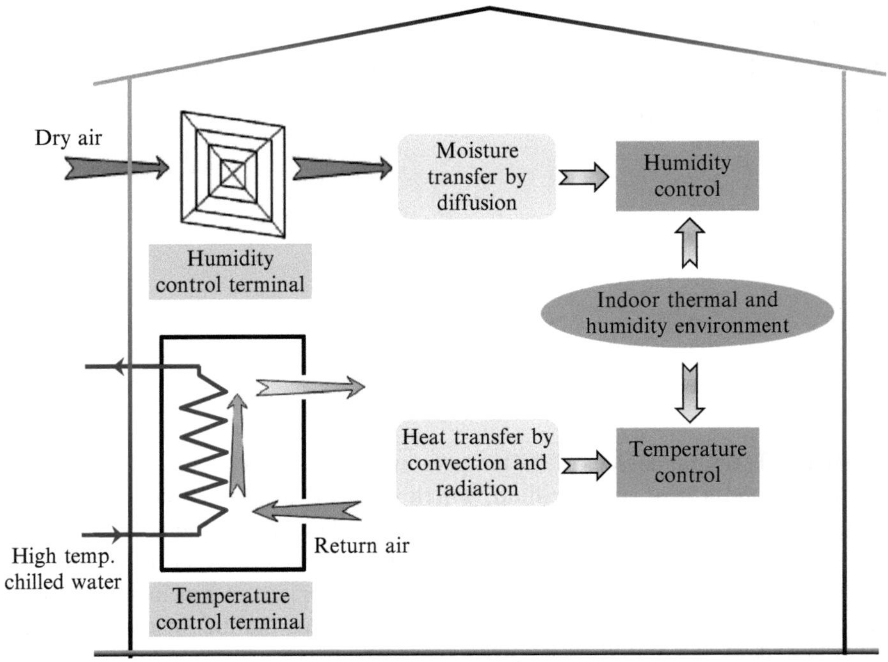

Fig. 3.1 Indoor terminal devices of the THIC air-conditioning system

3.1 Radiant Panels

3.1.1 Heat Transfer Process of Radiant Terminal Devices

In radiant cooling systems, chilled water is supplied by embedded circular pipes in a concrete layer or extruded sheet, which forms a low-temperature surface on the radiant panel. Then, the panel surface exchanges heat with the indoor environment. The cold media include chilled water and refrigerant. According to the location of the radiant terminal device, there can be radiant ceiling cooling systems, floor cooling systems, wall cooling systems, etc. According to the structure, the radiant terminal devices can be divided into two categories (as shown in Fig. 3.2): water pipe-embedded radiant floor/ceiling panels and metal or plastic radiant panels.

The radiant panel surface exchanges heat with the indoor air, internal wall surfaces, occupants, equipment, etc., through convection and radiation, as shown in Fig. 3.3.

As the surface of the radiant panel can transfer heat directly with other internal wall surfaces by longwave radiation, it reduces the forced air circulating loop, which is the major difference between radiant panels and the ordinary convective terminals in conventional air-conditioning systems. In addition, shortwave radiation (such as solar radiation) reaching the surface of the radiant panel could be

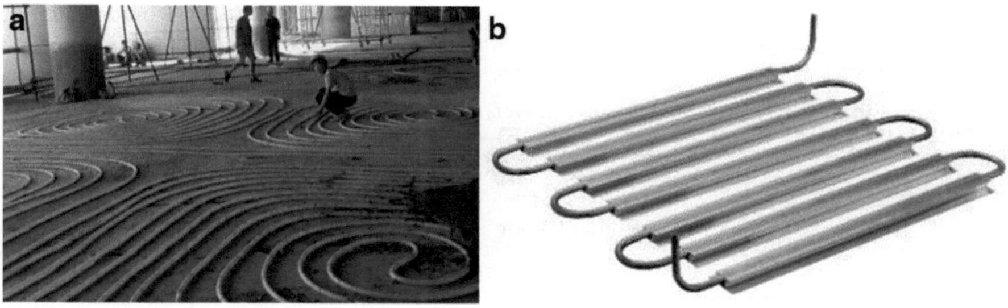

Fig. 3.2 Two types of radiant terminal devices: (**a**) concrete radiant floor and (**b**) metal radiant panel

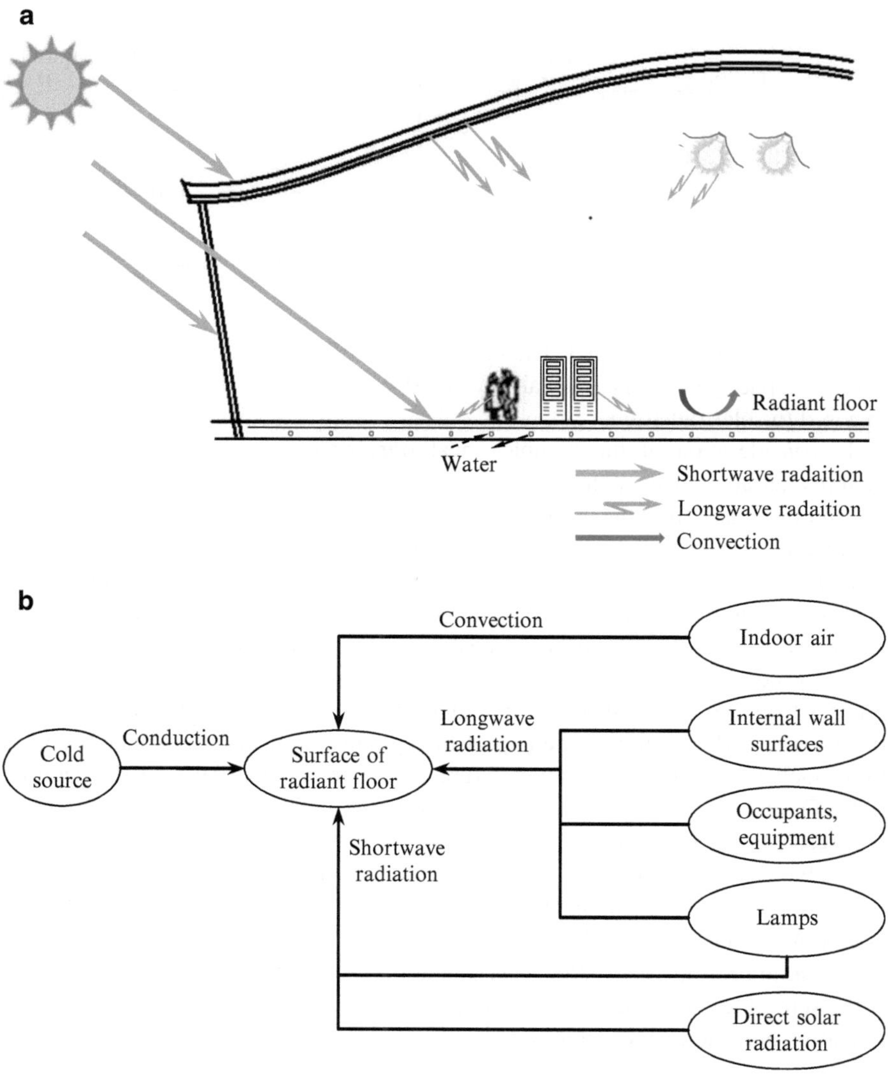

Fig. 3.3 Heat transfer between the radiant floor and its surroundings: (**a**) schematic diagram and (**b**) heat transfer process

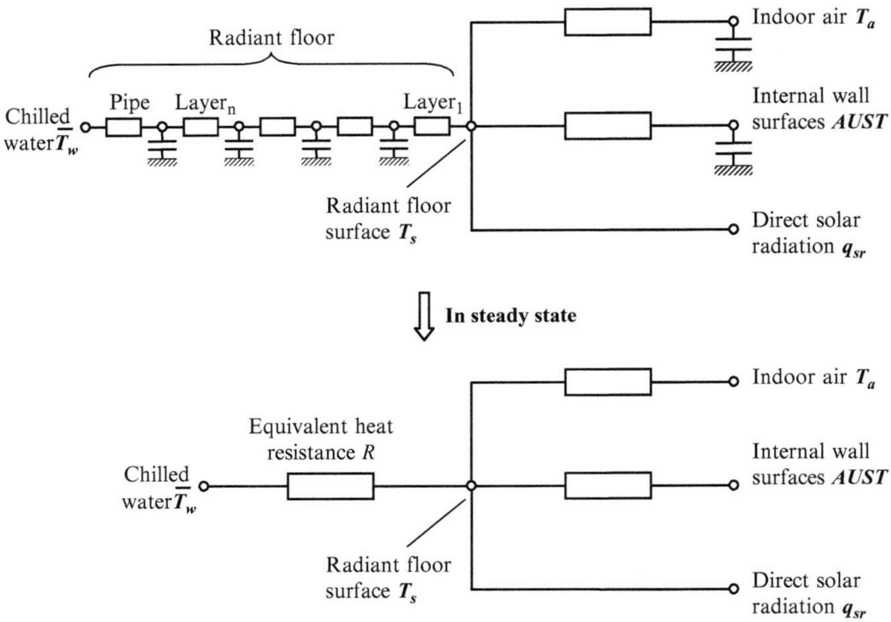

Fig. 3.4 Simplified analytical method of the radiant panel with thermal resistance

absorbed directly by the panel. Therefore, instead of a fixed value, the radiant panel cooling capacity can vary from about 30–60 W/m^2 to more than 100 W/m^2 when exposed to solar radiation.

In consideration of the complex heat transfer processes in radiant cooling systems, an analytical method based on thermal resistance is proposed. The heat transfer processes from the indoor space to the chilled water are depicted in Fig. 3.4 with the classical equivalent circuit model. In steady state, the thermal mass effect is negligible, and the cooling capacity of the radiant device is decided by the indoor heat source temperature/intensity, chilled water temperature, and heat transfer resistance at each link. The heat transfer process from the indoor space to the chilled water can be divided into two parts. The first part refers to the heat transfer process from the indoor space to the radiant panel surface, which is dependent on the average surface temperature (T_s) of the radiant panel, as well as indoor conditions such as the indoor air temperature (T_a), internal wall surface temperature (AUST), and shortwave radiation absorbed by the radiant panel (q_{sr}). The second part refers to the heat transfer process from the surface of the panel to the chilled water, which is determined by the construction parameters of the radiant panel and the difference between the average surface temperature (T_s) of the panel and the chilled water temperature ($\overline{T}_w$).

The total heat flux between the surface of the radiant panel and the indoor space is the sum of the heat flux through the convective heat transfer process, longwave radiation, and shortwave radiation:

$$q = q_c + q_{lr} + q_{sr} \tag{3.1}$$

where q is the total heat flux and q_c, q_{lr}, and q_{sr} are the heat fluxes of convective heat transfer, longwave radiation, and shortwave radiation, respectively.

Hereinto, the convective heat flux (q_c) and the longwave radiation (q_{lr}) can be achieved when the convective heat transfer coefficient (h_c), the longwave radiant heat transfer coefficient (h_{lr}), and the corresponding indoor conditions are known. Hence, the total heat flux in Eq. (3.1) can be written as Eq. (3.2):

$$q = h_c(T_a - T_s) + h_{lr}(\text{AUST} - T_s) + q_{sr} \tag{3.2}$$

The total heat flux between the chilled water and the panel surface can be calculated according to the difference between the average temperature of the chilled water ($\overline{T}_w$) and the average surface temperature (T_s) of the radiant panel, divided by the equivalent heat resistance of the radiant floor (R):

$$q = \frac{T_s - \overline{T}_w}{R} \tag{3.3}$$

Based on this analysis, the total heat flux into the radiant panel in steady state can be acquired by solving Eqs. (3.2) and (3.3):

$$q = \frac{h_c \cdot T_a + h_{lr} \cdot \text{AUST} + q_{sr} - (h_c + h_{lr}) \cdot \overline{T}_w}{1 + (h_c + h_{lr}) \cdot R} \tag{3.4}$$

The average temperature of the radiant panel surface can then be obtained:

$$T_s = \frac{h_c \cdot T_a + h_{lr} \cdot \text{AUST} + q_{sr} + \overline{T}_w/R}{h_c + h_{lr} + 1/R} \tag{3.5}$$

For the chilled water, the heat flux can be calculated by

$$q = \frac{\rho c_p \dot{g} (T_h - T_g)}{1 + r} \tag{3.6}$$

where T_g and T_h are the supply and return water temperatures of the chilled water, respectively; $\dot{g}$ is the water flow rate, $m^3/(s \cdot m^2)$; and r is the downward dimensionless heat transfer loss rate.

Therefore, according to the procedure shown in Fig. 3.4, it is most convenient to calculate the cooling capacity and the surface temperature of the radiant panel when the cooling source, indoor environment, and structure of the radiant panel are known. This analytical method is suitable not only for radiant floor cooling systems but also for radiant ceiling and wall cooling/heating systems.

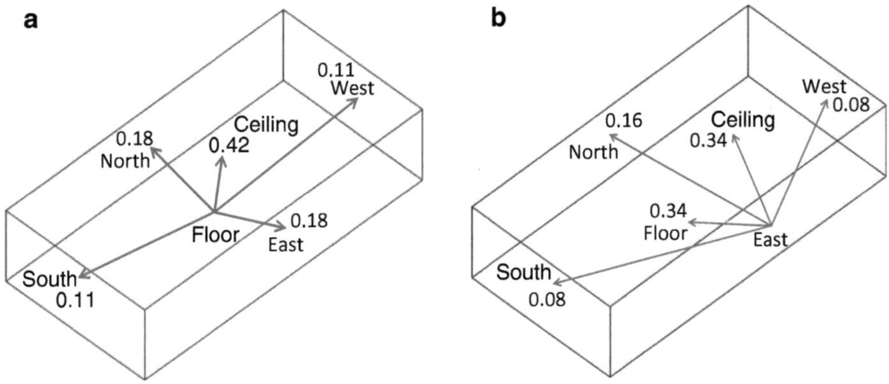

Fig. 3.5 View factors between the radiant panel surface and other internal wall surfaces: (**a**) radiant floor and (**b**) radiant wall

3.1.2 Key Parameters of Radiant Panels

In this simplified analytical method, there are several key parameters, including the convective heat transfer coefficient (h_c), longwave radiant heat transfer coefficient (h_{lr}), corresponding reference envelope surface temperature (AUST), and equivalent thermal resistance of the radiant floor (R).

3.1.2.1 Longwave Radiant Heat Transfer

Using the linear approximation, the heat flux (q_{lr}) can be expressed as the product of the temperature difference (AUST – T_s) and the corresponding radiant heat transfer coefficient (h_{lr}):

$$q_{lr} = h_{lr}(\text{AUST} - T_s) \tag{3.7}$$

The average uncooled surface temperature (AUST) can be calculated according to the area-weighted average value, or it can be calculated more precisely with view factors, as shown in Eq. (3.8):

$$\text{AUST} = \sqrt[4]{\sum_{j=1}^{n}\left(F_{s-j}(T_j + 273.15)^4\right)} - 273.15 \tag{3.8}$$

where F_{s-j} is the view factor between the radiant panel surface and internal wall surface j and T_j is the temperature of internal wall surface j.

Figure 3.5 shows the view factors between the radiant panel surface and other internal wall surfaces in a room with a size of 12 m × 6 m × 3 m. Figures 3.6 and 3.7 show some thermal imaging results for indoor surface temperatures measured at

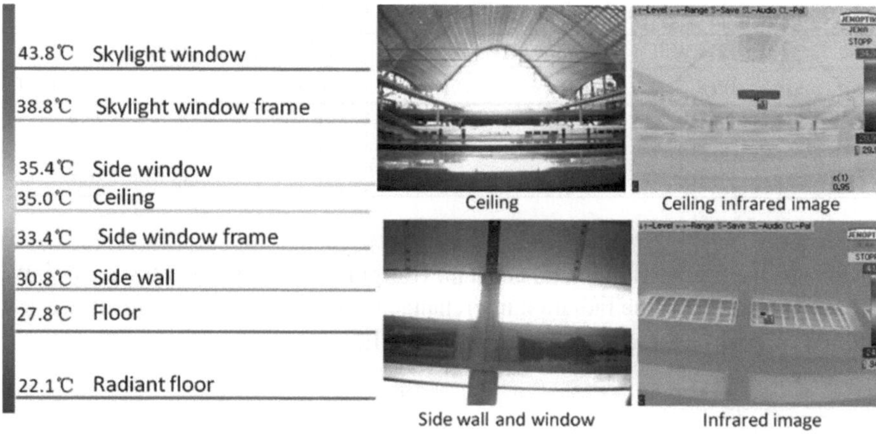

Fig. 3.6 Indoor wall surface temperature distribution in large space buildings

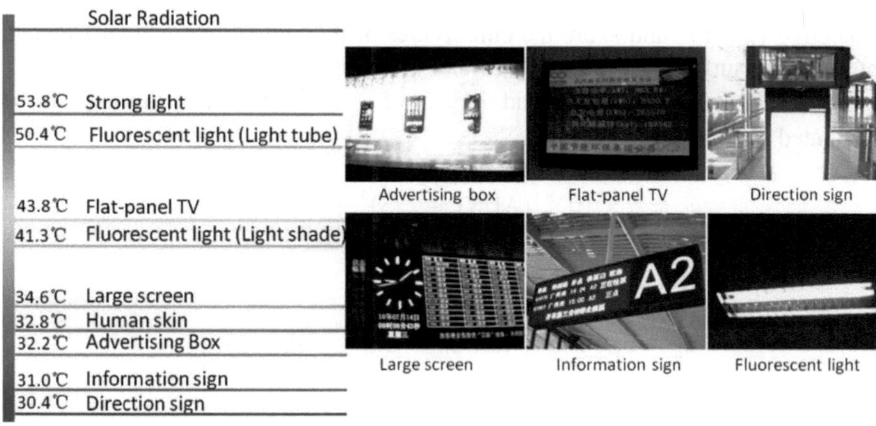

Fig. 3.7 Surface temperature distribution of indoor heat sources

several railway stations in China (Zhao et al. 2013). The temperatures of skylights and windows can reach 35–45 °C, while other opaque wall surfaces are around 30–35 °C in summer. Hence, the AUST values of these large space buildings are usually higher than 30 °C and can even reach 35–40 °C.

In common office buildings, AUST is close to the indoor air temperature. Taking the case in Fig. 3.5a as an example, the AUST calculation results for different internal wall surface temperatures are listed in Table 3.1.

Longwave radiation heat transfer coefficient h_{lr} is calculated by (Lu 2008):

$$h_{\mathrm{lr}} = \frac{\sigma \sum_{j=1}^{n} F_{\varepsilon_{s-j}} \left((T_j + 273)^4 - (T_s + 273)^4 \right)}{\mathrm{AUST} - T_s} \tag{3.9}$$

Table 3.1 AUST calculation results for different internal wall surface temperatures

Wall	East wall	South wall	West wall	North wall	Roof	
View factor	0.18	0.11	0.18	0.11	0.42	AUST (°C)
1	28	26	26	26	26	26.4
2	30	30	26	26	26	27.2
3	30	26	26	26	35	30.6

where σ is the Stefan-Boltzmann constant (Bergman et al. 2011), 5.67×10^{-8} W/(m$^2 \cdot$ K^4), and $F_{\varepsilon_{s-j}}$ is the radiation interchange factor between the floor surface and internal wall surface j, calculated according to Eq. (3.10):

$$F_{\varepsilon_{s-j}} = \frac{1}{[(1-\varepsilon_s)/\varepsilon_s] + (1/F_{s-j}) + (A_s/A_j)[(1-\varepsilon_j)/\varepsilon_j]} \tag{3.10}$$

where A_s and A_j are the areas of the radiant panel surface and internal wall surface j, respectively, and ε_s and ε_j are the emissivities of the radiant terminal surface and internal wall surface j, respectively, usually ranging from 0.90 to 0.95.

Hence, $F_{\varepsilon_{s-j}}$ is about 0.87, and $\sigma F_{\varepsilon_{s-j}} \approx 5 \times 10^{-8}$. And heat flux q_{lr} can be calculated as follows:

$$q_{\mathrm{lr}} \approx 5 \times 10^{-8}\left[(\mathrm{AUST}+273)^4 - (T_s+273)^4\right] \tag{3.11}$$

The radiation heat transfer coefficient (h_{lr}) is

$$h_{\mathrm{lr}} \approx 5 \times 10^{-8} \cdot [(\mathrm{AUST}+273) + (T_s+273)] \cdot \left[(\mathrm{AUST}+273)^2 + (T_s+273)^2\right] \tag{3.12}$$

As for the indoor cooling and heating system, the surface temperature of the radiant panel is usually 16–22 °C (cooling) and 25–35 °C (heating). Therefore, h_{lr} is almost a constant value, 5.5 W/(m^2 ·°C) for most indoor space, with an error of less than 4 % (Olesen and Michel 2000). Figure 3.8 shows the variance of heat flux q_{lr} according to the change in temperature difference between AUST and the radiant panel surface temperature. If low-emissivity material is adopted on a large area, h_{lr} is lower. For instance, low-E glass with emissivity of 0.1–0.15 is employed as the wall surfaces (F_{s-j} ranges from 0.25 to 0.15 for different scale buildings); the h_{lr} is about 4.5–4.8 W/(m^2 ·°C).

For the surface temperature of the radiant panel, there are temperature limits for the radiant floor, ceiling, and walls in the standards from different countries and regions. For instance, in the European standard EN1264:

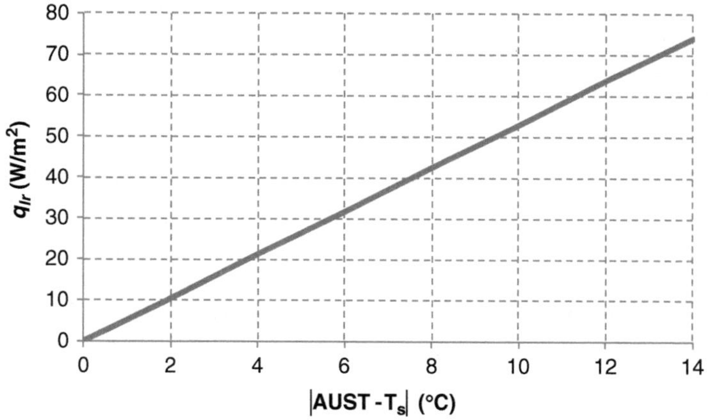

Fig. 3.8 Heat flux q_{lr} according to the change in temperature difference (AUST–T_s)

- Heating in winter: $T_s \leq 29$ °C in occupied zones and $T_s \leq 35$ °C in others.
- Cooling in summer: $T_s \geq 20$ °C in occupant seating zones and $T_s \leq 35$ °C in gyms; surface temperature should be higher than the dew point temperature of the surrounding air.
- Heating and cooling: within the limit of asymmetric thermal radiation.

3.1.2.2 Convective Heat Transfer Coefficient

The convective heat transfer coefficient between the surface of the radiant panel and the indoor air (h_c) is closely related to the difference between the indoor air temperature (T_a) and the floor surface temperature (T_s):

$$q_c = h_c(T_a - T_s) \tag{3.13}$$

The convective heat transfer coefficient is closely related to the location of and the air flow around the radiant panel and the cooling/heating pattern (Koschenz and Dorer 1999; Novoselac et al. 2006; Francesco et al. 2009). Table 3.2 summarizes the empirical formulas of the natural convective heat transfer coefficient. The natural convective heat transfer coefficients are ordered as follows: floor heating and ceiling cooling > wall heating and cooling > ceiling heating and floor cooling. Thus, if the surface temperature of the radiant panel and the indoor temperature are the same, the order of the cooling and heating capacities will be as follows: ceiling cooling > wall cooling > floor cooling and floor heating > wall heating > ceiling heating. It should be mentioned that the data in this table were obtained with only natural convection (i. e., no forced convection). If there is forced air ventilation, convective heat transfer coefficient h_c will increase. Figure 3.9 shows the convective heat transfer of the radiant panel with the temperature changes of the radiant panel surface and indoor air (the convective heat transfer coefficient is cited from BS EN1264-5).

Table 3.2 Natural convection heat transfer coefficient h_c

	Formula or value	Source	Explanation
Floor cooling and ceiling heating	$0.134 \cdot (\Delta T)^{0.25}$	Min et al. (1956) Lu (2008)	When $\Delta T = 5 \sim 8$ °C, $h_c = 0.20 \sim 0.23$ W/(m^2 · °C)
	$0.87 \cdot (\Delta T)^{0.25}$	Lu (2008)	During ceiling heating with intervals between adjacent radiant panels, the effect of the natural convection will be enhanced; when $\Delta T = 5 \sim 8$ °C, $h_c = 1.3 \sim 1.5$ W/(m^2 · °C)
	$\frac{0.704}{D^{0.601}} \cdot (\Delta T)^{0.133}$	Awbi and Hatton (1999)	Only heating: when $\Delta T = 5 \sim 8$ °C and $D = 5$ m, $h_c = 0.33 \sim 0.35$ W/(m^2 · °C)
	1.0	EN-1264	Estimated as $h_{lr} = 5.5$ W/(m^2 · °C)
Floor heating and ceiling cooling	$2.13 \cdot (\Delta T)^{0.31}$	Min et al. (1956) Lu (2008)	When $\Delta T = 5 \sim 8$ °C, $h_c = 3.5 \sim 4.1$ W/(m^2 · °C)
	$\frac{2.175}{D^{0.076}} \cdot (\Delta T)^{0.308}$	Awbi and Hatton (1999)	Only heating: when $\Delta T = 5 \sim 8$ °C and $D = 5$ m, $h_c = 3.2 \sim 3.7$ W/(m^2 · °C)
	$10.8 \cdot (\Delta T)^{0.1} - 5.5$	EN-1264	Estimated as $h_{lr} = 5.5$ W/(m^{2}°C); when $\Delta T = 5 \sim 8$ °C, $h_c = 7.2 \sim 7.8$ W/(m^2 · °C)
Wall heating and cooling	$1.78 \cdot (\Delta T)^{0.32}$	Lu (2008)	When $\Delta T = 5 \sim 8$ °C, $h_c = 3.0 \sim 3.5$ W/(m^2 · °C)
	$\frac{1.823}{D^{0.121}} \cdot (\Delta T)^{0.293}$	Awbi and Hatton (1999)	Only heating: when $\Delta T = 5 \sim 8$ °C and $D = 5$ m, $h_c = 2.4 \sim 2.8$ W/(m^2 · °C)
	2.5	EN-1264	Estimated as $h_{lr} = 5.5$ W/(m^2 · °C)

Notes: ΔT refers to the temperature difference between the radiant panel surface and the indoor air; $D = \frac{4 \times \text{Area}}{\text{Perimeter}}$ is the characteristic diameter of the radiant panel

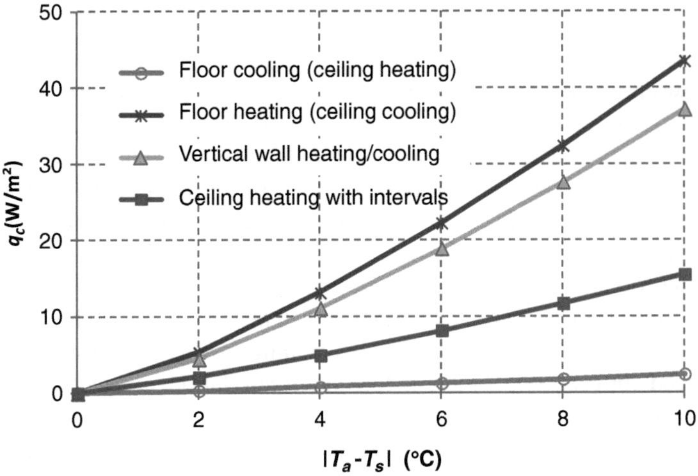

Fig. 3.9 Convective heat transfer of the radiant panel according to the variance of (T_a–T_s)

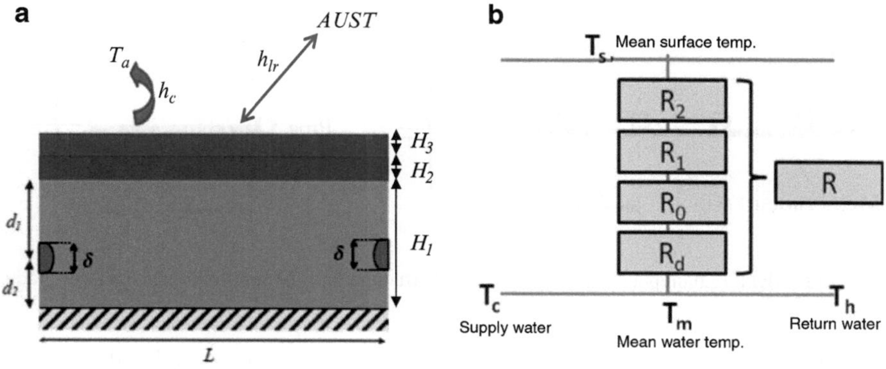

Fig. 3.10 Schematic of radiant panel structure and thermal resistance: (**a**) structure and (**b**) thermal resistance

3.1.2.3 Equivalent Thermal Resistance of the Radiant Panel

A radiant floor usually consists of a covering layer (e.g., marble, plastic, or a wood floor), a weight-bearing thermal diffusion layer with small pipes embedded in screed or prefabricated panels, and a thermal insulation layer on the base of the building structure. The structure of a typical radiant floor is shown in Fig. 3.10. At each transverse of this structure, the cooling capacity is transferred from the chilled water to the floor surface by heat conduction in both horizontal and vertical directions (Zhang et al. 2012). For two-dimensional heat conduction among multiple layers, it is convenient to define an equivalent thermal resistance (R) to represent the heat transfer characteristic for engineering applications:

$$R = \frac{T_\mathrm{s} - \overline{T}_\mathrm{w}}{q} \tag{3.14}$$

The equivalent thermal resistance R can be divided into two parts: (1) thermal resistance from the surface of the panel to the internal surface of pipes and (2) the thermal resistance resulting from the heat transfer between the chilled water and water pipes.

Thermal Resistance from the Surface of the Panel to the Internal Pipe Surface (R')

Thermal resistance R' of the radiant panel is decided by its structure and internal structural parameters. When the radiant panel structure is given, thermal resistance R' is nearly unaffected by the temperature in the high-temperature cooling and low-temperature heating operation conditions. By solving the heat conduction equation, the analytical expression of the thermal resistance of the radiant panel can be acquired.

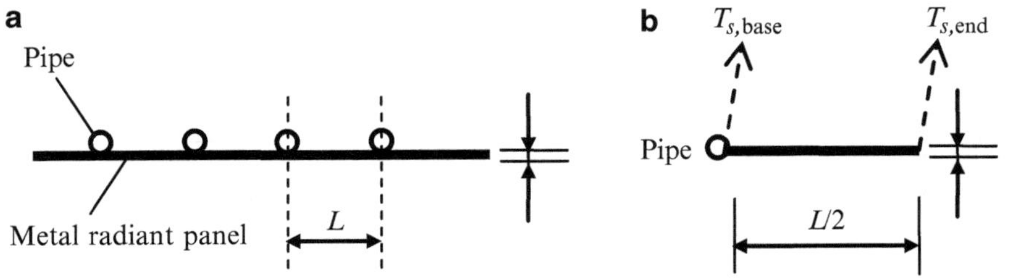

Fig. 3.11 Metal radiant panel structure: (**a**) panel structure and (**b**) simplified heat transfer model

For the actual radiant floor shown in Fig. 3.10, which is composed of three layers and a water pipe, the thermal resistance of each layer and the water pipe can be expressed as follows:

$$R' = \frac{L}{2\pi k_1}\left[\ln\left(\frac{L}{\pi\delta}\right) + \frac{2\pi d_1}{L} + \sum_{s=1}^{\infty}\frac{G(s)}{s}\right] + \frac{H_2}{k_2} + \frac{H_3}{k_3} + \frac{H_d}{k_d}\cdot\frac{L}{\pi\delta} \tag{3.15}$$

where L is the interval between adjacent water pipes, m; δ is the diameter of the water pipe, m; k is the thermal conductivity, W/(m · °C); H is the thickness of the layer, m; $G(s)$ is calculated according to Eq. (3.16); and h_z is the combined heat transfer coefficient, $h_z = h_c + h_{lr}$, W/(m^2 · °C):

$$G(s) = \frac{\frac{\text{Bi}+2\pi s}{\text{Bi}-2\pi s}e^{-\frac{4\pi s}{L}d_2} - 2e^{-\frac{4\pi s}{L}(d_1+d_2)} - e^{-\frac{4\pi s}{L}d_1}}{\frac{\text{Bi}+2\pi s}{\text{Bi}-2\pi s} + e^{-\frac{4\pi s}{L}(d_1+d_2)}}, \quad \text{where Bi} = h_z L/k \tag{3.16}$$

For examples:

- For a concrete radiant floor structure of pea gravel concrete (70 mm), cement mortar (25 mm), and granite (25 mm), with a pipe diameter of 20 mm and an interval between neighboring pipes of 150 mm, radiant floor heat resistance R' is 0.098 (m^2 · °C)/W.
- For a plastered capillary radiant panel with plastering with a thermal conductivity of 0.45 W/(m^2 · °C) and a thickness of 20 mm, a pipe diameter of 3.35 mm, and an interval between neighboring pipes of 15 mm, thermal resistance R' is 0.046 (m^2 · °C)/W.

For the metal radiant panel shown in Fig. 3.2b, the heat transfer process from the panel base (in contact with the water pipes) to the plate end can be simplified to the form shown in Fig. 3.11. The thickness (δ) and length (L) of the metal radiant panel are usually within the range of 0.5–2.0 mm and 70–200 mm, respectively. The temperature distribution along the thickness direction is almost uniform due to the high thermal conductivity of the metal material. Therefore, the heat transfer process for the radiant panel could be simplified as a one-dimensional heat transfer process along the direction of length L, with the temperature relation of the base temperature

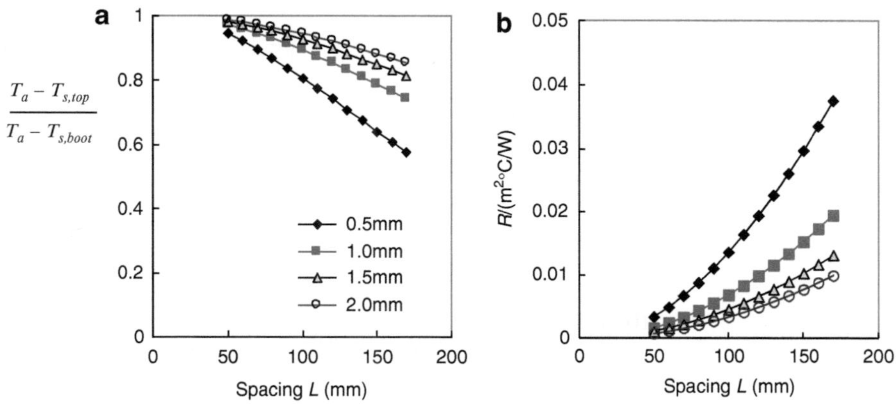

Fig. 3.12 Aluminum radiant panel end portion temperature and R': (**a**) temperature difference ratio at the end portion and the base portion and (**b**) thermal resistance R'

and the end portion, where T_z is the combined indoor temperature considering convection and radiation, $T_z = (h_c T_a + h_{lr}\text{AUST})/(h_c + h_{lr})$, °C:

$$T_{s,\text{end}} = T_z - (T_z - T_{s,\text{base}})/\text{ch}\left(\frac{L}{2}\sqrt{\frac{2h_z}{\lambda\delta}}\right) \tag{3.17}$$

Heat flux is

$$q = h_z \cdot (T_z - T_{s,\text{base}}) \cdot \text{th}\left(\frac{L}{2}\sqrt{\frac{2h_z}{\lambda\delta}}\right) / \left(\frac{L}{2}\sqrt{\frac{2h_z}{\lambda\delta}}\right) \tag{3.18}$$

Therefore, the heat resistance (R') of the metal radiant panel is

$$R' = \frac{1}{h_z}\left[\left(\frac{L}{2}\sqrt{\frac{2h_z}{\lambda\delta}}\right)/\text{th}\left(\frac{L}{2}\sqrt{\frac{2h_z}{\lambda\delta}}\right) - 1\right] \tag{3.19}$$

Figure 3.12 shows the end portion temperature of the aluminum radiant panel and thermal resistance R' according to the variation in thickness and spacing between the radiant panels ($h_c = 4.5$ W/(m^2 · °C), $h_{lr} = 5.5$ W/(m^2 · °C)). As the ratio of the temperature difference between the air and the radiant panel at the end portion and the base portion varies from 0.8 to 1.0, the radiant panel thermal resistance (R') is less than 0.015 (m^2 · °C)/W.

Thermal Resistance from the Internal Pipe Surface to the Chilled Water (R'')

Convective heat transfer coefficient h_w between the chilled water and the internal pipe surface can be calculated in accordance with the following equations (Zhang and Ren 2001):

$$Re \leq 2,300 \quad h_w = 1.86(Re \cdot Pr)^{1/3}\left(\frac{d}{l}\right)^{1/3}\frac{\lambda}{d} \tag{3.20}$$

$$2,300 < Re \leq 10,000$$
$$h_w = 0.012(Re^{0.87} - 280)Pr^{0.4}\left(1 + \left(\frac{d}{l}\right)^{2/3}\right)\frac{\lambda}{d} \tag{3.21}$$

$$Re > 10,000 \quad \begin{matrix} h_w = 0.023\mathrm{Re}^{0.8}Pr^{0.4}\dfrac{\lambda}{d} \quad \text{(cooling)} \\ h_w = 0.023\mathrm{Re}^{0.8}Pr^{0.3}\dfrac{\lambda}{d} \quad \text{(heating)} \end{matrix} \tag{3.22}$$

where Re and Pr are the Reynolds number and Prandtl number, respectively; λ is the thermal conductivity, W/(m · °C); d is the inner diameter of the water pipe, m; and l is the pipe length, m.

Common convective heat transfer coefficient h_w in radiant panel water pipes is shown in Fig. 3.13, where l is 5 m. The convective heat transfer coefficient under laminar flow conditions ($Re \leq 2{,}300$) is much smaller than that in the transition state (from $2{,}300 < Re \leq 10{,}000$) and during strong turbulent flow ($Re > 10{,}000$).

The convective heat transfer coefficient (h_w) can be translated into an equivalent heat transfer resistance R'' in accordance with the ratio of the radiant panel surface area to the inner area of the pipe:

$$R'' = \frac{1}{h_w} \cdot \frac{L}{\pi d} \tag{3.23}$$

Thermoplastic pipes and metal pipes are always used as the pipes in radiant panels, including plastic pipes with diameters of $\Phi 20$ and $\Phi 25$ and metal pipes with diameters of $\Phi 10$ and $\Phi 15$. For the thin and light radiant panel, a plastic pipe with a diameter of $\Phi 7$ is adopted, with a chilled water velocity of 0.5–1.0 m/s. In addition, a plastic tube with a diameter of $\Phi 3.35$ is used in the capillary radiant panel, with a chilled water velocity of 0.05–0.2 m/s. Common convective heat transfer coefficient h_w and the equivalent heat transfer resistance (R'') are listed in Table 3.3. The higher the water velocity is, the greater the convective heat transfer coefficient (h_w) will be, and, correspondingly, the smaller the equivalent heat transfer resistance (R'') will be.

For concrete structures and thin and light radiant panels, the ratio of R'' to R' is generally less than 5 %, so the thermal resistance from the internal surface of pipes to the chilled water (R'') could be ignored in the design calculations for engineering applications. For the metal radiant panel ($R' \approx 0.02$ (m^2 · °C)/W) and capillary radiant panel ($R' \approx 0.01$ (m^2 · °C)/W), the ratio of R'' to R' is in the range of 10–25 %, which needs to be taken into consideration.

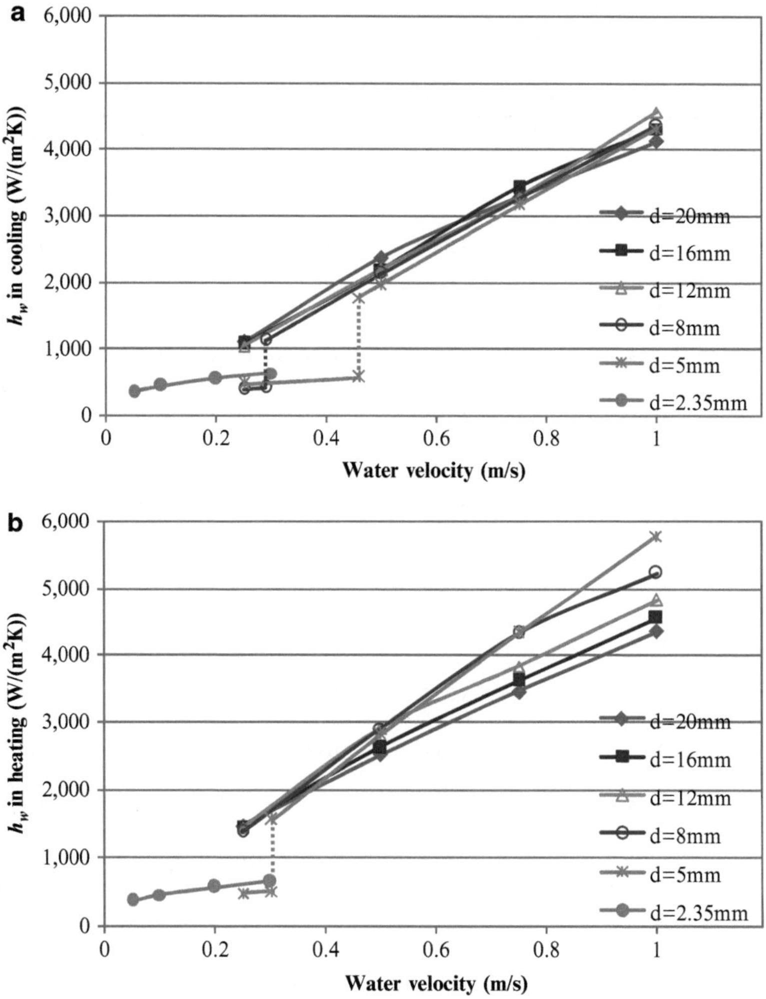

Fig. 3.13 Convective heat transfer coefficient h_w inside the pipes: (**a**) cooling mode with a water temperature of 20 °C and (**b**) heating mode with a water temperature of 40 °C

3.1.2.4 Surface Temperature Distribution of the Radiant Panel

Uniform Indoor Heat Sources

Apart from the mean surface temperature and heat flux, the surface temperature distribution is also an important aspect of the radiant panel, especially the lowest temperature of the panel surface in the cooling condition. The limitation of the lowest temperature for the panel surface is the dew point of the air near the panel.

Table 3.3 Thermal resistance from the internal surface of pipes to chilled/hot water (R'')

Pipe outer/inner diameter	Water flow rate	Pipe spacing	Cooling (chilled water: 20 °C)		Heating (hot water: 40 °C)	
			h_w	R''	h_w	R''
mm	m/s	mm	W/(m² · °C)	(m² · °C)/W	W/(m² · °C)	(m² · °C)/W
25/20	0.25	300/200	1,097	0.0044/0.0029	1,468	0.0033/0.0022
	0.5		2,371	0.0020/0.0013	2,504	0.0019/0.0013
	0.75		3,280	0.0015/0.0010	3,463	0.0014/0.0009
	1.0		4,128	0.0012/0.0008	4,360	0.0011/0.0007
20/16	0.25	300/200	1,076	0.0055/0.0037	1,463	0.0041/0.0027
	0.5		2,198	0.0027/0.0018	2,618	0.0023/0.0015
	0.75		3,429	0.0017/0.0012	3,621	0.0016/0.0011
	1.0		4,317	0.0014/0.0009	4,559	0.0013/0.0009
16/12	0.25	200/100	1,030	0.0052/0.0026	1,441	0.0037/0.0018
	0.5		2,191	0.0024/0.0012	2,902	0.0018/0.0009
	0.75		3,275	0.0016/0.0008	3,836	0.0014/0.0007
	1.0		4,572	0.0012/0.0006	4,829	0.0011/0.0005
10/8	0.25	150/75	392	0.0152/0.0076	1,368	0.0044/0.0022
	0.5		2,134	0.0028/0.0014	2,902	0.0021/0.0010
	0.75		3,272	0.0018/0.0009	4,335	0.0014/0.0007
	1.0		4,361	0.0014/0.0007	5,236	0.0011/0.0006
7/5	0.25	150/75	459	0.0208/0.0104	476	0.0201/0.0100
	0.5		1,963	0.0049/0.0024	2,814	0.0034/0.0017
	0.75		3,168	0.0030/0.0015	4,332	0.0022/0.0011
	1.0		4,321	0.0022/0.0011	5,783	0.0017/0.0008
3.35/2.35	0.05	20/10	345	0.0078/0.0039	358	0.0076/0.0038
	0.1		435	0.0062/0.0031	451	0.0060/0.0030
	0.2		548	0.0049/0.0025	568	0.0048/0.0024
	0.3		627	0.0043/0.0022	650	0.0042/0.0021

Concrete Radiant Floors

For the concrete radiant floor shown in Fig. 3.10, the temperature distribution of the floor surface is more uniform than the temperature distribution at the bottom of the floor; attenuation coefficient S is defined in Eq. (3.24), and it is a number usually smaller than 1. In this equation, $T_{s,max}$ and $T_{s,min}$ are the highest temperature and the lowest temperature, respectively, of the surface; T_g and T_h are the supply and return water temperatures, respectively:

$$S = \frac{T_{s,\max} - T_{s,\min}}{T_g - T_h} \tag{3.24}$$

According to the simplified analysis of the heat transfer process of the radiant panel, in engineering applications, attenuation coefficient S can be simplified as

$$\begin{aligned} S &\approx \frac{1}{[(h_z/k\beta_1 + 1)e^{\beta_1 H}/2]}, \quad \text{where} \quad \beta_1 = \frac{L}{\pi}, \quad H = H_1 + H_2 + H_3, \\ k &= \frac{k_1 e^{\beta_1 H_1} + k_2 e^{\beta_1 H_2} + k_3 e^{\beta_1 H_3}}{e^{\beta_1 H_1} + e^{\beta_1 H_3} + e^{\beta_1 H_3}} \end{aligned} \tag{3.25}$$

And the lowest surface temperature is

$$T_{s,\min} = T_s - \frac{S}{2}\left(T_g - T_h\right) \tag{3.26}$$

If the indoor heat sources are uniform, the surface temperatures are well distributed (the temperature difference between the mean and lowest points is less than 0.5 °C) and distinct from those of the supply chilled water for concrete structures.

Metal Radiant Panel

For the metal radiant panel shown in Fig. 3.2b, the lowest surface temperature is at the base of the panel ($T_{s,min} = T_{s,base}$), which is almost equal to the inlet temperature of the chilled water; the average surface temperature of the radiant panel can be calculated by

$$T_s = T_z - (T_z - T_{s,\text{base}}) \cdot \text{th}\left(\frac{L}{2}\sqrt{\frac{2h_z}{\lambda\delta}}\right) \Big/ \left(\frac{L}{2}\sqrt{\frac{2h_z}{\lambda\delta}}\right) \tag{3.27}$$

where T_z is the indoor integrated temperature considering heat convection and radiation, °C, and h_z is the combined heat transfer coefficient, $h_z = h_c + h_{lr}$, W/(m^2 · °C).

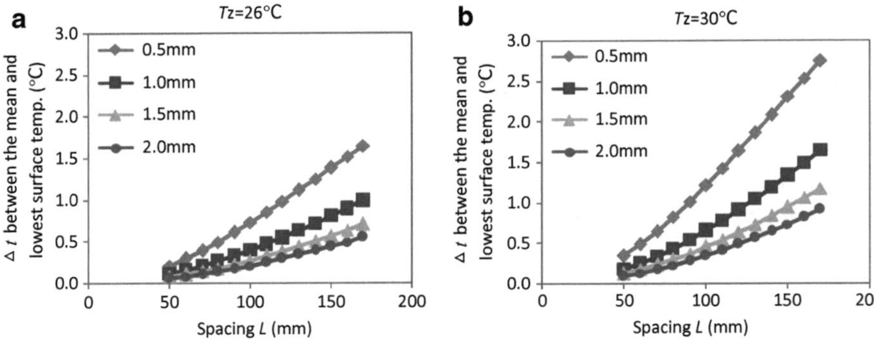

Fig. 3.14 Temperature difference between mean and lowest point of the metal radiant panel: (**a**) T_z is 26 °C and (**b**) T_z is 30 °C

Fig. 3.15 End portion temperature of the metal radiant panel

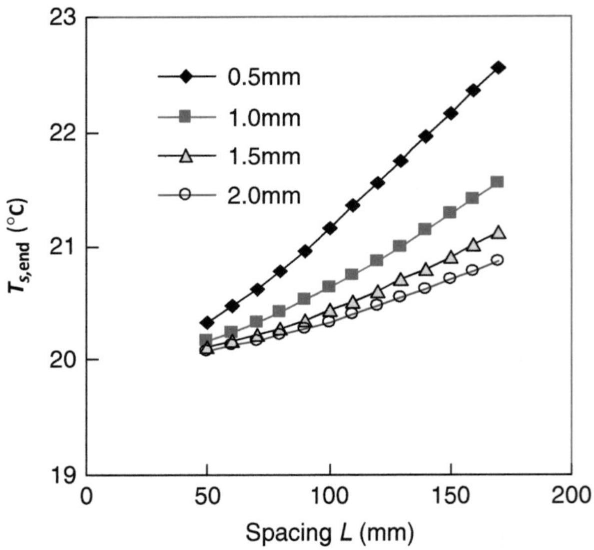

The temperature difference between the mean and lowest points is

$$T_s - T_{s,\text{base}} = (T_z - T_{s,\text{base}}) \cdot \left(1 - \text{th}\left(\frac{L}{2}\sqrt{\frac{2h_z}{\lambda\delta}}\right) \Big/ \left(\frac{L}{2}\sqrt{\frac{2h_z}{\lambda\delta}}\right)\right) \tag{3.28}$$

For metal radiant panels, the surface temperatures are hardly uniform, with a temperature difference of 1.5–3 °C as indicated in Fig. 3.14, and its minimum temperature is close to the chilled water temperature.

For the aluminum radiant panel shown in Figs. 3.11 and 3.12, the end portion temperature of the panel ($T_{s,\text{end}}$) according to the variance in panel thickness and spacing is shown in Fig. 3.15 ($T_z = 26$ °C, $T_{s,\text{base}} = 20$ °C):

- $L = 70$ mm: If the thickness of the radiant panel is 0.5, 1.0, 1.5, or 2.0 mm, the temperature difference between the end and the base portions is 0.6 °C, 0.3 °C, 0.2 °C, and 0.2 °C, respectively; as the thickness is 0.5 mm, the surface temperature is rather uniform.
- $L = 140$ mm: If the thickness of the radiant panel is 0.5, 1.0, 1.5, or 2.0 mm, the temperature difference between the end and the base portions is 2.0 °C, 1.1 °C, 0.8 °C, and 0.6 °C, respectively; as the thickness is 0.5 mm, the surface temperature is hardly uniform.

The cooling capacity of the radiant panel is closely related to its surface temperature. However, the lowest surface temperature must be higher than the dew point temperature of the surrounding air in order to avoid moisture condensation. Hence, a radiant panel with a uniform temperature distribution just above the dew point has the highest cooling capacity with a certain surrounding environment.

Nonuniform Indoor Heat Sources or Shading by Furniture

When the cold/heat transfer fluid temperature is given, the surface temperature of the radiant panel is closely related to the situation of the indoor heat sources. According to Eqs. (3.2) and (3.3), the average surface temperature of the radiant panel can be calculated by Eq. (3.29):

$$T_s = \frac{h_c \cdot T_a + h_{lr} \cdot \mathrm{AUST} + \overline{T}_w/R + q_{sr}}{h_c + h_{lr} + 1/R} \tag{3.29}$$

According to Eq. (3.29), the surface temperature difference of the radiant panel (ΔT_s) can be calculated as

$$\Delta T_s = \alpha_1 \cdot \Delta T_a + \alpha_2 \cdot \Delta\,\mathrm{AUST} + \alpha_3 \cdot \Delta \overline{T}_w + \beta \cdot \Delta q_{sr} \tag{3.30}$$

where

$$\alpha_1 = \frac{h_c}{h_c + h_{lr} + 1/R}, \quad \alpha_2 = \frac{h_{lr}}{h_c + h_{lr} + 1/R}, \quad \alpha_3 = \frac{1/R}{h_c + h_{lr} + 1/R}, \quad \beta = \frac{1}{h_c + h_{lr} + 1/R} \tag{3.31}$$

When only the changes of the internal wall surface temperature (AUST) and solar radiation q_{sr} are given, Eq. (3.30) can be simplified as

$$\Delta T_s = \alpha_2 \cdot \Delta\mathrm{AUST} + \beta \cdot \Delta q_{sr} \tag{3.32}$$

The values of factors α_2 and β for different values of radiant panel thermal resistance R are shown in Fig. 3.16. For instance, factors α_2 and β of a radiant panel

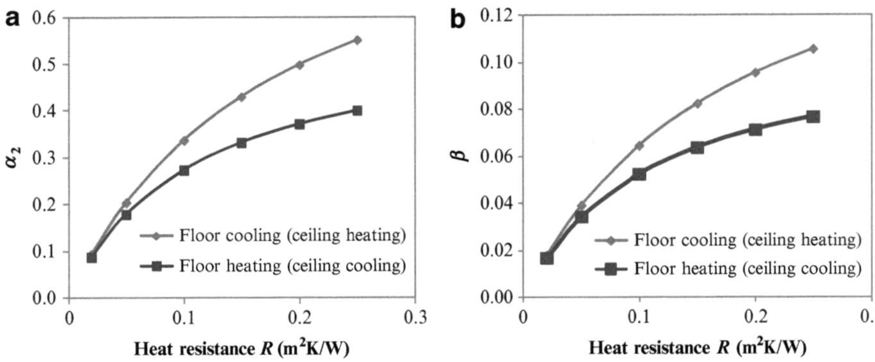

Fig. 3.16 Impact factors with different radiant panel thermal resistances: (**a**) indoor surface temperature factor α_2 and (**b**) solar radiation factor β

with an R of 0.1 (m^2 · °C)/W are 0.33 and 0.061, respectively. This means that the change in surface temperature (ΔT_s) is 1.7 °C when ΔAUST is 5 °C, and ΔT_s is 3.1 °C when Δq_{sr} is 50 W/m^2. Moreover, factors α_2 and β increase with an increase in the thermal resistance (R) of the radiant panel. In other words, surface temperatures of the radiant panels with higher R values are more susceptible to changes in shortwave solar radiation and indoor surface temperature AUST.

Several typical cases are shown in Fig. 3.17, where the chilled water temperature is 18 °C and the indoor air temperature (T_a) is 26 °C:

- For a radiant panel with a thermal resistance of 0.05 (m^2 · °C)/W, the surface temperature of the radiant panel (T_s) is about 22.1 °C, as AUST is 28 °C and q_{sr} is 50 W/m^2. However, it is only 20.1 °C if the radiant panel is shaded by furniture (AUST $= T_a$), resulting in a temperature difference of about 2 °C.
- For a radiant panel with a thermal resistance of 0.2 (m^2 · °C)/W, the surface temperature of the radiant panel (T_s) is about 28.0 °C, as AUST is 28 °C and q_{sr} is 50 W/m^2. However, it is only 23.2 °C if the radiant panel is shaded by furniture (AUST $= T_a$), resulting in a temperature difference of about 5 °C.

The higher the thermal resistance (R) of the radiant panel, the greater the difference between the surface temperatures with and without shading. By adopting this method, the lowest temperature of the radiant floor (i.e., under benches or in other places shaded by the indoor decoration) can be measured, and the radiant floor can be protected from moisture condensation through either controlling humidity or raising the chilled water temperature.

By analyzing of the unevenness of the surface temperature as a function of thermal resistance, it can be seen that:

- For radiant panels with higher R values (concrete structures, plastered capillary radiant panels), surface temperatures are well distributed and are distinct from the supply chilled water when the indoor heat sources are uniform. However,

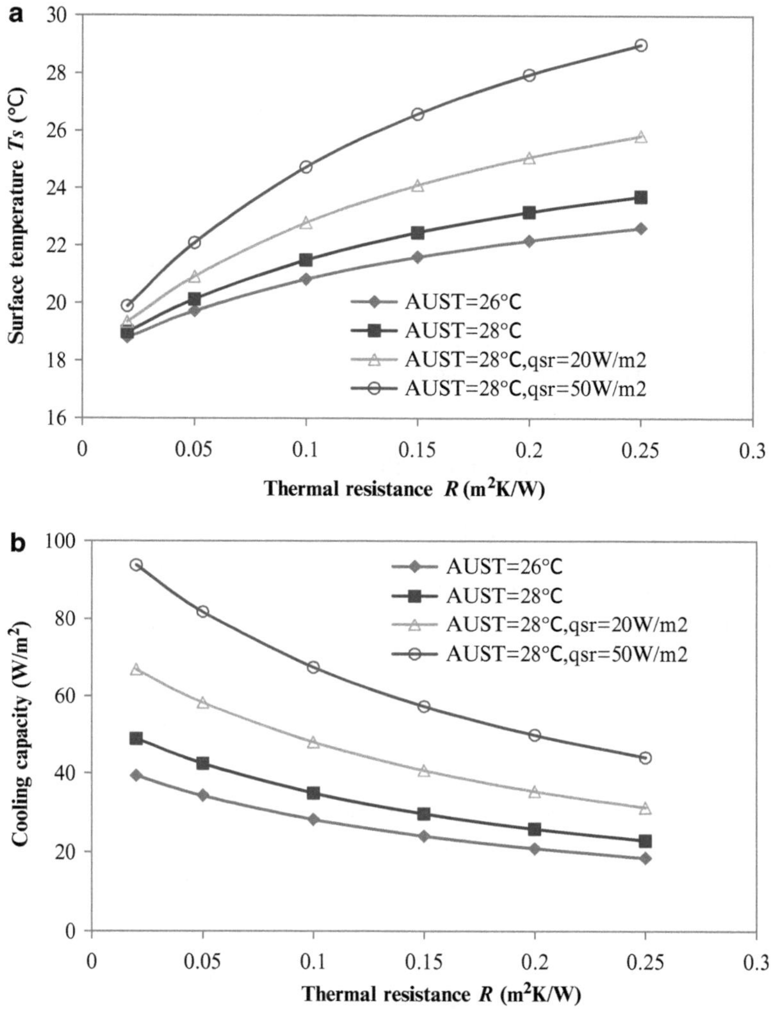

Fig. 3.17 Performance of radiant panels with variance in indoor environment and thermal resistance R: (**a**) surface temperature and (**b**) cooling capacity

surface temperatures of the radiant panels at different locations are significantly different; the surface temperature of radiant panels under furniture (e.g., chairs) will be the lowest and should therefore be protected from moisture condensation.

- For radiant panels with lower R values (e.g., metal radiant panels), surface temperatures are hardly uniform, and the minimum temperature is close to the chilled water temperature. However, the surface temperature is relatively uniform when it comes to nonuniform indoor heat sources or shading by furniture.

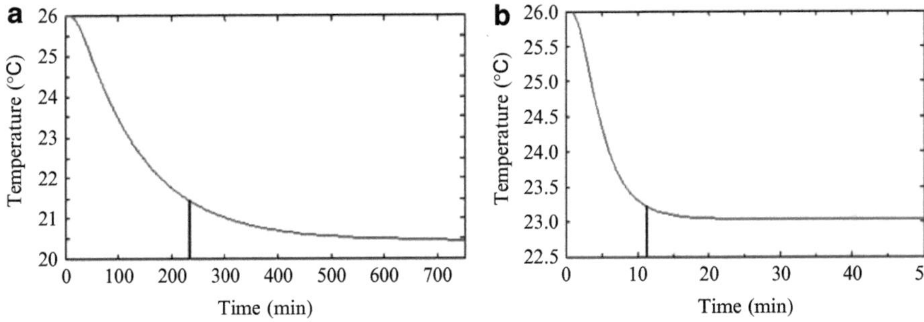

Fig. 3.18 Surface temperature and time constant of typical radiant panels: (**a**) concrete radiant floor and (**b**) capillary radiant panel with plaster

3.1.2.5 Thermal Inertia of Radiant Panels

Here, the concept of the "time constant" of automatic control is used to characterize the thermal inertia of the radiant terminals. The time constant is used as a time metric to measure how long the radiant terminal is required to approach the steady state in the heat transfer process. First, an energy equation is used to describe the dynamic heat transfer process of the radiant panel:

$$C\frac{\mathrm{d}T_{\mathrm{s}}}{\mathrm{d}\tau} = \frac{1}{R}\left(\overline{T}_{\mathrm{w}} - T_{\mathrm{s}}\right) + h_{\mathrm{c}} \cdot (T_{\mathrm{a}} - T_{\mathrm{s}}) + h_{\mathrm{lr}} \cdot (\mathrm{AUST} - T_{\mathrm{s}}) + q_{\mathrm{sr}} \tag{3.33}$$

where C is the heat capacity of the radiant panel filler (such as the concrete), J/(m · °C). When there is no shortwave radiation, and the wall temperature and the air temperature are the same, the time constant (T) of the radiant panel can be expressed as

$$T = \frac{C}{1/R + h_{\mathrm{c}} + h_{\mathrm{lr}}} = \frac{\sum_i c_{p,i}\rho_i\delta_i}{1/R + h_{\mathrm{c}} + h_{\mathrm{lr}}} \tag{3.34}$$

where $c_{p,i}$, ρ_i, and δ_i are the specific heat (kJ/(kg · °C)), density (kg/m^3), and thickness (m) of each layer of the radiant panel.

The time constant can be used as the metric of the thermal inertia of the radiant panel. Figure 3.18 shows the variance of the surface temperature of the radiant panels, including the concrete structure radiant floor and the plastered capillary radiant panel, where the time constant calculated according to Eq. (3.34) is plotted as the vertical line. As indicated by the figure, the time constant for the capillary radiant panel with plastering is around 10 min, while that for the concrete radiant floor is around 4 h. Thus, the smaller the thermal inertia of the radiant panel is, the faster the panel achieves a relatively steady state for cooling/heating in practical applications.

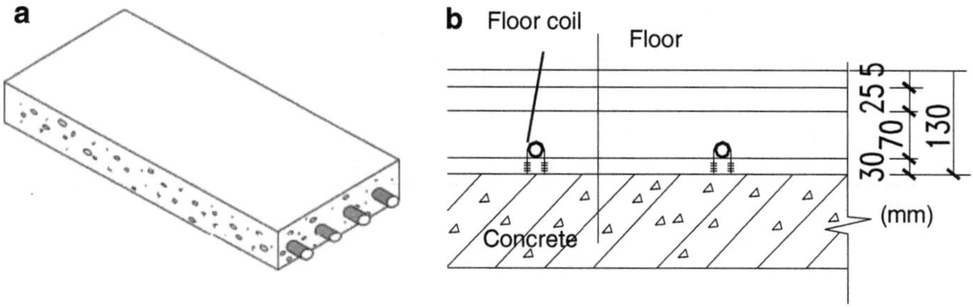

Fig. 3.19 Typical structure and form of concrete radiant panels: (**a**) schematic diagram; and (**b**) concrete radiant floor

3.1.3 Performances of Different Types of Radiant Panels

3.1.3.1 Concrete Radiant Floors

The concept of the concrete core follows that of the heating floor, which fixes plastic or stainless steel pipes with concrete reinforcing bars into poured concrete. This structure is widely used in Sweden and has also been used in the Beijing Tiptop International Apartments (Fig. 3.19). This kind of radiant panel has a mature structure technology and a low cost. On one hand, the large thermal mass of the concrete floor slab allows for adequate thermal storage. On the other hand, the large thermal inertia of the floor heating/cooling system means that the start-up time is long and the dynamic response is slow, leading to difficulties in both regulation and control.

Table 3.4 lists the thermal resistance and time constant of the concrete radiant floor with typical structures. Thermal resistance of the granite is around 0.1 ($m^2 \cdot$ °C)/W, while that of the plastic is around 0.15–0.2 ($m^2 \cdot$ °C)/W. The time constant of the radiant panel is around 3–4 h. From the calculated results of the thermal inertia for radiant panels, it can be seen that the time constant of this kind of radiant panel is large: it takes a long period of time to reach a stable cooling/heating effect. Compared to FCUs, concrete radiant floors have greater heat inertia if used for heating/cooling. For FCUs, heat exchange between the air and the water is realized through forced convection of fans. For radiant panel systems, heat is transferred from the cooling/heating fluid to the panel surface through heat conduction. The thermal inertia problem of radiant floors is significant because of the thickness of the material. If applied in airports and railway stations, where 24-h operation of air-conditioning systems is required, the effects of thermal storage are not obvious. However, these problems should be taken into consideration when applied in buildings with part-time operation.

3.1.3.2 Thin and Light Radiant Floors

The thin and light radiant floor has a small thermal inertia. In contrast to concrete radiant floors with a large thermal inertia, this kind of radiant floor is implemented

Table 3.4 Thermal resistances and time constants of concrete radiant floors with typical structures

	Configuration (from bottom to top)	Diameter of water pipe, mm	Space between adjacent pipes, mm	R (m^2 · °C/W)	Time constant, *h*	
					Floor cooling	Floor heating
Structure I	Pea gravel concrete (70 mm), cement mortar (25 mm), granite (25 mm)	20	150	0.098	3.8	3.2
Structure II	Pea gravel concrete (70 mm), cement mortar (25 mm), granite (25 mm)	20	200	0.116	4.2	3.4
Structure III	Pea gravel concrete (70 mm), cement mortar (25 mm), granite (25 mm)	20	250	0.138	4.6	3.7
Structure IV	Pea gravel concrete (70 mm), cement mortar (25 mm), granite (25 mm)	25	200	0.107	4.0	3.3
Structure V	Pea gravel concrete (50 mm), cement mortar (25 mm), granite (25 mm)	25	200	0.098	3.2	2.6
Structure VI	Pea gravel concrete (70 mm), cement mortar (25 mm), plastic (3 mm)	25	200	0.160	3.7	2.9
Structure VII	Pea gravel concrete (70 mm), cement mortar (25 mm), plastic (5 mm)	25	200	0.200	4.1	3.1

Notes: The thermal resistance of the radiant panels in the table is the same as $R'(R = R' + R'')$, because R'' is much smaller compared to R' (referring to Table 3.3). The thermal resistance for each material is as follows: pea gravel concrete is 1.84 W/(m · °C), cement mortar is 0.93 W/(m · °C), granite is 3.93 W/(m · °C), PERT plastic pipe is 0.4 W/(m · °C), and plastic floor is 0.05 W/(m · °C)

by putting pipes directly into insulation board with grooves on them. The pipes and insulation plates can be integrated together and then covered with decorative floors. Figure 3.20 illustrates a form of the thin and light radiant floor without the need for concrete. The pipes are laid in grooves of foam plastic insulation board and covered with metal plants or metal membranes to make the surface temperature evenly distributed (the pipes are the same size and height as the grooves). Thin and light radiant floors have been developing rapidly in recent decades, and they have been adopted in multiple residential buildings in China.

Table 3.5 lists the thermal resistance and time constant of thin and light radiant floors with typical structures; the thermal resistance is around 0.2–0.3 (m^2 · °C)/W. The time constant of this kind of radiant floor is small, which means stable

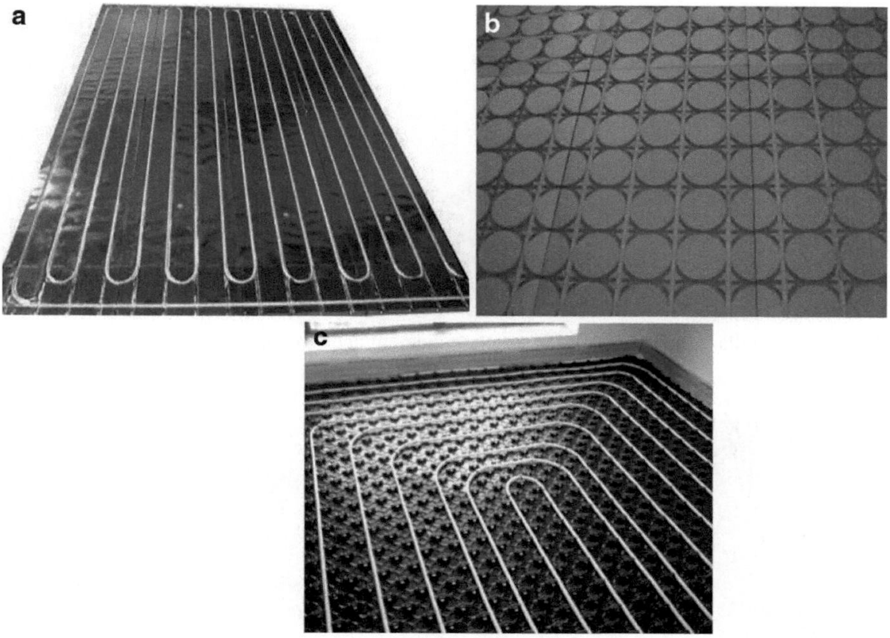

Fig. 3.20 Typical insulation plates with grooves in thin and light radiant floors: (**a**) type I, (**b**) type II, and (**c**) type III

Table 3.5 Thermal resistance and time constant of thin and light radiant floors with typical structures

	Configuration (from bottom to top)	Diameter of water pipe, mm	Space between adjacent pipes, mm	R ($m^2 \cdot °C$)/W	Time constant, *h*	
					Floor cooling	Floor heating
Structure I	Insulation plate (30 mm), aluminum foil (0.1 mm), wooden floor (10 mm)	20	250	0.340	0.32	0.25
Structure II	Insulation plate (25 mm), aluminum foil (0.1 mm), wooden floor (10 mm)	16	150	0.239	0.27	0.22
Structure III	Insulation plate (25 mm), cement mortar screed and brick (30 mm)	16	150	0.204	1.17	0.96
Structure IV	Insulation plate (12 mm), aluminum foil (0.12 mm), wooden floor (10 mm)	7	75	0.272	0.25	0.20

Notes: The thermal resistance of the radiant panels in the table is the same as R' ($R = R' + R''$), because R'' is much smaller compared to R' (referring to Table 3.3)

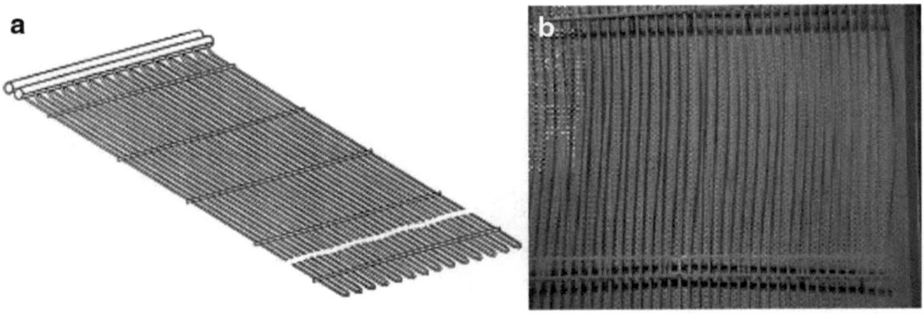

Fig. 3.21 Capillary radiant panel: (**a**) structure diagram and (**b**) sample

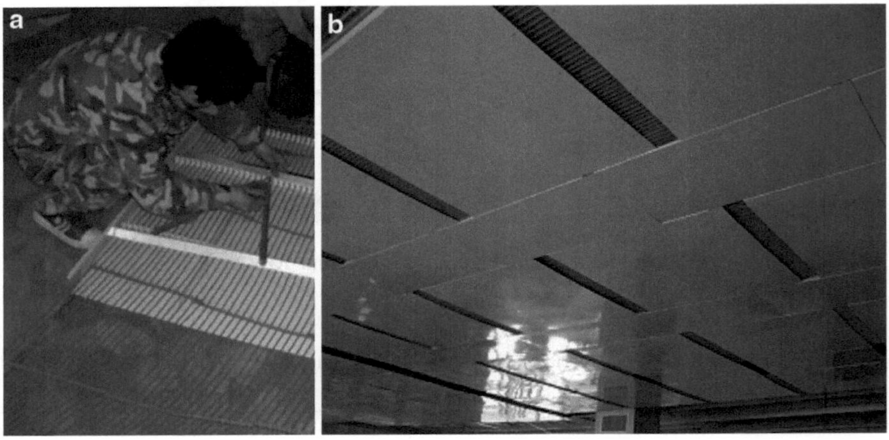

Fig. 3.22 Radiant panels combining capillaries and metal plates: (**a**) fixing pipes onto the metal plate and (**b**) photograph after setup

cooling/heating conditions can be obtained in a short time. The reduction of the thermal inertia indicates a shorter start-up time and a faster dynamic response.

3.1.3.3 Capillary Radiant Panels

Capillary radiant panels are made of densely covered plastic pipes with a small diameter (2–3 mm) and a small interval between neighboring pipes (10–20 mm). The pipes connect the distributor pipe on one side and the collector pipe on the other side to form a grid structure, as shown in Fig. 3.21. Water flow velocity is slow in the pipes; it is around 0.05–0.2 m/s, which is similar to that of blood flow in the capillaries of the human body. This structure can be combined with metal panels or concrete slabs and is thus widely used in practice in retrofitting projects.

When the capillary radiant panel is combined with a metal plate, good contact should be guaranteed in order to achieve a better heat transfer effect. Figure 3.22

shows a kind of capillary radiant panel where the pipe and the metal plate are not glued together. This leads to a high heat transfer resistance, and its cooling capacity is much lower compared to structures without metal plates.

Table 3.6 lists the thermal resistances of plastered capillary radiant panels with typical structures, where the capillary diameter is 3.35 mm. The thermal resistance and time constant of the capillary radiant panels, which are around 0.02–0.06 ($m^2 \cdot$ °C)/W and 5–15 min, respectively, are obviously lower than those of the concrete radiant floor.

3.1.3.4 Flat Metal Radiant Ceilings

The sandwich structure of the flat metal radiant ceiling mainly consists of metal, such as copper, aluminum, or steel. From the cross-sectional diagram, it can be seen that the pipe is in the middle, while the insulation material and the cover plate are on the top and the backing strap is on the bottom, as shown in Fig. 3.23a, b. As this type of structure includes decorative features, it is the most widely used structure. The indoor appearance after setup can be seen in Fig. 3.23c.

Because of the metal components in this structure, the mass is usually large, leading to a higher cost. Moreover, the fin efficiency is usually low due to the influence of the thickness of the panel. Figure 3.24 displays measurements made with an infrared thermometer, and it can be seen that the surface temperature is not evenly distributed.

According to Figs. 3.11 and 3.12, the thermal resistances of radiant panels with different thicknesses and intervals between neighboring pipes can be calculated. The results indicate that, if the ratio of temperature difference between the air and the radiant panel at the end portion to that at the base portion is within 0.8–1.0 (e.g., air temperature is 26 °C, panel surface temperature at the base portion is 20 °C, and panel surface temperature at the end portion is lower than 21.2 °C), the thermal resistance of the metal radiant panel is less than 0.015 ($m^2 \cdot$ °C)/W. Table 3.7 lists the thermal resistance and time constant of metal panels with typical structures. The time constants of the metal panels are quite small, i.e., less than 1 min.

3.1.3.5 Suspended Metal Radiant Ceilings with Enhanced Heat Convection

When adopting radiant panels for cooling, the surface temperature is required to be higher than the surrounding air dew point temperature to avoid condensation on the panel surfaces, which restricts the cooling capacity of the panels per unit area. For a certain installation area of radiant panels, strengthening convection is an effective way to improve cooling capacity. Figure 3.25 illustrates an actual device and the installation appearance of a type of suspended metal radiant ceiling with enhanced convection (Zhang et al. 2013). For this new type of radiant panel, the

Table 3.6 Thermal resistance and time constant of plastered capillary radiant panels with typical structures

	Packing layer material	Packing layer thickness, mm	Space between adjacent pipes, mm	R' (m^2 · °C)/W	R'' (m^2 · °C)/W	R (m^2 · °C)/W	Time constant, min		
							Ceiling cooling	Vertical heating or cooling	Ceiling heating
Structure I	Parget ($\lambda = 0.45$ W/(m · °C))	20	15	0.046	0.004–0.007	0.050–0.053	11	12	13
Structure II	Parget ($\lambda = 0.87$ W/(m · °C))	20	15	0.025	0.004–0.007	0.029–0.032	7	8	8
Structure III	Parget ($\lambda = 0.45$ W/(m · °C))	20	30	0.063	0.008–0.013	0.071–0.076	13	15	16
Structure IV	Parget ($\lambda = 0.45$ W/(m · °C))	10	15	0.024	0.004–0.007	0.028–0.031	4	4	4
Structure V	Cement mortar ($\lambda = 1.5$ W/(m · °C))	20	15	0.015	0.004–0.007	0.019–0.022	8	9	9

Notes: R'' is related to water velocity inside the pipes, which is taken as 0.05–0.2 m/s; the higher the velocity, the larger the R'' value

Fig. 3.23 Flat metal radiant ceiling panel: (**a**) sample, (**b**) top view, and (**c**) indoor appearance after setup

heat convection area is about 1.2–1.6 times greater than the area for radiation, indicating that about 20–60 % more area is added for heat convection.

Problems similar to those for flat metal radiant ceilings exist for this type of suspended metal radiant ceiling, e.g., the metal consumption should be taken into consideration and the surface temperature distribution is hardly uniform. Figure 3.26 shows the schematic diagram analyzing the surface temperature uniformity of this suspended radiant panel. The structure of this new kind of metal radiant ceiling is similar to that of the flat metal radiant ceiling. The major difference is that in order to enhance the heat convection per unit of projected area, there is a certain angle of slant in the installation of the new kind of metal radiant ceiling. However, the same method could be adopted for these two types of radiant panels. The influences of thickness and interval between neighboring pipes on the surface temperature uniformity are shown in Figs. 3.12 and 3.15, respectively. The thermal resistance and time constant for this kind of radiant panel are listed in Table 3.7.

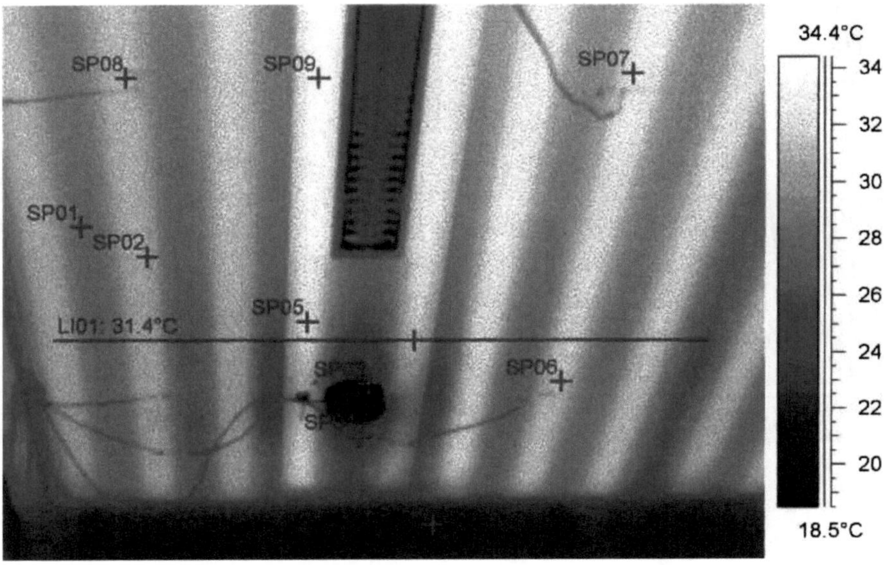

Fig. 3.24 Temperature distribution of the radiant surface measured by infrared thermometer

Table 3.7 Thermal resistance and time constant of metal (aluminum) panels with typical structures

	Thickness of metal δ, mm	Space between adjacent pipes L, mm	R' ($m^2 \cdot$ °C/W)	R'' ($m^2 \cdot$ °C/W)	R ($m^2 \cdot$ °C/W)	Time constant, min	
						Ceiling cooling	Ceiling heating
Structure I	0.5	80	0.009	0.001–0.002	0.010–0.012	0.2	0.2
Structure II	0.5	100	0.014	0.001–0.002	0.015–0.016	0.2	0.2
Structure III	1.0	100	0.007	0.001–0.002	0.008–0.009	0.3	0.3
Structure IV	1.0	140	0.013	0.001–0.003	0.014–0.016	0.5	0.5
Structure V	1.5	140	0.009	0.001–0.004	0.010–0.013	0.5	0.5

Notes: R'' is related to water velocity inside the pipes, which is taken as 0.4–0.8 m/s; the higher the velocity, the larger the value of R''

3.1.3.6 Performance Summary of Different Radiant Panels

Table 3.8 gives the performance comparison of the five types of radiant panels described in this section. For the flat metal radiant ceiling and the suspended metal radiant ceiling, the thermal resistance along the direction of plate thickness δ is relatively small; the main thermal resistance is along the direction of interval

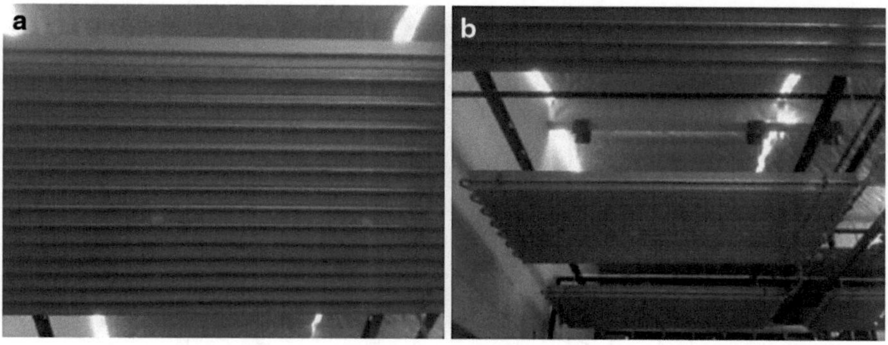

Fig. 3.25 Suspended metal radiant ceiling with strengthened heat convection: (**a**) an actual device and (**b**) its installation appearance

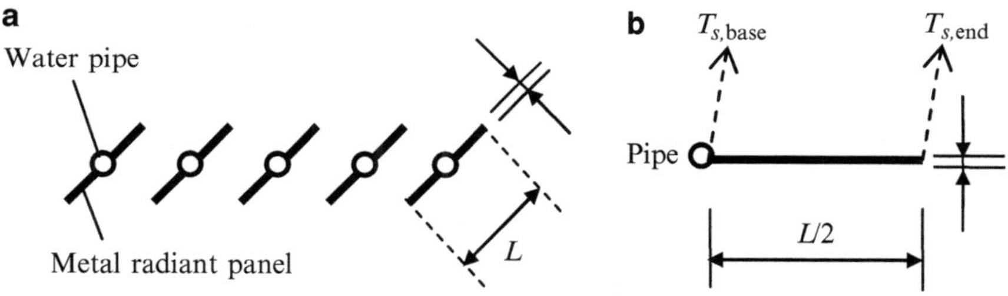

Fig. 3.26 Simplified model for the heat transfer of the suspended metal radiant panel: (**a**) schematic diagram of the panel and (**b**) simplified model

L between neighboring pipes. Thus, it is easy for the surface temperature distribution to be nonuniform, and the lowest surface temperature of the metal radiant panel approaches the supply chilled water temperature. For the concrete radiant floor and the capillary radiant ceiling with plastering, the thermal resistance is mainly along the direction of the plate thickness, which helps to make the surface temperature distribution more uniform. However, the temperature difference between the surface temperature (or the lowest surface temperature) and the supply water temperature is relatively larger compared to the metal type.

3.1.4 Performance in Summer

For radiant panels with different installations (ceiling type, floor type, or vertical wall type), if the indoor environmental parameters are the same and the average surface temperatures are also identical, the heat transfer rates between the panels and the environment will be the same for the different radiant panels (Jin et al. 2010; Shi and Zhang 2012). The major discrepancy for these radiant panels lies in

Table 3.8 Performance comparison between different radiant panels

	$R = R' + R''$			Surface temperature distribution		
	R' $(m^2 \cdot °C)/W$	R'' $(m^2 \cdot °C)/W$	Time constant	Uniform indoor heat sources and no shelter	Nonuniform indoor heat sources and partly sheltered[a]	Δt Between the surface and the supply water
Concrete radiant floor	~0.1 (granite)	<0.005	3–4 h	Surface temperature distribution is more uniform, and S is quite small	Nonuniform $\alpha_2 \approx 0.25–0.4$ $\beta \approx 0.05–0.08$	Δt is relatively larger
Light radiant floor	0.2–0.3	<0.005	15–20 min		Nonuniform $\alpha_2 \approx 0.4–0.6$ $\beta \approx 0.07–0.11$	
Capillary radiant ceiling with plastering	0.02–0.06	~0.005	5–15 min		A bit more uniform $\alpha_2 \approx 0.1–0.25$ $\beta \approx 0.02–0.05$	Large
Flat metal radiant ceiling	<0.02	<0.005	<1 min	Nonuniform surface temperature distribution and $S \approx 1$	A more uniform surface temperature distribution	Approaching the supply water temperature
Suspended metal radiant ceiling	<0.02	<0.005	<1 min		$\alpha_2 \approx 0.04–0.08$ $\beta \approx 0.01$	

[a]*Notes*: Variance of the mean surface temperature refers to Eq. (3.30); units: α_2 – dimensionless; β – $(m^2 \cdot °C)/W$

the different temperature requirements of the supply and return chilled or hot water. The smaller the thermal resistance of the radiant panel itself, the closer the panel surface temperature will be to the chilled water temperature. Table 3.9 lists the cooling capacities of radiant panels with different types of installations, where the average surface temperature and the temperature of the indoor environment are the same for all types. It can be seen that with the increase of the temperature difference between the indoor air temperature (T_a or AUST) and the surface temperature of the radiant panel (T_s), the cooling capacity of the radiant panel per unit area increases. And the radiant panel heat flux also increases with the increase of the shortwave radiation reaching the surface of the radiant panel.

For example:

- If $T_s = 20$ °C and $T_a =$ AUST $= 26$ °C, the radiant panel heat flux (q) is 34.3 W/m^2 if there is no shortwave radiation (i.e., solar radiation) directly reaching the surface. For radiant floors with a thermal resistance of 0.1 (m^2 · °C)/W, the required mean temperature of the chilled water is 16.6 °C.
- If $T_s = 20$ °C and $T_a =$ AUST $= 26$ °C, the radiant panel heat flux (q) is then 74.3 W/m^2 if there is shortwave radiation (i.e., solar radiation) of about 40 W/m^2 directly reaching the surface. Thus, the cooling capacity of the radiant panels increases significantly when there is shortwave irradiation. And for radiant floors with a thermal resistance of 0.1 (m^2 · °C)/W, the required mean temperature of the chilled water will be 12.6 °C.

3.1.4.1 Concrete Radiant Floors

Table 3.10 lists the operating performances of concrete radiant floors for cooling. Due to a relatively large thermal resistance of this kind of radiant panel, the surface temperature distribution is almost uniform if there is no partially shaded area. Thus, in this table, only the variances of the surface mean temperature and the cooling capacity of the radiant panels are listed. For the same indoor environment, if the thermal resistance of the radiant floor is larger, the required mean temperature of the supply and return chilled water will be lower for the same surface temperature of the radiant panel. For example, if $T_a =$ AUST $= 26$ °C and the shortwave radiation $q_{sr} = 0$, in order to achieve a surface temperature of 21 °C for the radiant floor, the required mean temperature of the supply and return chilled water ($\overline{T}_w$) is about 18 °C for Structure II, with a thermal resistance of 0.116 (m^2 · °C)/W. For Structure VII with a thermal resistance of 0.2 (m^2 · °C)/W, the required $\overline{T}_w$ is as low as 16 °C.

The cooling capacity of the radiant floor per unit area is significantly influenced by the operating environment (i.e., the environmental parameters including air temperature, surrounding wall temperatures, solar radiation) and the temperatures of the supply and return water. Taking Structure I of the concrete radiant floor as an example, with a structure of 70-mm-thick pea gravel concrete + 25-mm-thick cement mortar + 25-mm-thick granite, an external diameter of 20 mm for the

Table 3.9 Influence of installation and indoor environment on the cooling capacities of different radiant panels

	T_s, °C	T_a, °C	AUST, °C	q_{sr}, W/m^2	q, W/m^2	Required average chilled water temperature, °C			
						$R = 0.01$ (m^2 · °C)/W	$R = 0.05$ (m^2 · °C)/W	$R = 0.1$ (m^2 · °C)/W	$R = 0.2$ (m^2 · °C)/W
Ceiling cooling	20	26	26	0	55.3	19.4	17.2	–	–
	18	26	26	0	76.5	17.2	14.2	–	–
	22	26	26	0	35.1	21.6	20.2	–	–
	20	26	28	0	66.3	19.3	16.7	–	–
	20	26	26	10	65.3	19.3	16.7	–	–
	20	26	26	20	75.3	19.2	16.2	–	–
Vertical wall cooling	20	26	26	0	51.9	19.5	17.4	14.8	–
	18	26	26	0	71.7	17.3	14.4	10.8	–
	22	26	26	0	33.1	21.7	20.3	18.7	–
	20	26	28	0	62.9	19.4	16.9	13.7	–
	20	26	26	20	71.9	19.3	16.4	12.8	–
Floor cooling	20	26	26	0	34.3	–	18.3	16.6	13.1
	22	26	26	0	22.8	–	20.9	19.7	17.4
	20	26	28	0	45.3	–	17.7	15.5	10.9
	20	26	26	20	54.3	–	17.3	14.6	9.1
	20	26	28	20	65.3	–	16.7	13.5	6.9
	20	26	26	40	74.3	–	16.3	12.6	5.1

Notes: $h_{lr} = 5.5$ W/(m^2 · °C), and h_c is from Lu (2008); all types of radiant panels are suitable

Table 3.10 Operating performances of concrete radiant floors with different structures for cooling

			$T_a = 26$ °C, AUST = 26 °C, $q_{sr} = 0$		$T_a = 26$ °C, AUST = 28 °C, $q_{sr} = 0$		$T_a = 26$ °C, AUST = 26 °C, $q_{sr} = 50$ W/m^2	
	R (m^2 · °C)/W	$\overline{T}_w$, °C	T_s, °C	q, W/m^2	T_s, °C	q, W/m^2	T_s, °C	q, W/m^2
Structure I	0.098	16	19.5	35.4	20.1	42.2	22.7	68.1
		18	20.8	28.3	21.4	35.1	24.0	61.0
		20	22.1	21.2	22.7	28.0	25.3	53.9
Structure II	0.116	16	19.9	33.3	20.6	39.7	23.4	64.0
		18	21.1	26.6	21.8	33.0	24.6	57.3
		20	22.3	20.0	23.1	26.4	25.9	50.7
Structure III	0.138	16	20.3	31.0	21.1	37.0	24.2	59.6
		18	21.4	24.8	22.2	30.8	25.4	53.4
		20	22.6	18.6	23.4	24.6	26.5	47.2
Structure IV	0.107	16	19.7	34.3	20.4	40.9	23.1	66.0
		18	20.9	27.4	21.6	34.0	24.3	59.1
		20	22.2	20.6	22.9	27.2	25.6	52.2
Structure V	0.098	16	19.5	35.4	20.1	42.2	22.7	68.1
		18	20.8	28.3	21.4	35.1	24.0	61.0
		20	22.1	21.2	22.7	28.0	25.3	53.9
Structure VI	0.160	16	20.6	29.0	21.5	34.6	24.9	55.8
		18	21.7	23.2	22.6	28.8	26.0	50.0
		20	22.8	17.4	23.7	23.0	27.1	44.2
Structure VII	0.200	16	21.2	26.0	22.2	31.0	26.0	50.0
		18	22.2	20.8	23.2	25.8	27.0	44.8
		20	23.1	15.6	24.1	20.6	27.9	39.6

supply and return water pipe, a distance of 150 mm between neighboring pipes, and $\overline{T}_w = 16$ °C:

- If the indoor air temperature and the surrounding wall temperatures are all 26 °C, the floor surface mean temperature is 19.5 °C and the radiant floor heat flux (q) is 35.4 W/m^2 when there is no shortwave radiation (i.e., solar radiation) directly reaching the surface.
- If the surrounding wall temperatures are 28 °C and other conditions are the same as above, the mean surface temperature for the radiant floor increases to 20.1 °C and the heat flux is then about 42.2 W/m^2, representing an increase of about 20 %.
- If the indoor air temperature and the surrounding wall temperatures are all 26 °C and there is solar radiation of 50 W/m^2 directly reaching the floor surface, the mean surface temperature for the radiant floor goes up to 22.7 °C. The radiant floor heat flux increases to 68.1 W/m^2, which is almost twice as large as the heat flux with no solar radiation.

Table 3.11 Operating performances of capillary radiant ceilings with plastering

			$T_a = 26$ °C, AUST $= 26$ °C, $q_{sr} = 0$		$T_a = 26$ °C, AUST $= 28$ °C, $q_{sr} = 0$	
	R (m^2 · °C)/W	$\overline{T}_w$, °C	T_s, °C	q, W/m^2	T_s, °C	q, W/m^2
Structure I	0.051	16	19.2	62.0	19.5	68.8
		18	20.5	48.6	20.8	55.6
		20	21.8	35.6	22.2	42.6
Structure II	0.030	16	18.2	72.3	18.4	80.2
		18	19.7	56.5	19.9	64.5
		20	21.2	41.2	21.5	49.3
Structure III	0.073	16	19.9	54.0	20.4	60.0
		18	21.1	42.5	21.5	48.6
		20	22.3	31.2	22.7	37.4
Structure IV	0.029	16	18.1	72.9	18.3	80.8
		18	19.7	57.0	19.9	65.0
		20	21.2	41.5	21.4	49.7
Structure V	0.020	16	17.6	78.6	17.7	87.2
		18	19.2	61.3	19.4	70.0
		20	20.9	44.6	21.1	53.3

3.1.4.2 Capillary Radiant Ceilings with Plastering

Table 3.11 lists the operating performances of the capillary radiant ceilings with plastering for cooling. It can be seen that similar conclusions can be drawn for this type of radiant panel as for the concrete radiant floors described above. For the same indoor environment, the required mean temperature of the supply and return chilled water will be lower if the thermal resistance of the radiant floor is larger for the same average surface temperature of the radiant panel. In addition, the cooling capacity of the capillary radiant ceiling per unit area is significantly influenced by the operating environment (i.e., air temperature, surrounding wall temperature, solar radiation) and the temperature of the supply and return water.

Comparing the cooling capacities of the capillary radiant ceilings with plastering listed in Table 3.11 and the concrete radiant floors listed in Table 3.10, it can be seen that the cooling capacities of the capillary radiant ceilings with plastering are higher than those of the concrete radiant floors (with the same mean temperature of the supply and return water and the same indoor environment). This performance discrepancy is mainly due to two key factors: first, the convective heat transfer coefficient for the radiant cooling ceiling is significantly higher than that of the radiant cooling floor (as indicated in Table 3.2), and, second, the thermal resistances of the capillary radiant ceilings are considerably lower than those of the concrete radiant floors.

3.1.4.3 Metal Radiant Ceilings

Table 3.12 shows the performance comparison between the flat metal radiant panels and the suspended metal radiant panels for cooling, where the angle of slant for the

Table 3.12 Cooling performance comparison of flat and suspended metal radiant ceilings (T_a = AUST = 26 °C, q_{sr} = 0)

	Inlet and outlet temp. of chilled water, °C	Flat metal radiant panel		Suspended metal radiant panel	
		T_s, °C	q, W/m^2	T_s, °C	q, W/m^2
Structure I	18, 21	20.0	52.5	20.3	86.1
	18, 23	20.9	44.4	21.1	72.9
Structure II	18, 21	20.2	50.5	20.6	81.0
	18, 23	21.1	42.7	21.4	68.6
Structure III	18, 21	19.9	53.3	20.1	88.3
	18, 23	20.8	45.1	21.0	74.7
Structure IV	18, 21	20.2	50.6	20.6	81.3
	18, 23	21.1	42.8	21.4	68.8
Structure V	18, 21	20.0	52.4	20.3	85.9
	18, 23	20.9	44.3	21.2	72.7

installation of the suspended metal type is 45°, indicating that its area for heat convection is about 2.8 times larger than that of the flat metal radiant panel with the same projected area. The cooling capacities listed in this table are calculated with the same supply water temperature for the radiant cooling panels; the lowest temperature is about 18 °C for all surfaces. The cooling capacity of the metal radiant panel increases with the decrease of the temperature difference between the supply and return water (in the condition when the lowest surface temperature remains the same). Taking Structure I of the flat metal radiant roof as an example, the cooling capacities of the radiant panels are 44.4 and 52.5 W/m^2 when the supply/return water temperatures are 18 °C/23 °C and 18 °C/21 °C, respectively. Therefore, the increase of the cooling capacity with a lower water temperature difference is about 18 %. This is why the operating mode of "higher water flow rate and lower temperature difference" is usually chosen for radiant panels in the condition of no condensation.

Compared with the flat metal radiant panel, the suspended metal radiant panel effectively increases the area for heat convection with the same projected area; thus, its cooling capacity increases significantly. The results shown in Table 3.12 indicate that the increasing percentage of the cooling capacity is about 60–65 %. The cooling capacity for the suspended metal radiant roof per unit of projected area is 68–75 W/m^2 if the supply and return water temperatures are 18 °C/23 °C and increases to 81–88 W/m^2 if the water temperatures are 18 °C/21 °C. The cooling capacity of the suspended metal radiant ceiling is sufficient for the cooling requirement in most commercial buildings.

3.1.5 Performance in Winter

Tables 3.13 and 3.14 list the heating performances of the concrete radiant floor and the capillary radiant ceiling with plastering in winter, respectively, where the indoor

Table 3.13 Heating performances of typical concrete radiant floors in winter

		Performance with hot water inlet/outlet temperature of 40 °C/30 °C		Performance with hot water inlet/outlet temperature of 35 °C/30 °C	
	R, (m^2 · °C)/W	T_s, °C	q, W/m^2	T_s, °C	q, W/m^2
Structure I	0.098	27.9	72.2	26.7	59.4
Structure II	0.116	27.3	66.4	26.2	54.7
Structure III	0.138	26.6	60.5	25.6	49.9
Structure IV	0.107	27.6	69.2	26.4	57.0
Structure V	0.098	27.9	72.2	26.7	59.4
Structure VI	0.160	26.1	55.6	25.2	45.9
Structure VII	0.200	25.3	48.4	24.5	40.0

Table 3.14 Heating performances of typical capillary radiant ceilings with plastering in winter

		Performance with hot water inlet/outlet temperature of 40 °C/30 °C		Performance with hot water inlet/outlet temperature of 35 °C/30 °C	
	R, (m^2 · °C)/W	T_s, °C	q, W/m^2	T_s, °C	q, W/m^2
Structure I	0.051	31.6	67.1	29.6	55.9
Structure II	0.030	32.8	74.1	30.6	61.8
Structure III	0.073	30.5	61.1	28.8	50.9
Structure IV	0.029	32.8	74.5	30.7	62.1
Structure V	0.020	33.4	78.0	31.2	65.0

air temperature and the surrounding wall temperatures are 20 °C. The figures show that the heating performance discrepancy for the two types of radiant panels is not as significant as that in summer. Although the thermal resistances of the concrete radiant floors are considerably higher than those of the capillary radiant ceilings with plastering, the convective heat transfer coefficient of the radiant floor is higher than that of the radiant ceiling, so the performance discrepancy is not significant. Moreover, as the supply and return hot water temperatures are 35 °C/30 °C, the radiant panels can satisfy the heating requirement for most buildings.

3.1.6 Self-Regulating Property of Radiant Panels

In addition to the cooling/heating fluids' temperature levels, the radiant panel cooling/heating capacity is closely related to the indoor thermal conditions (i.e., air temperature, wall temperature, and solar radiation); this is quite different from convective heat exchangers (such as FCUs). Equation (3.30) quantitatively describes the surface temperature of the radiant panel with the variances of the indoor thermal conditions, the heat transfer coefficient, and the thermal resistance of the radiant panels. Performances of the radiant panels in summer and winter conditions are described in separate examples:

3.1.6.1 Radiant Panel for Cooling

A radiant floor with a thermal resistance R of 0.1 ($m^2 \cdot °C$)/W and an average chilled water temperature of 16 °C are employed to maintain an indoor temperature of 26 °C:

- Wall temperature AUST = 26 °C and no direct solar radiation: The average temperature of the radiant floor surface is 19.5 °C, and the cooling capacity is 35.4 W/m^2.
- Wall temperature AUST = 28 °C and no direct solar radiation: The average temperature of the radiant floor surface increases to 20.1 °C, and the cooling capacity is 42.2 W/m^2, 20 % higher than in the previous example.
- Wall temperature AUST = 26 °C and solar radiation q_{sr} = 50 W/m^2: The average temperature of the radiant floor surface increases to 22.7 °C, and the cooling capacity is 68.1 W/m^2, two times higher than in the condition without solar radiation.

3.1.6.2 Radiant Panel for Heating

A radiant floor with a thermal resistance R of 0.1 ($m^2 \cdot °C$)/W and hot water temperatures of 40 °C/30 °C are employed to maintain an indoor temperature of 20 °C:

- No direct solar radiation: The average temperature of the radiant floor surface is 27.9 °C, and the heating capacity is 70.6 W/m^2.
- Solar radiation q_{sr} = 30 W/m^2: The average temperature of the radiant floor surface increases to 29.5 °C, and the heating capacity is 54.8 W/m^2, 22 % lower than in the previous example.
- Solar radiation q_{sr} = 50 W/m^2: The average temperature of the radiant floor surface increases to 30.6 °C, and the heating capacity is 44.2 W/m^2, 40 % lower than in the condition without solar radiation.

From the above analysis of these cases, it can be seen that the cooling/heating capacity of radiant panels is significantly influenced by the indoor environment. If there is direct solar radiation reaching the surface of the radiant panel, it will be absorbed by the panel and transferred to the cooling/heating fluid, thereby increasing the radiant panel cooling capacity in summer (reducing the radiant panel heating capacity in winter). Wall surface temperature has the same impact on the performance of the radiant panel. Hence, the radiant panel can self-modify its cooling/heating capacity based on the variance of the indoor thermal environment: as the cooling/heating load increases (i.e., solar radiation or wall temperature changes), the cooling/heating capacity of the radiant panel also increases. This "adaptive" characteristic of radiant panels makes it possible to eliminate indoor temperature fluctuations to a certain extent. For example, if there is direct solar radiation coming into the indoor space in winter, some buildings may appear to overheat, but radiant floors actually stabilize the indoor air temperature to a greater extent than conventional systems.

Table 3.15 Operative temperatures in large space buildings

	View factor	No radiant floor cooling (°C)	Radiant floor cooling (°C)	
			Same air temperature	Same operative temperature
Roof	0.09	38.0	38.0	38.0
Wall	0.42	28.0	28.0	28.0
Floor	0.29	26.0	21.0	21.0
Shaded floor	0.20	26.0	26.0	26.0
Air temperature, T_c		26.0	26.0	28.6
Equivalent mean radiant temperature, T_r		27.9	26.5	26.5
Operative temperature, T_{OP}		27.2	26.3	27.2

Table 3.16 Operative temperatures in office rooms

	View factor	No radiant floor cooling (°C)	Radiant floor cooling (°C)	
			Same air temperature	Same operative temperature
Roof	0.31	26.0	26.0	26.0
Wall	0.33	28.0	28.0	28.0
Floor	0.18	26.0	21.0	21.0
Shaded floor	0.18	26.0	26.0	26.0
Air temperature, T_c		26.0	26.0	27.6
Equivalent mean radiant temperature, T_r		26.7	25.8	25.8
Operative temperature, T_{OP}		26.4	25.8	26.4

3.1.7 Impact on Indoor Thermal Comfort

Radiant panels exchange heat with the human body through radiation. The operative temperature (ASHRAE Handbook) is usually used to evaluate human thermal comfort, defined as Eq. (3.35):

$$T_{OP} = \frac{h_c T_c + h_r T_r}{h_c + h_r} \tag{3.35}$$

where h_c is the convective heat transfer coefficient, W/(m^2 · °C); T_c is the ambient air temperature for occupants, °C; h_{lr} is the radiation heat transfer coefficient, W/(m^2 · °C); and T_r is the equivalent mean radiant temperature, °C.

Equivalent mean radiant temperature is determined by the view factors and wall temperatures. Usually, there is a large view factor between the cooling floor and the occupants; hence, a floor with lower surface temperature can reduce the operative temperature effectively. Tables 3.15 and 3.16 list the operative temperatures of a typical large space building (100 m × 30 m × 20 m) and an office room (5 m × 5m × 3 m), respectively. As radiant floor cooling is adopted, the operative temperature is lower than that of an ordinary convective terminal-based air-conditioning system with the same indoor air temperature; the set value of the indoor air temperature could also be raised to achieve the same operative temperature.

3.2 Dry Fan Coil Units (FCUs)

3.2.1 Differences Between Dry FCUs and Wet FCUs

In conventional air-conditioning systems using condensing dehumidification, chilled water supplied into the FCU is usually around 7 °C, with a temperature low enough for both sensible cooling and dehumidification. Condensate water from moist air accumulates in the condensate water pan and then is drained, as shown in Fig. 3.27.

This kind of FCU with a condensate water pan is referred to as a wet FCU in the following analysis. The condensate water pan and corresponding condensate water pipe make the wet FCU's structure complex, and the condensate water pan is also likely to become a hotbed for the breeding of microorganisms. In THIC air-conditioning systems, FCUs are only used to remove indoor sensible heat. Assuming the task for temperature control only, the temperature of the supply chilled water can be increased to about 16–18 °C, which is higher than the indoor air dew point temperature. Furthermore, there will be no water condensation during the operation of the FCU. This kind of FCU operating with no condensate water is thus called a dry FCU.

Due to the increase of the supply chilled water temperature, the temperature difference between the high-temperature chilled water in the dry FCU and the indoor air is significantly lower compared to the conventional wet FCU. Taking the indoor temperature of 26 °C as an example, as the surface temperature of the dry

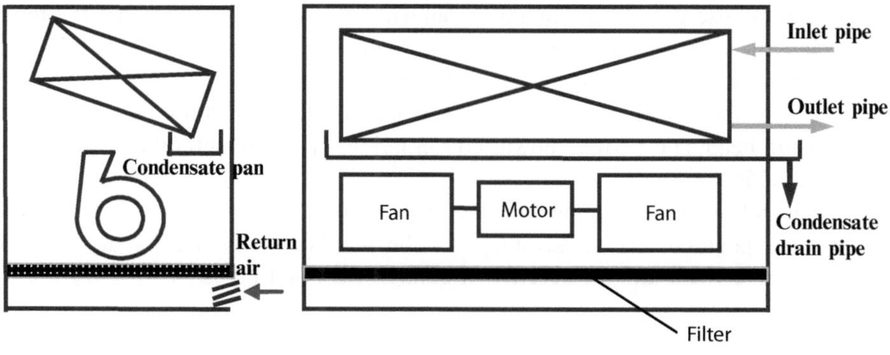

Fig. 3.27 Schematic of wet FCU with condensate water pan

FCU is 18 °C, the temperature difference between the surface of the dry FCU and the indoor air is 8 °C. If the conventional wet FCU (with 7 °C chilled water) corresponds to a surface temperature of 9 °C, the temperature difference between the surface and the indoor air is as high as 17 °C. It can be seen that the temperature difference between the surface of the dry FCU and the indoor air is only about half of that between the surface of the conventional wet FCU and the indoor air. As this temperature difference can be regarded as the driving force for the heat transfer process, decreasing the temperature difference leads to a significant cooling capacity drop of the dry FCU per unit area. Therefore, the key issue for designing dry FCUs is to find a way to achieve an efficient heat transfer process with a much lower temperature difference. Compared with the wet FCU, greater heat transfer area or air flow rate is needed for the dry FCU when assuming the same sensible cooling capacity.

For an FCU, fan power consumption can be expressed by Eq. (3.36):

$$P = \frac{\Delta p \cdot G}{\eta} \tag{3.36}$$

where P is the fan power consumption, W; Δp is the pressure drop of the FCU, Pa; G is the circulating air flow rate, m^3/s; and η is the fan efficiency.

An increased air flow rate in the heat transfer process does not necessarily result in an increase of fan energy consumption. Since there is no condensate water in the dry FCU, Δp could be lower compared to that of the wet FCU with condensate water. As indicated by Eq. (3.36), although the air flow rate increases for the dry FCU to achieve the same cooling capacity, the decrease of Δp can slow down a substantial increase in fan power consumption. In contrast to the conventional wet FCU, there is no requirement for considering the removal of condensate water for the dry FCU. The structure of the dry FCU could be greatly simplified, and new structures could be designed and developed. The typical ideas for designing and developing dry FCUs can be summarized as follows:

- Choose a larger heat transfer area and a lower row number of tubes to reduce the flow resistance in the air side.
- Select and use a new tube bundle arrangement, and try to ensure that the flow arrangement between the air and chilled water is a countercurrent flow; this will help improve the heat transfer effect as much as possible.
- Choose a through-flow fan or axial-flow fan with a larger flow rate, a smaller pressure head, and a lower power consumption, or utilize natural convection in the air side.
- Choose a flexible arrangement (e.g., ceiling fans) that can be installed in a corner or other position, taking advantage of the flexibility afforded by the lack of condensate water and the absence of condensate water pipes.

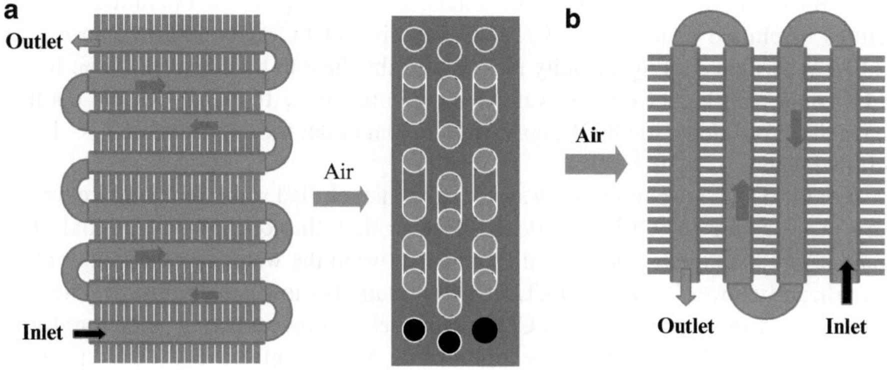

Fig. 3.28 Structure comparison of the FCU: (**a**) conventional FCU and (**b**) counterflow arrangement

Table 3.17 Performance comparison of FCU (FP-68) in wet and dry conditions (Zhang et al. 2011)

	Conventional FCU	Conventional FCU running in dry condition	Dry FCU
Inlet air dry-bulb temperature, °C	27.0	26.0	26.0
Inlet air wet-bulb temperature, °C	19.5	18.7	18.7
Inlet water temperature, °C	7.0	16.0	16.0
Outlet water temperature, °C	12.0	21.0	21.0
Outlet air dry-bulb temperature, °C	14.5	22.2	19.8
Outlet air wet-bulb temperature, °C	13.6	17.4	16.6
Water flow rate, m^3/h	0.685	0.181	0.294
Cooling capacity, W	3,826	989	1,604
Water pressure drop, kPa	18.1	3.2	11.4
Power of fan, W	61	63	61
Cooling capacity per air flow rate, $W/(m^3/h)$	6.0	1.3	2.2

3.2.2 Developed Dry FCU with a Similar Structure as a Wet FCU

A "dry FCU" and a "conventional FCU running in the dry condition" are two different concepts. The structure of the conventional wet FCU is shown in Fig. 3.28a, where the chilled water coils and the supply air constitute a cross-flow arrangement. Although a higher supply chilled water temperature is chosen, the conventional FCU with high-temperature chilled water still retains a certain degree of cooling capacity. Nevertheless, it has an obvious disadvantage: the conventional FCU is not specially designed to undertake the indoor sensible load. Taking a type FP-68 FCU as an example, Table 3.17 lists the performance of the conventional wet

FCU operating in the dry condition (Zhang et al. 2011). As the operating temperatures of chilled water are 7 °C/12 °C for this wet FCU, its cooling capacity is 3,826 W, with a cooling capacity ratio of sensible heat to total heat of 71 %. If this FCU runs with higher chilled water temperatures, e.g., 16 °C/21 °C, the cooling capacity drops down to 989 W, only 26 % of that under the condition of 7 °C/12 °C chilled water.

For the THIC air-conditioning system, the inlet chilled water temperature for the dry FCU (about 16 °C) is significantly higher than that of the conventional FCU (about 7 °C), so the temperature difference between the water and the indoor air is significantly lower for the dry FCU. It is important to know how to improve the heat transfer performance of the dry FCU, and the design processes (e.g., structure form, fin pitch, coil diameter) must be optimized. Accordingly, a dry FCU specially developed for the THIC system has been marketed in China. Tian et al. (2011) and Zhang et al. (2011) analyzed the applicability of conventional FCUs used in THIC systems, as well as the performance of a specially designed dry FCU. Table 3.17 shows the performance results of the newly developed dry FCU. The heat exchanger of the dry FCU adopts a new type of aluminum fin with a 3-row window; the corresponding fin width is 92 mm, the thickness is 0.115 mm, and the pitch is 1.6 mm. Φ7 brass pipes, with horizontal and vertical distances between neighboring pipes of 18.2 mm and 21 mm, respectively, and a counterflow arrangement (as shown in Fig. 3.28b) help to reduce the thermal resistance and enhance the heat transfer process. As indicated in Table 3.17, the cooling capacity of the newly developed dry FCU is as high as 1,604 W, achieving an increase of 62 % compared to that of the conventional FCU running in the dry condition.

3.2.2.1 Performance of Wet FCUs Based on China's Standard

Table 3.18 lists the input fan power, heating capacity, and cooling capacity in rated operating conditions for wet FCUs based on China's national standard GB/T 19232–2003. For the rated operating condition in summer, indoor dry-bulb temperature is 27.0 °C, with a corresponding wet-bulb temperature of 19.5 °C. The temperatures of inlet/outlet chilled water are 7 °C/12 °C, indicating that the temperature difference between the indoor dry-bulb temperature and the average temperature of inlet/outlet chilled water is 17.5°. The cooling capacity per air flow rate of the wet FCU is 5.3 W/(m^3/h); cooling capacity per input power is about 49–59 W/W (a corresponding cooling capacity for extracting sensible heat per input power is about 40 W/W, which accounts for 75 % of the total cooling capacity).

3.2.2.2 Performance of Dry FCUs Based on China's Standard

Table 3.19 gives the performance of dry FCUs according to an industry standard for dry fan coil units, JB/T 11524–2013. In the rated cooling condition, the dry-bulb temperature of the indoor air is 26.0 °C, and the corresponding wet-bulb

Table 3.18 Performance of wet FCUs (low static pressure units, GB/T 19232–2003)

Type	Rated air flow rate, m^3/ h	Input fan power, W	Rated cooling condition: Cooling capacity, W	Cooling capacity: Per air flow rate, W/(m^3/h)	Cooling capacity: Per input power, W/W	Rated heating condition: Heating capacity, W	Heating capacity: Per air flow rate, W/(m^3/h)	Heating capacity: Per input power, W/W
FP-34	340	37	1,800	5.3	49	2,700	7.9	73
FP-51	510	52	2,700	5.3	52	4,050	7.9	78
FP-68	680	62	3,600	5.3	58	5,400	7.9	87
FP-85	850	76	4,500	5.3	59	6,750	7.9	89
FP-102	1,020	96	5,400	5.3	56	8,100	7.9	84
FP-136	1,360	134	7,200	5.3	54	10,800	7.9	81
FP-170	1,700	152	9,000	5.3	59	13,500	7.9	89
FP-204	2,040	189	10,800	5.3	57	16,200	7.9	86
FP-238	2,380	228	12,600	5.3	55	18,900	7.9	83

Cooling condition: Dry-bulb temperature of the indoor air is 27.0 °C, the corresponding wet-bulb temperature is 19.5 °C, the chilled water inlet temperature is 7 °C, and the outlet temperature is 12 °C

Heating condition: Dry-bulb temperature of the indoor air is 21 °C, the hot water inlet temperature is 60.0 °C, and the flow rate of the hot water is equal to that of the chilled water in the cooling condition

temperature is 18.7 °C; chilled water inlet/outlet temperatures are 16 °C/21 °C, indicating that the temperature difference between the indoor dry-bulb temperature and the average temperature of the chilled water is 7.5 °C. As shown in this table, the cooling capacity per air flow rate of the dry FCU is 2.0 W/(m^3/h), and the cooling capacity per input power varies from 18 to 22 W/W. In other words, as the supply chilled water temperature increases, the cooling capacity per fan input power for dry FCUs decreases to only about 50 % of the cooling capacity of wet FCUs (40 W/W).

As for the standard for dry FCUs, chilled water inlet/outlet temperatures are 16 °C/21 °C (Zhang et al. 2009; SAC/TC238). The corresponding water temperature difference is 5 °C, which is consistent with the water temperature difference of conventional wet FCUs. Therefore, the power consumption of water distribution is not increased for FCUs operating in the dry condition.

According to the heating performance of FCUs in winter shown in Table 3.19, the heating performance of dry FCUs as a function of supply water temperature and indoor air temperature can be investigated. The calculated results with the same circulating water flow rate are listed in Table 3.20. Taking the typical model of FP-34 as an example, when indoor air temperature is 21 °C, heating capacity corresponding to 40 °C hot water is 1,490 W, and heating capacity corresponding to 35 °C hot water is 1,098 W. When the indoor temperature is 20 °C, the heating capacities corresponding to the 40 and 35 °C hot water temperatures are 1,568 W and 1,176 W, respectively. These results are much higher than the cooling capacity of 680 W in summer. Therefore, if dry FCUs are adopted as the sensible terminal, supply hot water with a temperature of 35 °C is sufficient for heating in winter.

Table 3.19 Performance of dry FCUs (low static pressure type, JB/T 11524–2013)

Type	Rated air flow rate, m^3/h	Input power, W	Rated cooling condition: Cooling capacity, W	Cooling capacity: Per air flow rate, W/(m^3/h)	Cooling capacity: Per input power, W/W	Rated heating condition: Heat capacity, W	Heating capacity: Per air flow rate, W/W	Heating capacity: Per input power, W/W
FP-34	340	37	680	2.0	18	1,490	4.4	40
FP-51	510	52	1,020	2.0	20	2,240	4.4	43
FP-68	680	62	1,360	2.0	22	2,990	4.4	48
FP-85	850	76	1,700	2.0	22	3,740	4.4	49
FP-102	1,020	96	2,040	2.0	21	4,500	4.4	47
FP-136	1,360	134	2,720	2.0	20	5,980	4.4	45
FP-170	1,700	152	3,400	2.0	22	7,480	4.4	49
FP-204	2,040	189	4,080	2.0	22	8,970	4.4	47
FP-238	2,380	228	4,760	2.0	21	10,470	4.4	46

Cooling condition: Dry-bulb temperature of the indoor air is 26.0 °C, the corresponding wet-bulb temperature is 18.7 °C, the chilled water inlet temperature is 16 °C, and the outlet temperature is 21 °C

Heating condition: Dry-bulb temperature of the indoor air is 21.0 °C, the hot water inlet temperature is 40.0 °C, and the flow rate of the hot water is equal to that of the chilled water in the cooling condition

Table 3.20 Heating capacity of dry FCUs in winter

Type	Heating capacity with 40 °C supply water, W: Indoor temperature 21 °C	Indoor temperature 20 °C	Indoor temperature 19 °C	Heating capacity with 35 °C supply water, W: Indoor temperature 21 °C	Indoor temperature 20 °C	Indoor temperature 19 °C
FP-34	1,490	1,568	1,647	1,098	1,176	1,255
FP-51	2,240	2,358	2,476	1,651	1,768	1,886
FP-68	2,990	3,147	3,305	2,203	2,361	2,518
FP-85	3,740	3,937	4,134	2,756	2,953	3,149
FP-102	4,500	4,737	4,974	3,316	3,553	3,789
FP-136	5,980	6,295	6,609	4,406	4,721	5,036
FP-170	7,480	7,874	8,267	5,512	5,905	6,299
FP-204	8,970	9,442	9,914	6,609	7,082	7,554
FP-238	10,470	11,021	11,572	7,715	8,266	8,817

3.2.2.3 Performance Comparison Between Wet and Dry FCUs Based on China's Standard

For the test results in Tables 3.18 and 3.19, the testing conditions of wet FCUs and dry FCUs are different in winter and summer. Thus, it is unsuitable to use heating capacity or cooling capacity to make the comparison. Instead, a method to compare the heat transfer ability (*KF*) between wet and dry FCUs is chosen: *KF* of the FCUs

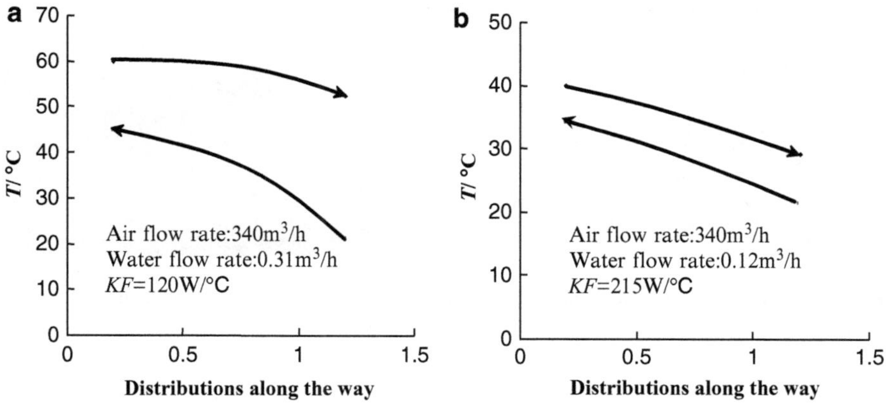

Fig. 3.29 Performances of wet and dry FCUs: (**a**) wet FCU with a supply water temperature of 60 °C and (**b**) dry FCU with a supply water temperature of 40 °C

can be determined by the heating capacity and supply/return water temperatures. The calculation of the heat transfer ability is as follows:

$$KF = \frac{Q_{\rm h}}{\Delta t_{\rm m,h}} \tag{3.37}$$

where Q_h is the heating capacity in rated conditions, W; K is the heat transfer coefficient, W/(m^2 · °C); F is the heat transfer area, m^2; and $\Delta t_{m,h}$ is the logarithmic mean temperature difference in the heating condition, °C. Figure 3.29 shows a comparison of the heat transfer performance between conventional wet FCUs and dry FCUs based on China's standards. Taking an air flow rate of 340 m^3/h for the FCUs as an example, KF of the conventional wet FCU is 120 W/°C, while that of the dry FCU is 215 W/°C. As a result, optimizations such as adjusting the flow structure, the fin structure (slotted fin, fin density, etc.), and the water pipe diameter are required to enhance the heat transfer performance and meet the requirements for dry FCUs in the standard.

The problem that remains is that although the performance of the dry FCU is better than that of the conventional wet FCU running with high-temperature chilled water, cooling capacity per input power is only about 20 W/W (as COP of the chiller equals the cooling capacity divided by the chiller's input power, the transport coefficient for the FCU could be defined as the cooling capacity of the FCU divided by its fan power). While the transport coefficient of the wet FCU varies from 50 to 60 W/W, the corresponding transfer coefficient of the cooling capacity for extracting sensible heat is about 40 W/W. Therefore, how to take advantage of the absence of condensate water in dry FCUs to design a novel structure and decrease the transfer resistance is the key issue for further improvements.

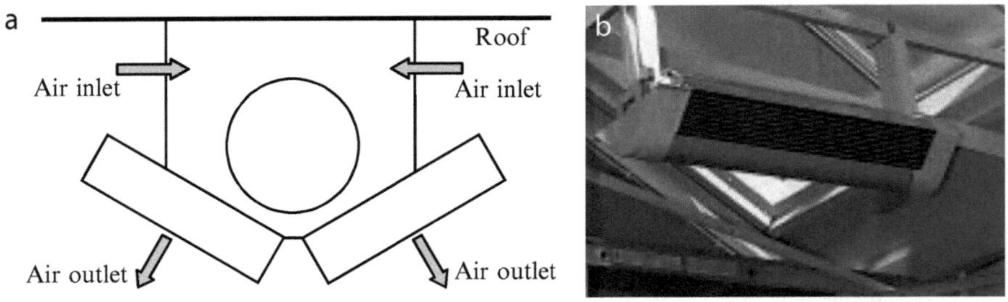

Fig. 3.30 "Imitation of ceiling fan" type dry FCU: (**a**) schematic diagram and (**b**) installation photo (Carnegie Mellon University)

3.2.3 Dry FCU with New Structures

As the temperature difference between the chilled water and the indoor air decreases, the heat transfer performance for FCUs in the dry condition decreases much more substantially than for FCUs in the wet condition with the same heat transfer area. Even though dry FCUs are based on the structure of wet FCUs, which have made great improvements to the flow arrangement between air and water, it must be noted that the transfer coefficient of the dry FCU is still significantly lower than that of the wet FCU. The chiller's COP for the system using dry FCUs is significantly higher than the system adopting wet FCUs (7 °C chilled water). However, the transfer coefficient of the dry FCU is relatively low, restricting the energy performance of the entire system to some extent. Thus, innovative ideas and new structures for dry FCUs are needed to fully utilize the advantage of the absence of condensate water. Here, two examples are introduced:

- "Imitation of Ceiling Fan" Type Dry FCU

An installed "imitation of ceiling fan" type dry FCU is shown in Fig. 3.30. The only change is to arrange coils in the air flow direction; this reduces the chances that anything might increase the resistance to air side flow and the required fan pressure head. At the same time, this type of installation can greatly cut down the cost of FCUs and installation fees, and ceiling space is no longer occupied.

- "Through-Flow" Dry FCU

Figures 3.31 and 3.32 show the cross-sectional structure and installation examples of dry FCUs (Danfoss Group 2004). The dry FCU shown in Fig. 3.31 has a modular design and is flexible in length so that it can easily fit the size of different buildings. A special material (VORTEX) is chosen for the area between the fan and the deflector, which can eliminate noise due to high wind speeds. The motor is a brushless DC (direct current) type with high efficiency. Moreover, the motor is continuously adjustable within a range of 400–3,000 rpm (revolutions per minute), and cooling capacity per input power is 50–60 W/W.

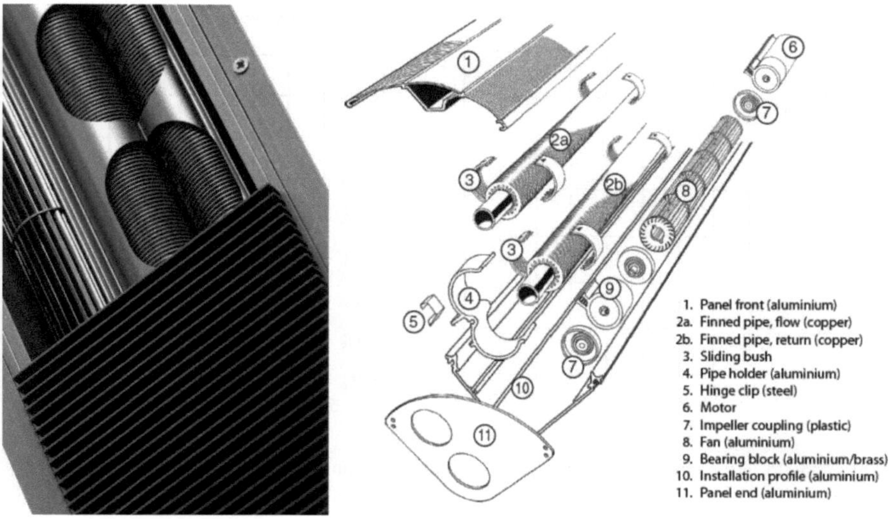

Fig. 3.31 Schematic of the compact dry FCU (Danfoss)

Fig. 3.32 Applications of the dry FCU units

Figure 3.33 shows the heating capacity of the through-flow dry FCU from Fig. 3.31 varying with the water temperature and wind speed. The rated heating condition is defined as supply/return water temperatures of 60 °C/50 °C, an indoor air temperature of 20 °C, and a fan speed of 1,500 rpm. The cooling capacity and heating capacity of the dry FCU provided by the manufacturer are shown in Table 3.21, where Δt is the temperature difference between the average water temperature and the indoor air temperature. From the data in Fig. 3.33 and Table 3.21, it can be seen that the increase of wind speed can significantly improve the output capability. Therefore, fan speed could be changed to regulate the cooling/heating capacity of FCUs, achieving continuous adjustment of the indoor air temperature.

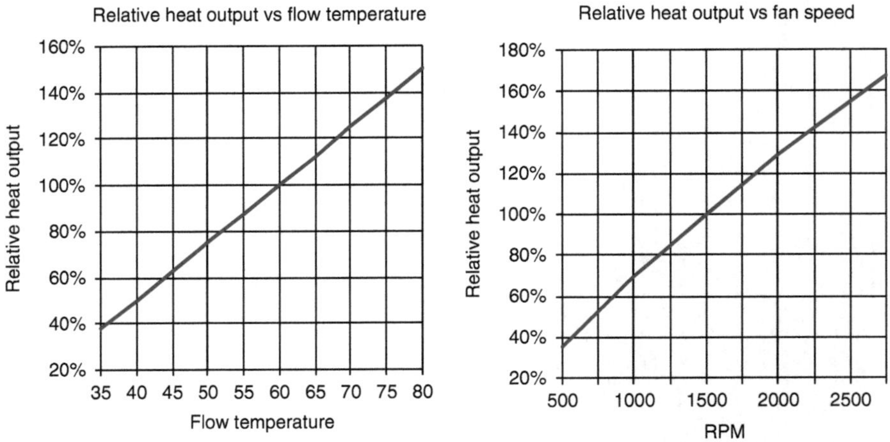

Fig. 3.33 Relative heat output varying with water temperature and wind speed of the dry FCU

Table 3.21 Cooling and heating capacity of the dry FCU

	Air velocity	
Output capacity of the dry FCU	1,500 rpm	2,800 rpm
$\Delta t = 35$ °C, output capacity per meter	400 W	730 W
$\Delta t = 9$ °C, output capacity per meter	110 W	190 W

The aforementioned through-flow dry FCU designed by Danfoss is not widely used in the Chinese market because of the noise caused by the eccentric bearing, but its transfer coefficient is as high as 50–60 W/W, which is the same as that of conventional wet FCUs. Therefore, in order to maximize the performance of dry FCUs, new structures must take full advantage of the absence of condensate water. So far, these new structures suitable for the dry condition have yet to be designed.

References

Awbi HB, Hatton A (1999) Natural convection from heated room surfaces. Energy Build 30:233–244

Bergman TL, Lavine AS, Incropera FP, Dewitt DP (2011) Fundamentals of heat and mass transfer, 7th edn. Wiley, New York

BS EN1264-5 2008 (2009) Water based surface embedded heating and cooling systems. Part 5: Heating and cooling surfaces embedded in floors, ceilings and walls–determination of the thermal output. British Standards Institution, London

Danfoss Group (2004) Danfoss Group. Samples of fan coil units, 2004. http://www.danfoss.com/

Francesco C, Stefano PC, Marco F, Bjarne O (2009) Experimental evaluation of heat transfer coefficients between radiant ceiling and room. Energy Build 41:622–628

Jin X, Zhang XS, Luo YJ, Cao RQ (2010) Numerical simulation of radiant floor cooling system: the effects of thermal resistance of pipe and water velocity on the performance. Build Environ 45(11):2545–2552

Koschenz M, Dorer V (1999) Interaction of an air system with concrete core conditioning. Energy Build 30:139–145

Lu YQ (2008) Practical design manual for heating and air-conditioning, 2nd edn. China Architecture & Building Press, Beijing (in Chinese)

Min TC, Schutrum LF, Parmelee GV, Vouris JD (1956) Natural convection and radiation in a panel-heated room. ASHRAE Transit 62:337–358

Novoselac A, Burley BJ, Srebric J (2006) New convection correlations for cooled ceiling panels in room with mixed and stratified airflow. HVAC&R Res 12:279–294

Olesen BW, Michel E (2000) Heat exchange coefficient between floor surface and space by floor cooling—theory or a question of definition. ASHRAE Trans 106:684–694

SAC/TC238. JB/T 11524–2013 Dried fan-coil unit (in Chinese)

Shi XM, Zhang X (2012) Effects of radiant terminal and air supply terminal devices on energy consumption of cooling load sharing rate in residential buildings. Energy Build 49:499–508

Tian XD, Zhang XP, Du LW, Song YQ, Wu JF (2011) Dry fan-coil units used in temperature and humidity independent control systems. Chin HV&AC 41:28–32 (in Chinese)

Zhang XM, Ren ZP (2001) Heat transfer, 4th edn. China Architecture & Building Press, Beijing (in Chinese)

Zhang XP, Pan YG, Tian XD, Du LW, Jia L, Feng XW, Song YQ (2009) Adaptability research on common fan-coil units used in temperature and humidity independent control system. Fluid Mach 37(1):72–76 (in Chinese)

Zhang XP, Xu BQ, Tian XD, Fan YG, Yao Y, Wu JF (2011) Research on nominal test conditions and the basic specification of the units in the standard "the dry fan coil units". Fluid Mach 39:59–63 (in Chinese)

Zhang L, Liu XH, Jiang Y (2012) Simplified calculation for cooling- heating capacity, surface temperature distribution of radiant floor. Energy Build 55:397–404

Zhang L, Liu XH, Jiang Y (2013) Experimental evaluation of a suspended metal ceiling radiant panel with inclined fins. Energy Build 62:522–529

Zhao K, Liu XH, Jiang Y (2013) Application of radiant floor cooling in a large open space building with high-intensity solar radiation. Energy Build 66:246–257

Chapter 4
Key Components of the THIC System: Outdoor Air Handling Methods

Abstract In the THIC system, outdoor air is usually handled to a state dry enough to remove indoor moisture for humidity control during cooling season. In this chapter, requirements for outdoor air handling devices within different climate regions are discussed, and the basic air handling devices for heat recovery, dehumidification, and humidification are introduced. Evaporative cooling method, which is feasible for outdoor air handling process in the dry region, the corresponding devices, as well as the analyzing method are introduced. As to the humid region in summer, dehumidification is required to handle the outdoor air to a state dry enough. Condensation dehumidification method and solid desiccant dehumidification method are introduced in this chapter.

4.1 Basic Outdoor Air Handling Devices

4.1.1 Requirements for Outdoor Air Handling Devices in Different Climate Regions

The requirements for outdoor air handling process in the THIC system have been examined in Sect. 2.4.2 of Chap. 2. The requirements for outdoor air handling devices can be determined according to the required parameters of the supplied air. Moreover, the outdoor air handling devices should satisfy the handling requirements for different outdoor conditions. The outdoor climate parameters for summer and winter in China have been analyzed in Sect. 2.4.3 in Chap. 2; the country can be divided into two regions (Zone I and Zone II) according to the different outdoor air handling requirements, as shown in Fig. 4.1.

X. Liu et al., *Temperature and Humidity Independent Control (THIC) of Air-conditioning System*, DOI 10.1007/978-3-642-42222-5_4,

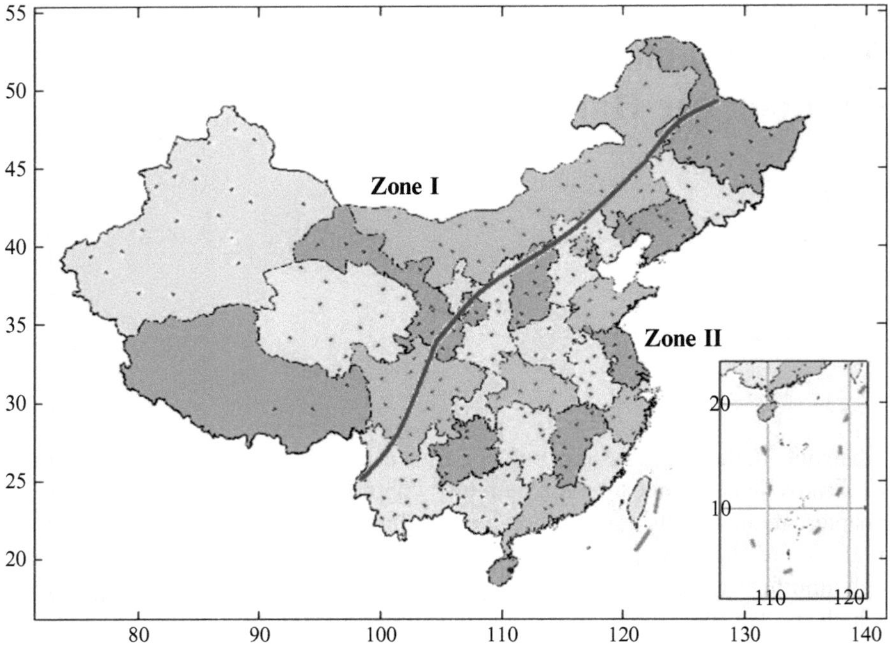

Fig. 4.1 Main climate regions of China

4.1.1.1 Dry Region in the West (Zone I)

In the northwest of China, the outdoor air humidity ratio is low enough in summer that only sensible cooling is required to handle the outdoor air before it is supplied to the indoor environment. The typical annual outdoor air parameters of Urumchi are listed in Fig. 4.2, and it can be seen that the outdoor air humidity ratio is low throughout the year. Because of the dry outdoor air humidity ratio, the evaporative cooling method can be adopted to cool the outdoor air in summer. Section 4.2 will introduce the different evaporative cooling solutions and devices in detail. In this region, the outdoor air temperature and humidity ratio are both quite low in winter, so there are both heating and humidification requirements in the outdoor air handling process.

4.1.1.2 Humid Region in the East (Zone II)

In the THIC system, dry air is supplied to the indoor environment to extract the indoor moisture load. The outdoor climate is hot and humid during summer in the southeast of China. The typical annual outdoor climate parameters of Beijing and Shanghai are shown in Figs. 4.3 and 4.4, respectively. The mean outdoor humidity ratios in July and August approach 17 g/kg for Beijing and 19 g/kg for Shanghai. Thus, the outdoor air has to be dehumidified before being supplied to the indoor environment.

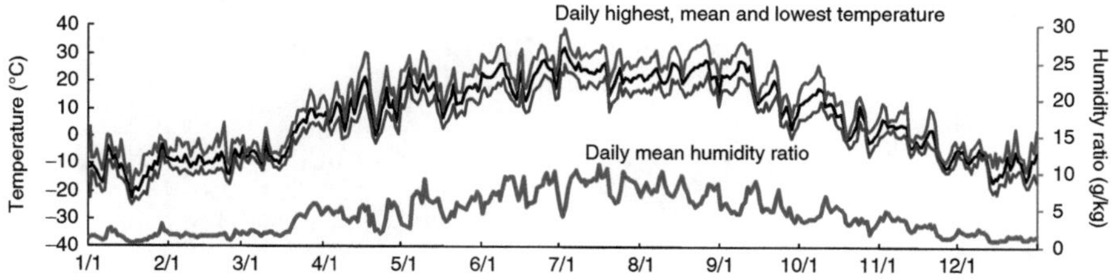

Fig. 4.2 Annual outdoor climate of Urumchi

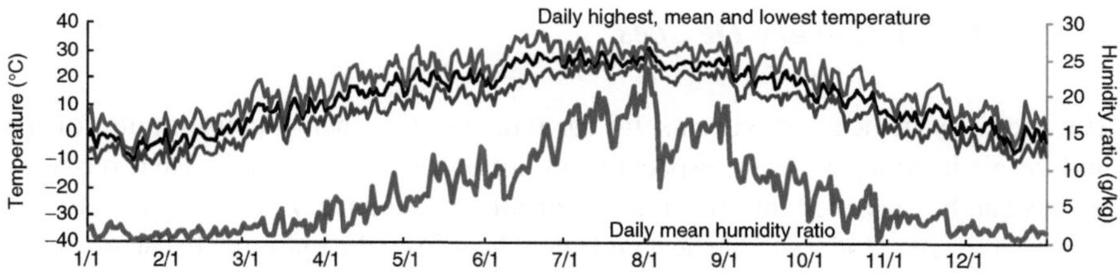

Fig. 4.3 Annual outdoor climate of Beijing

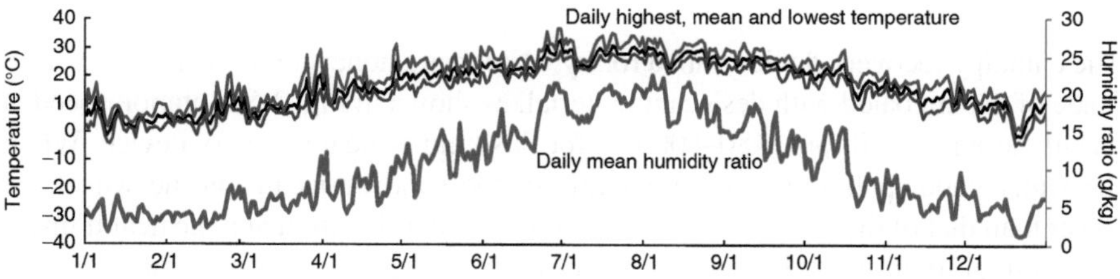

Fig. 4.4 Annual outdoor climate of Shanghai

According to the analysis of the requirements for outdoor air handling processes from Sect. 2.4.2 in Chap. 2, as well as the performance requirements for outdoor air handling devices in different climate regions of China discussed above, various kinds of outdoor air handling solutions and devices can satisfy these requirements. The following subsection will introduce the basic handling devices in the outdoor air handling process.

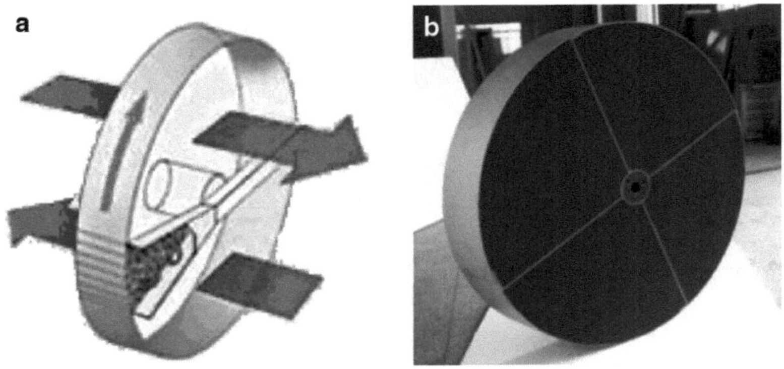

Fig. 4.5 Desiccant wheel used for heat recovery: (**a**) schematic diagram and (**b**) rotary wheel

4.1.2 Heat Recovery Devices

Heat recovery devices are very useful for reducing the energy consumption of the outdoor air handling process, especially in hot and humid climates. Heat recovery devices can be classified into two types: sensible heat recovery devices and enthalpy recovery devices. Both sensible heat and latent heat from the indoor exhaust air can be recovered by enthalpy recovery devices, which demonstrate greater performance than sensible heat recovery devices. There are many different configurations of heat recovery devices, as described in detail in various handbooks (Lu 2008, China Institute of Building Standard Design & Research 2006). In this section, only enthalpy recovery devices using rotary wheels and those using liquid desiccant are examined.

4.1.2.1 Enthalpy Recovery Device Using a Rotary Wheel

The enthalpy recovery device with a rotary wheel has a honeycomb cylinder structure made of paper coated with desiccant material, as shown in Fig. 4.5. Rotation speed is usually around 8–10 rpm (480–600 r/h) for this kind of heat recovery device. When the water vapor pressure and temperature of the solid desiccant on the wheel are lower than that of the outdoor air, the moisture transfer direction and the heat transfer direction are both from the outdoor air to the desiccant, which makes air dehumidification possible. With the rotation of the wheel, the fully adsorbed desiccant rotates to the indoor exhaust air zone and is dried and cooled by the exhaust air. In this process, exhaust air and outdoor air flow through the top and bottom sections of the wheel, respectively, with a counterflow configuration to achieve enthalpy recovery.

4.1.2.2 Enthalpy Recovery Device Using Liquid Desiccant

The heat recovery device using liquid desiccant is also an option for recovering energy from the indoor exhaust air. Packed-bed towers are common handling

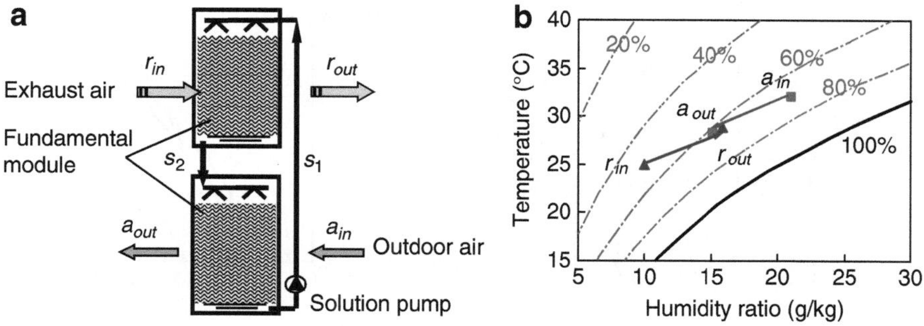

Fig. 4.6 Single-stage enthalpy recovery device using liquid desiccant: (**a**) operating schematic and (**b**) air handling process in psychrometric chart

devices in liquid desiccant systems, and the packing in the tower can increase the contact area between the solution and the air, thereby enhancing the effectiveness of the heat and moisture transfer process. Figure 4.6 shows a typical cross-flow single-stage enthalpy recovery device using liquid desiccant, where the indoor exhaust air is used to preprocess the outdoor air. The device consists of two packed towers and a circulating solution pump: the top module is the indoor exhaust air handling module, and the bottom module is for handling the outdoor air. The outdoor air and the exhaust air are represented by a and r, respectively, and the solution state is represented by s, as shown in Fig. 4.6.

For the operating process in summer, the solution pump transports the liquid from the bottom of the solution tank in the bottom module to the top of the top module, and then the solution is sprayed from the top to wet the packing. The coupled heat and mass transfer process proceeds between the indoor exhaust air and the solution. The exhaust air is heated and humidified by the solution and then flows out of the module. The cooled and concentrated solution mixes together and flows out of the top module. The solution is then sprayed from the top of the bottom module and is evenly distributed to the bottom packing. The outdoor air flows into the bottom module and is cooled and dehumidified, as the solution temperature or vapor pressure in the packing is lower than the air. Then the solution is diluted and flows to the bottom of the bottom module, completing the cycle. In this enthalpy recovery device, the outdoor air is cooled and dehumidified, and the indoor exhaust air is heated and humidified with the help of the solution cycle, achieving energy recovery. The operation in winter is similar to that in summer, with only the direction for energy transfer, which indicates that the outdoor air is heated and humidified, being reversed.

The multistage device is composed of a series of the single-stage devices shown in Fig. 4.6. Figure 4.7 illustrates the operating schematic of a three-stage enthalpy recovery device using liquid desiccant, with the outdoor air and the exhaust air flowing through the modules in reverse order. The temperature and concentration of the solution sprayed in each stage are determined by the parameters of the outdoor air and the exhaust air. The air handling process is shown in Fig. 4.7b.

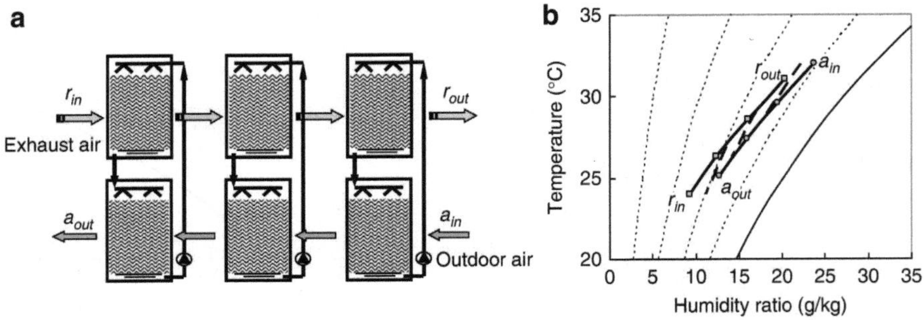

Fig. 4.7 Three-stage enthalpy recovery device using liquid desiccant: (**a**) operating schematic and (**b**) air handling process in psychrometric chart

4.1.3 Dehumidification Devices

4.1.3.1 Condensation Dehumidification Methods

Condensation dehumidification uses chilled water or refrigerant with a temperature low enough to cool the humid air; as the air temperature drops below its dew point temperature, the moisture condenses and the air is dehumidified. The operating schematic of the condensation dehumidification method is demonstrated in Fig. 4.8.

The relatively low-temperature cooling medium flows into the cooling coil, and then the temperature of the humid air flowing through the coil decreases. When the saturation condition is achieved, the humid air is dehumidified, and the moisture continues to be condensed as the temperature continues to decrease. Figure 4.9 shows the reachable handling region (nearly a triangular region) of the outlet air state using the condensation dehumidification method. The temperature and humidity ratio of the processed outlet air are both lower than those of the inlet, and the state of the processed outlet air usually approaches the saturation state.

4.1.3.2 Dehumidification Methods Using Liquid Desiccant

The operating principle of the typical dehumidification-regeneration cycle using liquid desiccant is shown in Fig. 4.10. The left side of this figure is the air dehumidification process, and the right side is the desiccant regeneration process. In the air dehumidification process, moisture transfers from the gas phase (air) to the liquid phase (solution) because the partial pressure of water vapor in the air is greater than that in the solution. With the help of the mass transfer process, the humidity ratio of the moist air is reduced, i.e., the air is dehumidified, and the solution is diluted due to moisture absorption. Thus, the partial pressure of the water vapor in the solution is gradually increased, so that the water vapor pressure difference between the solution and the air is gradually reduced. As a result, the

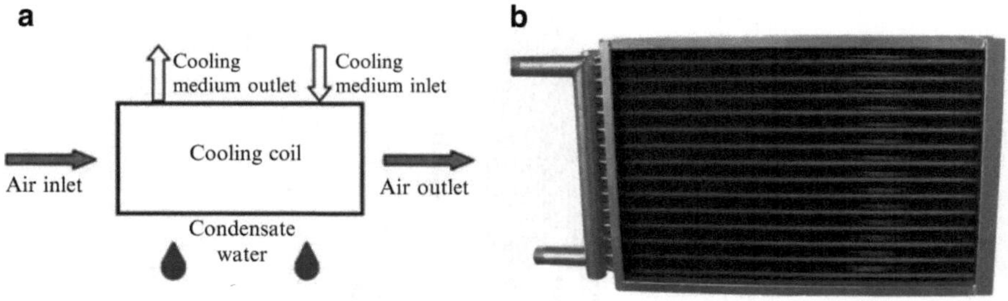

Fig. 4.8 Operating schematic of the condensation dehumidification method: (**a**) operating principle and (**b**) picture of the cooling coil

Fig. 4.9 Reachable air outlet region of the condensation dehumidification method

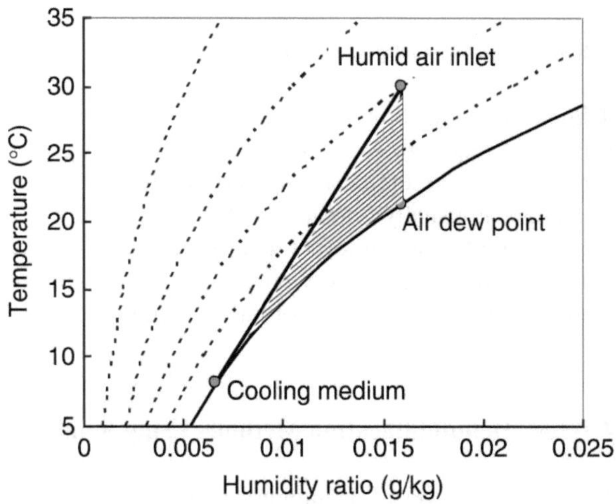

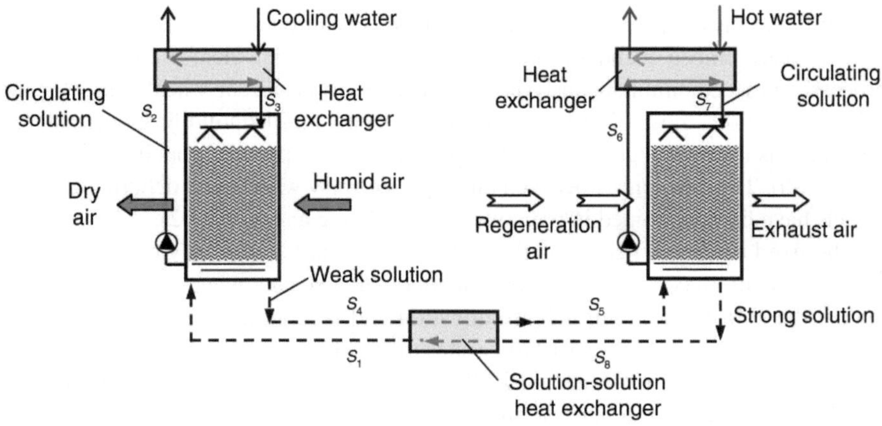

Fig. 4.10 Principle of the typical liquid desiccant dehumidification-regeneration cycle

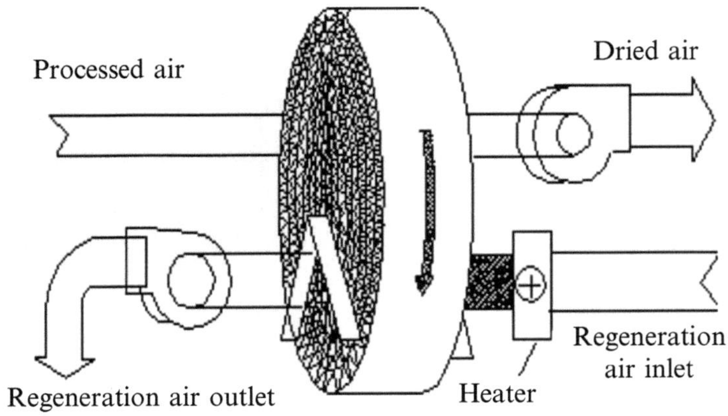

Fig. 4.11 Operating principle of the desiccant wheel used for dehumidification

dehumidification ability of the solution is weakened, and regeneration of the diluted solution is required. As shown in the right part of Fig. 4.10, hot water provides the heat required for the desiccant regeneration process. A solution-to-solution heat exchanger is usually adopted in this air handling system to precool the solution flowing into the dehumidifier and to preheat the solution flowing into the regenerator. Thus, the heat recovery process between solutions of different temperatures is realized, and the energy performance is improved.

4.1.3.3 Dehumidification Methods Using Solid Desiccant

There are two types of dehumidification methods that use solid desiccant: those with rotary wheels and those with fixed desiccant beds. Desiccant wheels are widely used due to their ability to achieve continuous dehumidification and regeneration. The operating principle of a desiccant wheel is shown in Fig. 4.11. Dehumidification wheels are similar to the enthalpy recovery desiccant wheels introduced in Sect. 4.1.2, which have a honeycombed channel coated with desiccant material. Heat and mass transfer processes proceed simultaneously in the channels. In desiccant wheels used for air dehumidification, most of the surface area is used for dehumidification, and the remainder is used for regeneration. An optimal rotation speed is around 0.2–0.5 rpm (12–30 r/h). The differences between enthalpy recovery wheels and dehumidification wheels have been examined in previous studies (Zhang and Niu 2002).

The fixed desiccant bed is another kind of dehumidification device that utilizes solid desiccant. In contrast to the rotation mode of desiccant wheels, desiccant beds realize the shifting between dehumidification and regeneration modes by shifting the air flow direction. Figure 4.12 illustrates the working principle of a fixed descant bed (Suzuki and Oya 1983; Zhang 2005). During the first half of the cycle, the left desiccant bed works in regeneration mode, while the right bed works in dehumidification mode. Humid air enters the right bed, and cooling water enters the right

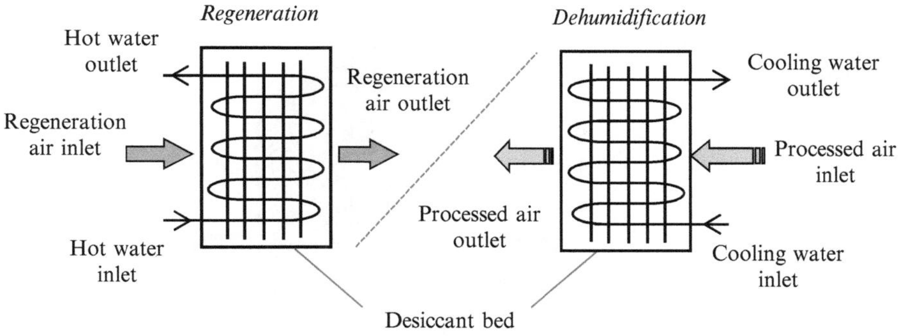

Fig. 4.12 Operating principle of the fixed desiccant bed (the first half of the cycle)

bed to take away the adsorption heat. Regeneration air enters the left bed, and hot water enters the left bed to provide desorption heat. During the second half of the cycle, the left bed works in dehumidification mode (with cooling water entering it), and the right bed works in regeneration mode. During the dehumidification period, the supplied air humidity ratio changes periodically due to the shifting of fan valves and water valves, and this instability of the supplied air parameters limits the use of fixed desiccant beds to some extent.

4.1.4 Humidification Devices

There are multiple methods for air humidification in air-conditioning systems. Common air humidification solutions adopted in air-conditioning systems (The 10th Design and Research Institute of Ministry of Electronics Industry 1995) include wet film evaporative humidification, dry steam humidification, electrode-type humidification, and ultrasonic humidification. The operating schematic of the wet film evaporative humidification method is shown in Fig. 4.13. The water distributed from the sprinkler is sprayed on the wet film. The dry air flows through the wet film, and the sprayed water turns into steam due to evaporation. The moisture then flows into the air, and the dry air is humidified. The humidification process of this type of wet film evaporative humidification method is similar to evaporative cooling. In the operating period during winter, the inlet dry air is usually preheated before flowing through the wet film to ensure the required humidification capacity.

The operating schematic of the dry steam humidification process is shown in Fig. 4.14. The saturated steam flows into the evaporation chamber through the bend. A baffle plate ensures the smooth flow of the steam flowing into the evaporation chamber. The saturated steam flowing out of the evaporation chamber then flows into the drying chamber and extracts all the water to ensure that the dry saturated steam flows out. A control valve is adopted to regulate the flow rate of the dry saturated steam flowing into the effuser. Then the dry saturated steam erupts from the effuser and diffuses uniformly along the width of the air duct, realizing the humidification of the dry air.

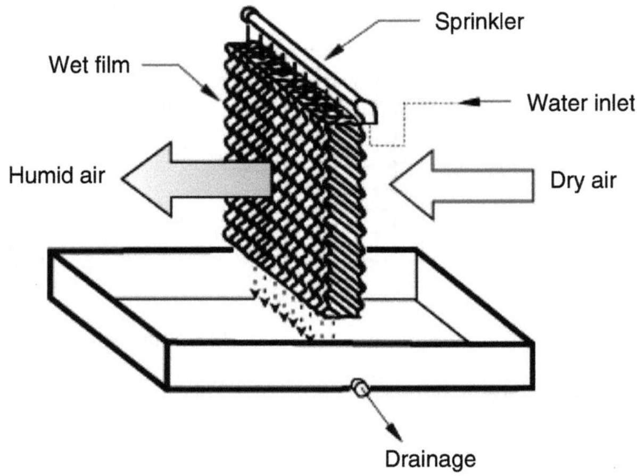

Fig. 4.13 Operating schematic of the wet film humidifier

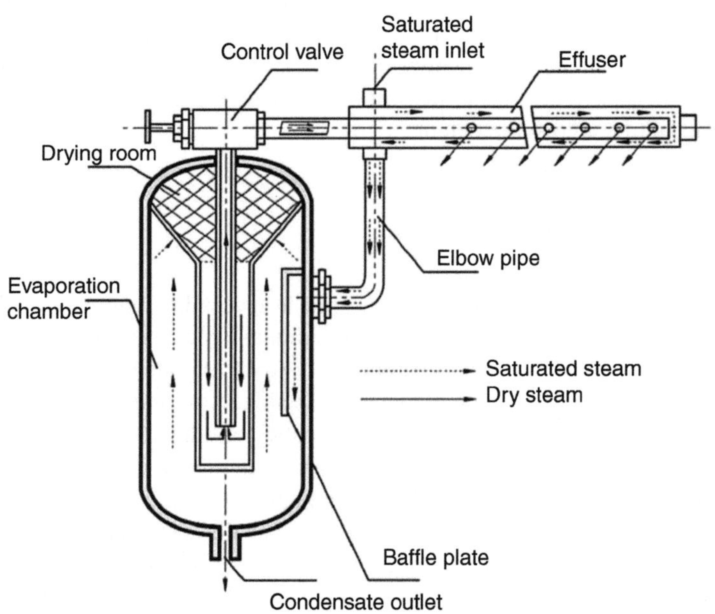

Fig. 4.14 Operating schematic of the dry steam humidifier

In the electrode-type humidification method, an electrode is immersed in the water, and the water is treated as the electrical resistance. The electrical voltage heats the water, and steam is produced, which is then distributed to the area where humidification is required.

However, the application conditions for the various humidification methods are different. If there is a steam source, then the dry steam humidification process should be given priority; this dehumidification method is often utilized in the operating rooms of hospitals. If no steam source can be adopted directly, the wet film evaporative humidification method or the electrode-type humidification method could be utilized.

4.2 Outdoor Air Handling Process in the Dry Region

For most parts of northwestern China, the humidity ratio of the outdoor air is seldom higher than 12 g/kg during summer. Thus, the dry outdoor air could be supplied to the indoor environment to remove indoor moisture. Therefore, the major task of handling the outdoor air is to cool it to a state with a suitable temperature. According to the outdoor air parameters, the air handling devices using direct or indirect evaporative cooling methods could be applied to handle the outdoor air to a state with a temperature of about 18–21 °C and a humidity ratio of about 8–10 g/kg. When the temperature of the handled outdoor air is a bit lower than the indoor temperature, the supplied outdoor air could also extract part of the sensible load, as well as the indoor moisture load. However, the supply air parameters for evaporative cooling rely on the local dry-bulb and wet-bulb temperatures. In designing an evaporative cooling outdoor air handling processor, the stage number and the handling process must both be determined to ensure that the outdoor air can effectively extract the indoor moisture load.

4.2.1 Outdoor Air Handling Process Using Evaporative Cooling

Evaporative cooling devices include direct evaporative cooling devices, indirect evaporative cooling devices, and devices combining direct and indirect evaporative cooling. The operating schematics and corresponding air handling processes for direct and indirect evaporative cooling devices are shown in Figs. 4.15 and 4.16, respectively. For the direct evaporative cooling process, the lowest temperature of the outlet air is equal to the wet-bulb temperature of the inlet air in theory. For the indirect evaporative cooling process, the lowest temperature of the outlet air is equal to the dew point temperature of the inlet air in theory (Huang 2010). In the indirect evaporative cooling process, only the temperature of the handled air decreases, while the humidity ratio stays constant, achieving a cooling process with a constant humidity ratio. The indirect evaporative cooling device shown in Fig. 4.16 is composed of a direct evaporative cooling device and a sensible heat exchanger. The indirect evaporative cooling device could also be constructed by combining a sensible heat exchange component inside a direct evaporative cooling device, which would approach the functionality of an internally cooled indirect evaporative cooling device.

A multistage evaporative cooling process can be constructed based on the direct and indirect evaporative cooling devices introduced above. Figure 4.17 shows a

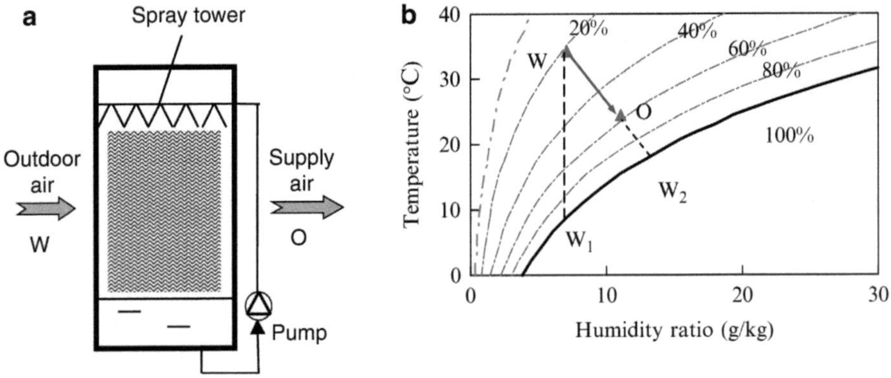

Fig. 4.15 Direct evaporative cooling device: (**a**) schematic diagram and (**b**) air handling process in psychrometric chart

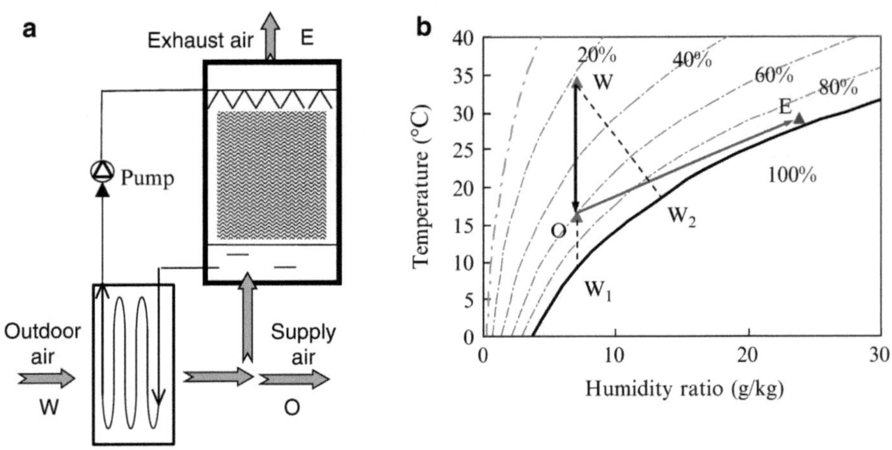

Fig. 4.16 Indirect evaporative cooling device: (**a**) schematic diagram and (**b**) air handling process in psychrometric chart

two-stage evaporative cooling device that includes two indirect evaporative cooling components (Xie and Jiang 2010). Figure 4.18 shows a two-stage air handling process that combines the direct and indirect evaporative cooling methods. In the indirect evaporative cooling process, the limiting temperature of the inlet air to be cooled is determined by the source of the secondary air (i.e., the air participating in the direct evaporative cooling process). If the secondary air is a part of the outlet primary air, as shown in Fig. 4.17, the limiting temperature of the outlet air for this indirect evaporative cooling process is the dew point of the inlet air. In the air handling process shown in Fig. 4.18, the outdoor air is first cooled at a constant humidity ratio by the indirect evaporative cooling method and then cooled with humidification in the direct evaporative cooling module. In practice, the limiting

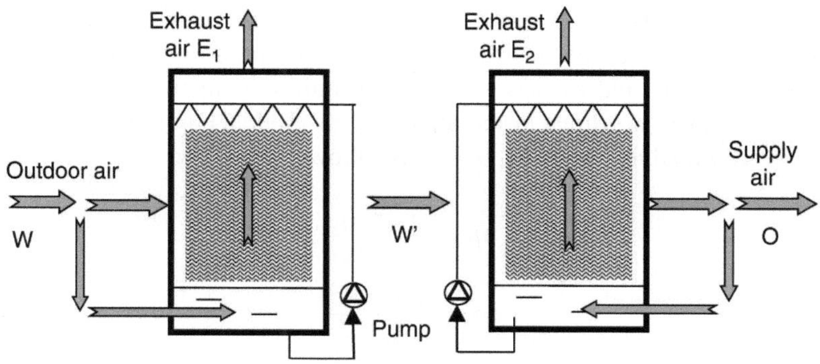

Fig. 4.17 A two-stage indirect evaporative cooling device

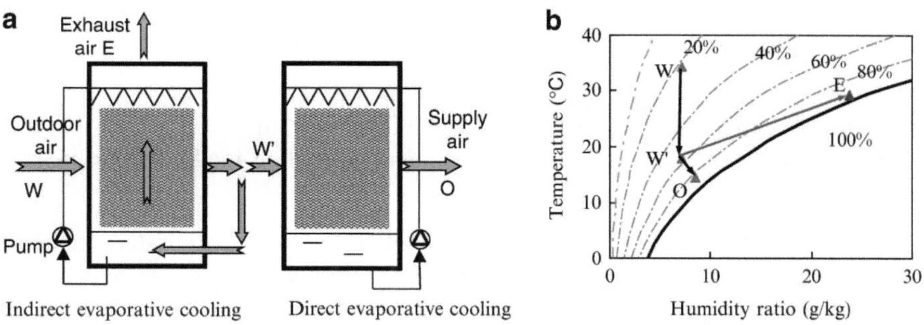

Fig. 4.18 Air handling process combining indirect and direct evaporative cooling: (**a**) operating principle and (**b**) air handling process in psychrometric chart

temperature of the air to be cooled is lower than the wet-bulb temperature of the outdoor air but higher than the dew point of the outdoor air.

In the handling processes using direct evaporative cooling, indirect evaporative cooling, or a combination of direct and indirect types, the outlet air parameters can be expressed by Eqs. (4.1) and (4.2), where W is the outdoor air state, W_1 is the relative dew point state, W_2 is the relative wet-bulb state, and O is the supplied air state after evaporative cooling (shown in Figs. 4.15, 4.16, 4.17 and 4.18, respectively). The temperature and humidity ratio of the supplied air are expressed as follows (Xie and Jiang 2010):

$$t_O = t_W - \eta_1 \cdot (t_W - t_{W1}) - \frac{r}{c_{p,m}}(d_O - d_W) \tag{4.1}$$

$$d_O = d_W + \eta_2 \cdot (1 - \eta_1) \cdot (d_{W2} - d_W) \tag{4.2}$$

where r is the latent heat of water vaporization, $c_{p,m}$ is the specific heat capacity of the humid air, η_1 is the device efficiency of the indirect evaporative cooling module (taking the outdoor dew point as the limiting temperature), and η_2 is the device efficiency of the direct evaporative cooling module humidifying the air. The definitions of $\eta_1$and η_2 can be expressed as follows:

$$\eta_1 = \frac{t_W - t_1}{t_W - t_{W1}} \tag{4.3}$$

$$\eta_2 = \frac{d_W - d_O}{d_W - d_2} \tag{4.4}$$

where t_1 is the dry-bulb temperature of the outlet air of the indirect evaporative cooling module and d_2 is the humidity ratio of the saturated air, with a temperature equal to the wet-bulb temperature of the inlet air in the direct evaporative cooling module. Thus, the values of the relative coefficients in Eqs. (4.1) and (4.2) can be determined for the three kinds of evaporative cooling devices:

- Direct evaporative cooling method: $\eta_1 = 0$ and $0 < \eta_2 \leq 1$
- Indirect evaporative cooling method: $\eta_2 = 0$ and $0 < \eta_1 < 1$
- Combination of direct and indirect evaporative cooling methods: $0 < \eta_1 < 1$ and $0 < \eta_2 < 1$

For internally cooled or externally cooled indirect evaporative cooling devices, dew point efficiency η_1 expressed by Eq. (4.3) is determined by the ratio between the primary air and the secondary air, the flow rate ratio between the air and the water in the direct evaporative cooling module, and the input heat transfer ability of the indirect evaporative cooling module. For handling processes combining direct and indirect evaporative cooling methods, the higher the efficiency (η_1) of the indirect evaporative cooling module, the lower the humidification efficiency of the air supplied by the device, as indicated in Eq. (4.2). The reason is that, as the efficiency increases for the indirect evaporative cooling module, the outlet air approaches the saturation line, leading to a decrease of the driving force for cooling and humidification in the direct evaporative cooling module.

Of all the current outdoor air handling devices that utilize evaporative cooling methods, the structure of the direct evaporative cooling type is the simplest, and its air humidification efficiency (η_2) can be about 90 % or even higher in some cases. For the indirect evaporative cooling type, the air cooling efficiency (η_1) based on the outdoor dew point temperature is usually 40–70 %, mostly depending on the structures and input parameters. Table 4.1 lists the outlet air parameters of different evaporative cooling devices in typical cities of Xinjiang province, with η_1 calculated as 60 % and η_2 calculated as 90 % to simplify the analysis.

Table 4.1 Outlet air states using evaporative cooling

Area	Atmospheric pressure (kPa)	Outdoor design parameters in summer t_{db} (°C)	t_{wb} (°C)	t_{dew} (°C)	ω (g/kg)	Outlet air of DEC device (one stage) t (°C)	ω (g/kg)	Outlet air of IEC device (one stage) t (°C)	ω (g/kg)	Outlet air of IDEC device (two stages) t (°C)	ω (g/kg)
Altay	93.4	30.6	18.7	12.6	9.9	26.3	14.2	19.8	9.9	18.1	11.6
Karamay	96.9	34.9	19.1	9.4	8.2	29.2	13.9	19.6	8.2	17.3	10.5
Gulja	94.2	32.2	21.4	15.7	12.9	28.2	16.9	22.3	12.9	20.7	14.5
Urumchi	91.8	34.1	18.5	7.5	8.5	28.5	14.1	18.1	8.5	15.9	10.8
Turpan	101.3	40.7	23.8	12.3	11.8	34.5	18.0	23.7	11.8	21.2	14.3
Kumul	93.1	35.8	20.2	11.3	9.9	30.1	15.6	21.1	9.9	18.8	12.2
Kashgar	87.2	33.7	19.9	13.4	11.4	28.6	16.5	21.5	11.4	19.5	13.4
Khotan	86.2	34.3	20.4	13.6	12.2	29.3	17.2	21.9	12.2	19.9	14.2

Notes: t temperature, ω humidity ratio, t_{db}, dry-bulb temperature, t_{wb} wet-bulb temperature, t_{dew}, *dew* point temperature, *DEC* direct evaporative cooling, *IEC* indirect evaporative cooling, *IDEC* combination of direct and indirect evaporative cooling

4.2.2 Outdoor Air Humidification in Winter

4.2.2.1 Operating Schematic of the Handling Device

In the dry region, the outdoor air temperature and humidity ratio are both relatively low in winter, indicating that the outdoor air has to be heated and humidified before being supplied to the indoor environment. Figure 4.19a, b illustrates the typical outdoor air handling device and the handling process, respectively, where the cold and dry outdoor air first flows through the heater to be preheated and then flows into the packed tower to be humidified to reach a satisfactory state for the indoor environment.

4.2.2.2 Discussion of the Position of the Heater

For these common air humidification processes, the water or the air is usually heated in the humidification process to achieve a better humidification result. The heater could be used to heat the inlet water or the inlet air. For air humidification processes that use a heater to heat the inlet air, there are some cases in practice when the humidification effect is not satisfactory or is lower than the requirement. What could account for these insufficient humidification results? And does the difference in inlet heating targets have any affect?

Figure 4.20 illustrates two humidification cases in an attempt to answer the above questions. A heater is adopted to heat the inlet air in scenario A and the inlet water in scenario B. Figure 4.21 illustrates the air handling processes that correspond to the two devices described above. The conditions of the inlet air (point a_I)

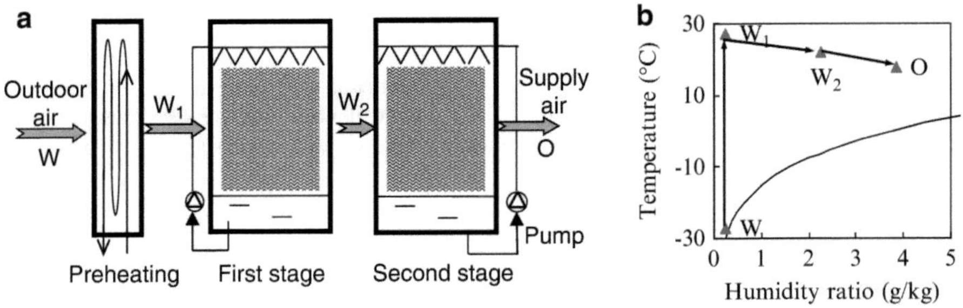

Fig. 4.19 Outdoor air humidification process in the dry region during winter: (**a**) operating principle and (**b**) air handling process in psychrometric chart

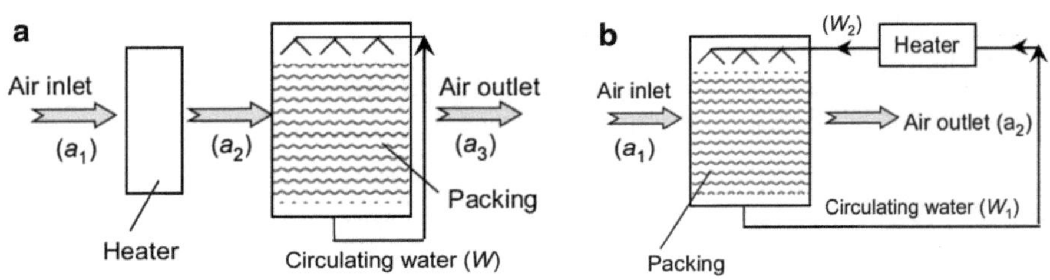

Fig. 4.20 Operating schematics of two air humidification processes: (**a**) heating the inlet air and (**b**) heating the inlet water.

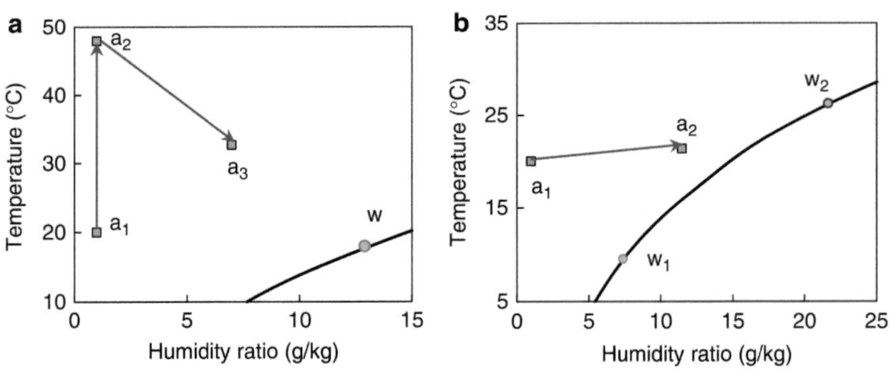

Fig. 4.21 Air handling processes of two different humidification processes in psychrometric chart: (**a**) heating the inlet air and (**b**) heating the outlet air

are 1 kg/s, 20 °C, and 1 g/kg, and the flow rate of the circulating water (point w) is 0.4 kg/s. The heater capacity is 28 kW, and the NTU_m value of the packed-bed humidifier is 1. The humidification performances of the two modes are shown in Fig. 4.21 and Table 4.2.

Table 4.2 Calculated results for different air humidification processes

	Inlet parameters of the packed module			Outlet parameters of the packed module			
Heating object	$t_{a,in}$ (°C)	$\omega_{a,in}$ (g/kg)	$t_{w,in}$ (°C)	$t_{a,out}$ (°C)	$\omega_{a,out}$ (g/kg)	$t_{w,out}$ (°C)	Humidification effect (g/kg)
Air	47.9	1.0	18.0	32.6	7.0	18.0	6.0
Water	20.0	1.0	26.2	21.4	11.5	9.5	10.5

Notes: t temperature, ω humidity ratio; subscripts: *a* air, *w* water, *in* inlet, *out* outlet

As indicated by the results listed in Table 4.2, when the heater is adopted to heat the inlet air in scenario A, the actual humidification effect of the outlet air is 6.0 g/kg; when the heater is adopted to heat the inlet water in scenario B, the actual humidification effect of the outlet air is as high as 10.5 g/kg. In other words, when inlet water is the heating target, 75 % more humidification is achieved compared to the scenario when inlet air is the heating target. Therefore, for the packed air-water device for humidification shown in Fig. 4.20, the inlet water should be heated, rather than air, in order to gain better humidification performance. Thus, by adjusting the heating target, the humidification effect can be improved significantly in the air-water packed device.

4.3 Outdoor Air Handling Process in the Humid Region

In this region, the outdoor air temperature and humidity ratio are both relatively high in summer, where the outdoor air has to be cooled and dehumidified before being supplied to the indoor environment. Therefore, maximizing the efficiency of the outdoor air handling process is the key issue. In this section, air dehumidification methods that utilize condensation dehumidification and solid desiccant are investigated. Dehumidification methods using liquid desiccant will be discussed in Chap. 5.

4.3.1 Condensation Dehumidification Method

Condensation dehumidification is widely utilized in conventional air-conditioning systems, and it can also be adopted in THIC systems. In conventional air-conditioning systems, chilled water with a temperature of around 7 °C is usually adopted to handle the outdoor air, with the temperature and humidity ratio of the outdoor air both decreasing. Taking a system utilizing an FCU with an outdoor air handling device as an example, the outdoor air is usually handled to a state with a similar humidity ratio as the indoor state before being supplied to the indoor environment. The indoor moisture load (e.g., the moisture generated from occupants) is then extracted by the

FCU operating in the wet condition. In the THIC system, where dry air is responsible for removing the indoor moisture, the outdoor air has to be handled to a state lower than the indoor humidity ratio and dry enough to extract the indoor moisture load. Compared with the conventional system, the required humidity ratio of the supplied outdoor air in the THIC system is lower, leading to a higher performance requirement for outdoor air handling processors.

As the indoor design parameters are 26 °C and 60 % (with a corresponding humidity ratio of 12.6 g/kg), if only the moisture generated from occupants is taken into account, then the humidity ratio of the handled outdoor air in the conventional system is about 12.6 g/kg. For the THIC system, the required humidity ratio of the handled outdoor air is about 9.6 g/kg, as calculated by Eq. (2.20), if the outdoor air flow rate is 30 m^3/h per person. The above parameters of the handled outdoor air in conventional and THIC systems are listed in Table 4.3. Taking the outdoor climate design parameters of Beijing (dry-bulb temperature of 33.2 °C, wet-bulb temperature of 26.4 °C, with a corresponding humidity ratio of 19.1 g/kg) as an example, the required parameters of cooling coils and chilled water flow rate can be determined. The inlet chilled water temperature is 7 °C, and the outdoor air flow rate is 12,000 m^3/h. The required rows of cooling coils and the water flow velocity are listed in Table 4.4, which shows that a 4-row cooling coil is sufficient for the conventional system. Due to its lower required humidity ratio, the THIC system requires more rows of cooling coils than the conventional system, which means a 6-row cooling coil is sufficient.

When the condensation dehumidification method is adopted for the outdoor air handling process, there is a significant temperature difference between the inlet outdoor air and the cooling source. Thus, this low-temperature cooling source leads to an obvious temperature mismatch for the total heat transfer process, resulting in heat transfer loss. To reduce this kind of heat transfer loss caused by the significant temperature difference between the inlet fluids, using an appropriate high-temperature cooling source to precool the outdoor air is a feasible solution, which results in a cascade process and improved energy performance. On the other hand, the handled outdoor air usually approaches the saturated state, and its temperature is too low to be supplied directly to the indoor environment, even though the humidity ratio is satisfactory for humidity control. Thus, reheating is needed to a certain extent. To avoid energy dissipation caused by the reheating process, the indoor exhaust air or the outdoor air itself could be used to reheat the handled air (MOHURD, China Institute of Building Standard Design & Research 2009). The following subsection focuses on common precooling methods and reheating solutions.

4.3.1.1 Precooling the Outdoor Air with Heat Recovery from the Indoor Exhaust Air

If the gas tightness of a building's windows and that of the building envelope are satisfactory, some indoor air should be extracted to maintain the air balance when the outdoor air is supplied to the conditioned space. By setting an appropriate indoor exhaust air system and venting the indoor exhaust air in an organized way,

Table 4.3 Supply air parameters after condensation dehumidification

Type	Dry-bulb temperature (°C)	Relative humidity (%)	Humidity ratio (g/kg)	Wet-bulb temperature (°C)	Specific enthalpy (kJ/kg)
Conventional system	18.5	95	12.6	17.9	50.6
THIC system	14.2	95	9.6	13.8	38.6

Table 4.4 Selected cooling coils for condensation dehumidification

Type	γ	Cooling coil type	Rows	Air velocity (m/s)	Water velocity (m/s)	η_h
Conventional system	2.14	JW20-4	4	1.78	1.01	0.56
THIC system	2.29	JW20-4	6	1.78	1.39	0.72

Notes: γ ratio of the total cooling capacity to the sensible cooling capacity, η_h heat exchange efficiency

the heat recovery device can be implemented between the indoor exhaust air and the outdoor air, where the energy can be recovered. Figure 4.22 illustrates the outdoor air handling process using condensation dehumidification with the enthalpy recovery module. It can be seen that the enthalpy of the outdoor air decreases after the enthalpy recovery process (from state W to W_1). Thus, the enthalpy difference required in the dehumidification process decreases, helping to reduce the energy consumption of the outdoor air handling process.

4.3.1.2 Precooling the Outdoor Air with High-Temperature Chilled Water

To improve the operating performance of the outdoor air handling process, the high-temperature chilled water (about 16–18 °C) of the THIC system could be adopted to precool the air. The lower-temperature chilled water could be used to dehumidify the air further, as shown in Fig. 4.23. The high-temperature chilled water could be directly obtained from natural cooling sources such as underground water and could also be available from the high-temperature water chiller. With the help of the precooling process, the outdoor air could be cooled from the hot and humid state to the saturated state (or approaching the saturated state). The major task of the precooling process is to cool the air but not to dehumidify it, i.e., from state W to state W_1, as shown in Fig. 4.23b. Low-temperature chilled water is then adopted to dehumidify the air from state W_1 to state O, satisfying the humidity ratio requirement of the supplied air. Moreover, using high-temperature chilled water for precooling takes full advantage of the energy efficiency of the high-temperature cooling source.

Based on the variances of the outdoor air parameters and the required supply air parameters, the air handling processor can meet the requirement of the supplied air

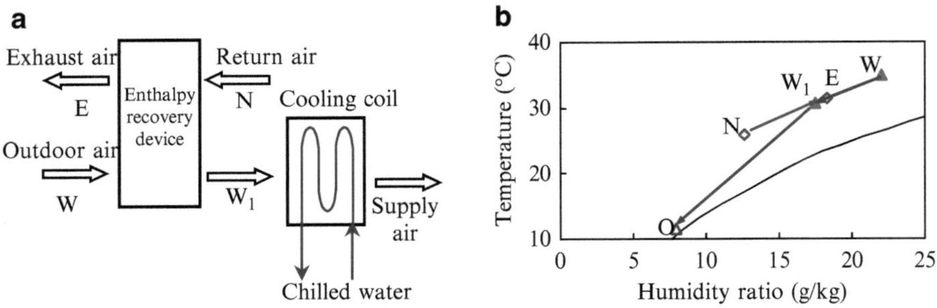

Fig. 4.22 Condensation dehumidification process with the enthalpy recovery module: (**a**) operating principle and (**b**) air handling process in psychrometric chart

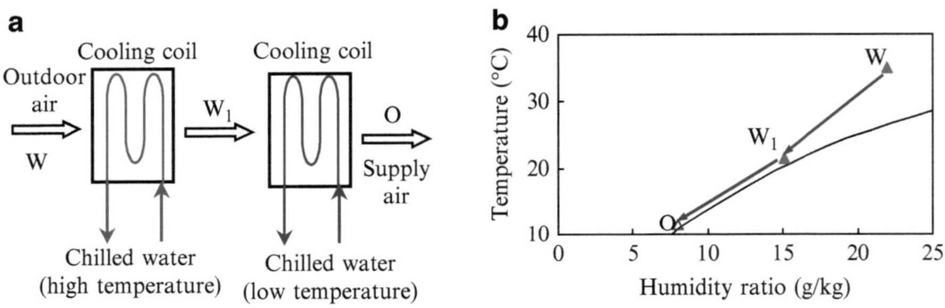

Fig. 4.23 Condensation dehumidification process using high-temperature chilled water for precooling: (**a**) operating principle and (**b**) air handling process in psychrometric chart

by regulating the flow rate of the low-temperature chilled water and ensuring the use of high-temperature chilled water as much as possible. Taking the required humidity ratio of 8 g/kg (the relative temperature is 11.5 °C with a relative humidity of 95 %) as an example, Table 4.5 lists the enthalpy differences using the high-temperature chilled water to precool the air as a function of the outdoor climate, where the high-temperature chilled water is 16 °C. As indicated by the results, the enthalpy difference of the precooling process is significantly higher as the outdoor temperature and humidity ratio are higher. The enthalpy difference of the precooling process accounts for about 50 % of the total enthalpy difference during the outdoor air handling process. Thus, precooling using high-temperature chilled water could undertake the load of handling the outdoor air effectively and is a feasible approach for improving the efficiency of the outdoor air handling process.

For the outdoor air handling process using high-temperature chilled water to precool the air in Fig. 4.23, low-temperature chilled water is adopted for further dehumidification, leading to two plumbing systems for chilled water of different temperatures. To make the air handling processor more flexible, some improvements are proposed for the condensation dehumidification outdoor air handling processes utilizing high-temperature chilled water for precooling. Figure 4.24 illustrates an

Table 4.5 Outdoor air enthalpy differences under typical conditions

Outdoor air parameters					
Temperature (°C)	Humidity ratio (g/kg)	Enthalpy (kJ/kg)	Δh_1 (kJ/kg)	Δh_2 (kJ/kg)	$\Delta h_1/(\Delta h_1 + \Delta h_2)$
35	22	91.6	32.7	27.2	54.7 %
30	22	86.4	29.1	25.6	53.2 %
35	16	76.2	22.0	22.5	49.4 %
30	16	71.0	18.3	21.0	46.7 %

Notes: Δh_1 is the enthalpy difference during precooling, Δh_2 is the enthalpy difference during dehumidification

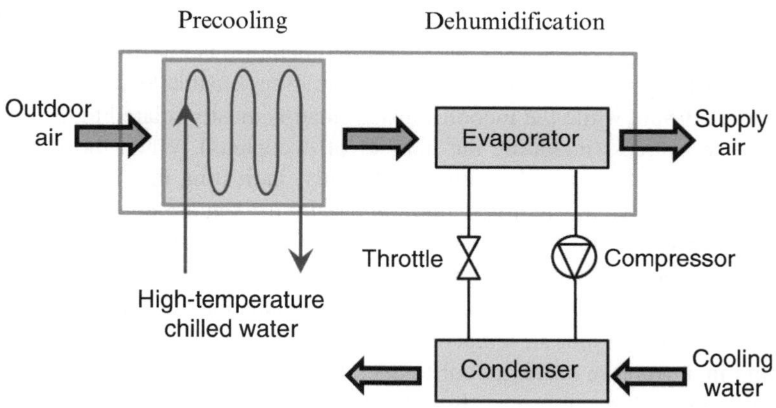

Fig. 4.24 Condensation dehumidification outdoor air handling process with a separate heat pump

improved outdoor air handling process using condensation dehumidification. In the handling process, the outdoor air is first precooled by the high-temperature chilled water (16–18 °C) from the cooling source. The air is dehumidified further by the evaporator of the separate heat pump cycle to meet the humidity ratio requirement. The condenser of the heat pump could be an air-cooled type utilizing indoor exhaust air or a water-cooled type utilizing cooling water. For the outdoor air handling process shown in Fig. 4.24, the refrigerant inside the evaporator evaporates directly, and the air is dehumidified by the heat transfer process between the refrigerant and the moist air. As the separately installed heat pump is responsible for dehumidification, only a single plumbing system for the high-temperature chilled water is required, resulting in a much simpler arrangement of the processors.

For the aforementioned outdoor air handling processes using condensation dehumidification, a common problem is that the supply air temperature is usually too low to be supplied to the indoor environment directly. As the humidity ratio is about 8–10 g/kg, the corresponding air temperature is about 11.5–14.8 °C. If air with such a low temperature is supplied to the conditioned space, the occupants could experience thermal discomfort. Thus, an air diffuser with good inductivity or

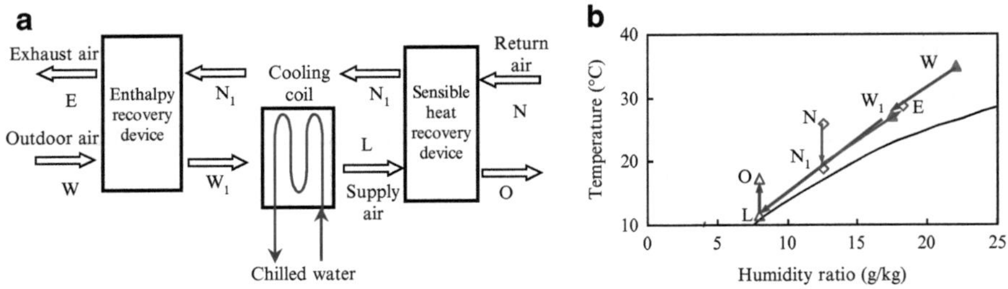

Fig. 4.25 Condensation dehumidification process using return air for precooling and reheating: (**a**) operating schematic and (**b**) air handling process in psychrometric chart

diffusivity is required, and the air distribution should be checked carefully. However, there are some cases when the indoor temperature is too low due to the low supplied air temperature. This is because the building sensible load is related to the outdoor condition, while the indoor moisture load is mostly related to the variance of moisture sources (including the number of occupants). When the variance of occupant number does not fluctuate significantly, indicating that the indoor moisture load is steady, the required humidity ratio for indoor humidity control is therefore steady. If the condensation dehumidification method is adopted, the required supplied air state is then fixed. As a result, during the partial sensible cooling load, the supplied outdoor air temperature can sometimes become too low because of lower outdoor air temperatures or insufficient solar radiation, inevitably resulting in a decrease of the indoor temperature.

If the dehumidified air is directly supplied to the indoor environment, it may lead to overcooling in the partial load. Therefore, the dehumidified air using condensation dehumidification should be reheated to reach an appropriate temperature before being supplied to the indoor environment. Common reheating methods include electrical reheating and steam reheating. However, these methods lead to additional energy dissipation and should be avoided in practice, except for certain special requirements. Instead, reheating the supplied air after dehumidification using the indoor exhaust air or the outdoor air is a more feasible solution, since it achieves the reheating effect while reducing unnecessary energy consumption.

4.3.1.3 Reheating the Handled Air with Indoor Exhaust Air

Figure 4.25 shows the operating schematic of a method for reheating the supply air. The sensible heat recovery process is conducted between the dehumidified outdoor air (L) and the indoor return air (N), realizing the reheating of the outdoor air (from L to O). The return air is cooled to state N_1 by the dehumidified air and then flows through the enthalpy recovery device. The outdoor air is first precooled by the indoor exhaust air before being dehumidified, from state W to state W_1. Thus, in this process, the indoor exhaust air is first used to reheat the dehumidified air. It then absorbs more heat from the enthalpy recovery module and precools the outdoor air.

Table 4.6 Performance of the condensation dehumidification process using exhaust air for precooling and reheating

State *W* (outdoor air)			State W_1			State *O*	State N_1	State *E*		
t	*ω*	*h*	*t*	*ω*	*h*	*t*	*t*	*t*	*ω*	*h*
(°C)	(g/kg)	(kJ/kg)	(°C)	(g/kg)	(kJ/kg)	(°C)	(°C)	(°C)	(g/kg)	(kJ/kg)
35.0	22.0	91.6	27.3	17.5	72.0	17.3	18.7	28.6	18.3	75.3
30.0	22.0	86.4	24.6	17.5	69.3	17.3	18.7	25.5	18.3	72.2
35.0	16.0	76.2	27.2	14.4	64.0	17.3	18.7	28.5	14.7	66.1
30.0	16.0	71.0	24.6	14.4	61.3	17.3	18.7	25.5	14.7	63.0

Notes: *t* temperature, *ω* humidity ratio, *h* enthalpy

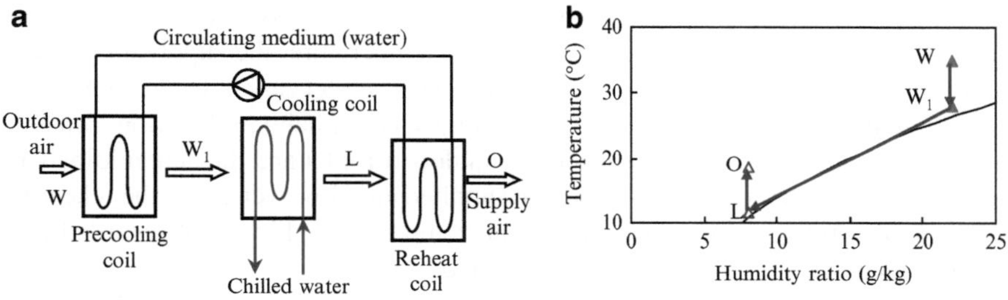

Fig. 4.26 Outdoor air handling process using intermediate for precooling and reheating: (**a**) operating schematic and (**b**) air handling process in psychrometric chart

Table 4.6 lists the operating performances of the sensible heat exchange module and the total heat exchange module of the outdoor air handling device shown in Fig. 4.25 that utilizes condensation dehumidification. The indoor design parameters are 26 °C with a relative humidity of 60 % (the corresponding humidity ratio is 12.6 g/kg). The flow rate ratio between the indoor exhaust air and the outdoor air is 0.8, and the dehumidified air state (*L*) is 11.5 °C with a relative humidity of 95 % (the corresponding humidity ratio is 8 g/kg). The supply air temperature could be around 17.3 °C using the indoor exhaust air to reheat the dehumidified air, which is suitable for the indoor environment. Based on these indoor parameters, it can be said that this reheating method is superior in relatively stable operating conditions.

4.3.1.4 Reheating the Handled Air with Outdoor Air

As the fluid intermediate can act as the medium for the energy exchange between the outdoor air and the dehumidified air, it could be adopted for precooling and reheating, as shown in Fig. 4.26. In contrast to the condensation dehumidification process shown in Fig. 4.25, this kind of precooling and reheating process uses the intermediate, which circulates in a closed loop. The intermediate's major task is to cool or reheat the air in the handling process undertaking the sensible load.

Table 4.7 Temperatures of the outdoor air handling process using intermediate for precooling and reheating (°C)

State *W* (outdoor air)	State W_1	State *L*	State *O* (supply air)	Intermediate
35.0	27.9	11.5	18.5	23.2
32.0	25.8	11.5	17.6	21.7
29.0	23.7	11.5	16.7	20.2
26.0	21.6	11.5	15.8	18.7

Table 4.7 lists the precooling and reheating process in typical conditions using Freon as the intermediate, where the dehumidified air state (*L*) is 11.5 °C with a relative humidity of 95 % (the corresponding humidity ratio is 8 g/kg). This table shows that the effect of the precooling and reheating process using the circulating fluid is affected by the variance of outdoor air temperature; the supply air temperature can fluctuate after reheating, and the reheating effect cannot be guaranteed. In some outdoor air handling devices using condensation dehumidification, sensible heat exchangers are installed between the outdoor air inlet and the dehumidified air and use the outdoor air itself to reheat the supplied air. However, since the reheating effect is also influenced by the outdoor air temperature, this is not regarded as a solution that can satisfy a steady reheating requirement.

Thus, for this kind of reheating process that uses the intermediate or the outdoor air itself, the influence of the variance of outdoor air parameters on the reheating effect should be taken into account, especially when the outdoor air temperature is low (e.g., during plum rain season in Shanghai), as the supplied air temperature after reheating cannot be guaranteed.

4.3.2 Solid Desiccant Dehumidification Method

There are two kinds of solid desiccant dehumidifiers: the rotary wheel type (continuous dehumidification process) and the fixed bed type (noncontinuous dehumidification process). In this section, a desiccant wheel dehumidifier and the DESICA fixed bed system (developed by Daikin Industries, Ltd.) are introduced.

4.3.2.1 Properties of Commonly Used Solid Desiccant Materials

Solid desiccant material can be used to adsorb water vapor molecules in the air to achieve dehumidification, and the water vapor pressure difference between the air and the desiccant surface is the driving force for this moisture transfer process. As humid air flows through the surface of the solid desiccant material, water vapor is adsorbed and the air is dehumidified. Air states approach the isenthalpic line, with the humidity ratio decreasing and the temperature increasing. Thus, the air temperature after dehumidification by solid desiccant is usually too high to be supplied

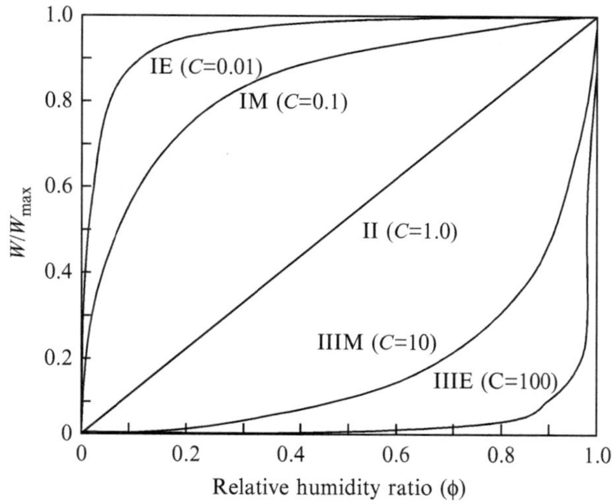

Fig. 4.27 Adsorption isotherms of ideal adsorbents

directly to the indoor environment. To realize a lower supplied air temperature, the dried air usually has to be cooled down by associated cooling sources. Furthermore, as the moist air is dehumidified, the water content of the desiccant material increases, and the equilibrium humidity ratio on the surface also increases. Desiccant material loses its adsorption ability when an equilibrium state is reached, at which point it has to be regenerated.

When the desiccant material reaches the equilibrium state, the mass of adsorbed water per unit mass of desiccant material, which is called the absorbed content, is an important index for evaluating the performance of the desiccant material. The water content at the equilibrium state can be described by Zhang et al. (2006):

$$\frac{W}{W_{\max}} = \frac{\phi}{C + (1 - C)\phi} \tag{4.5}$$

where W is water content, kg (water)/kg (desiccant material); $W_{\max}$ is the maximum water content if the moist air is of 100 % relative humidity, kg (water)/kg (desiccant material); C is a shape factor of the desiccant material; and ϕ is the relative humidity of the moist air.

According to different equilibrium water contents under variant relative humidity, adsorption isotherms of the solid desiccant can be obtained. Adsorption isotherms can be divided into different types according to various isotherm characteristics. Type I (including IE and IM), type II, and type III (including IIIE and IIIM) adsorption isotherms have corresponding shape factors of greater than 1, equal to 1, and lower than 1, respectively, as shown in Fig. 4.27. For dehumidification processes using desiccant wheels, desiccant materials with adsorption isotherms of types I and II are usually preferred.

Solid desiccant materials commonly used for dehumidification include silica gel, activated aluminum oxide, molecular sieves, calcium chloride, and zeolite.

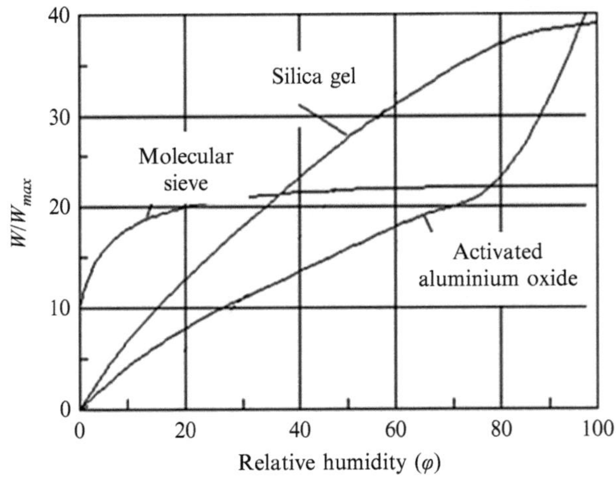

Fig. 4.28 Adsorption isotherms of commonly used adsorbents (Zhang 2005)

Figure 4.28 shows the adsorption isotherms of these desiccant materials, where the molecular sieve is type I, silica gel is type II, and activated aluminum oxide is type III.

Silica gel is a kind of nontoxic, odorless semitransparent crystalline, with a porosity greater than 70 % and W_{max} higher than 30 % (Suzuki and Oya 1983; Jia 2006). Some kinds of silica gel are macroporous, while others are microporous; some change color from blue to pink when adsorbing water because of the addition of cobalt chloride, while others do not. Macroporous silica gels can be saturated easily, while microporous silica gels can adsorb water vapor for a longer period of time. Silica gel is commonly used in air-conditioning systems. Hot air with a temperature between 100 and 150 °C is usually used for the regeneration of silica gel to remove the adsorbed water from the desiccant.

Activated aluminum oxide is highly attracted to water, and under certain circumstances, it can dry air to a dew point lower than −70 °C. Its regeneration temperature is much lower than that of a molecular sieve.

Calcium chloride is a salty, white porous crystal with strong adsorption ability and can deliquesce easily after adsorbing water. Calcium chloride solutions have strong corrosion effects on metal, making the substance inconvenient to use. Nevertheless, because it is cheap and can be easily regenerated, it is a commonly used desiccant material.

Zeolite is an aluminosilicate compound with a strong adsorption ability that is related to the ratio between Si and Al; a smaller ratio helps to increase the adsorption ability (Suzuki and Oya 1983). Synthetic zeolite (a molecular sieve) is of balanced microporosity, and it can be used to produce dehumidification products with different porosity characteristics.

When the solid desiccant and humid air reach equilibrium, their water vapor pressures are identical. In other words, the equivalent humidity ratio of the desiccant is the same as the humidity ratio of the air. The state of the desiccant can be represented in a psychrometric chart according to the equilibrium of the air state with the desiccant. Figure 4.29 depicts the states of commonly used desiccants in a

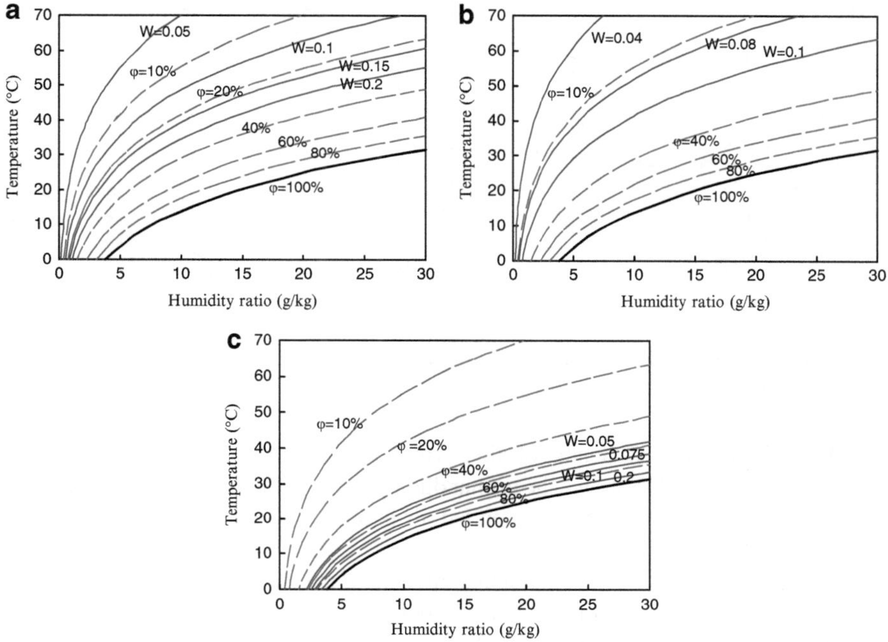

Fig. 4.29 States of commonly used desiccants in psychrometric chart: (**a**) RD-type silica gel, (**b**) molecular sieve, and (**c**) activated aluminum oxide

psychrometric chart. For example, the states of RD-type silica gel (Pesaran and Mills 1987) are shown in Fig. 4.29a, with *W* representing water content (kg water/kg desiccant). As indicated by this figure, the iso-water content lines of the silica gel coincide with the iso-relative humidity lines of the humid air.

It can be observed that, for the iso-water content line, lowering the temperature helps to lower the vapor pressure and strengthen its adsorption ability. In the process of dehumidification, the driving force for mass transfer is the vapor pressure difference between the desiccant and the air. Reducing the desiccant's temperature and water content helps to increase the mass transfer driving force. The regeneration process is similar to the dehumidification process but with an opposite mass transfer direction.

4.3.2.2 Dehumidification Process Using Desiccant Wheel

Figure 4.30 shows a dehumidification system using a desiccant wheel developed by the Swedish company Munters (2004), which uses silica gel as the desiccant material. As the processed air flow rate is 790 m^3/h, to achieve a dehumidification capacity of 3.7 kg/h, the regeneration air temperature needs to be around 130 °C with a flow rate of 220 m^3/h; the power consumption of the heater is 3.9 kW. Defining dehumidification efficiency as the latent heat difference of the processed air divided by the heater's power,

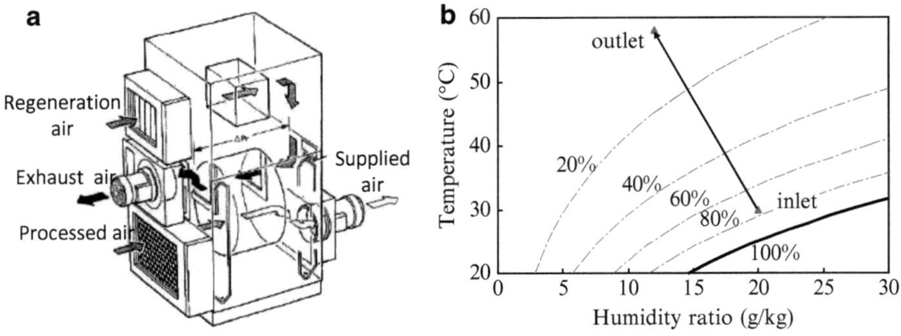

Fig. 4.30 Desiccant wheel developed by Munters, series SX: (**a**) schematic diagram and (**b**) air handling process in psychrometric chart

the dehumidification efficiency of this wheel is about 0.7 ((3.7/3,600 × 2,500)/3.9 = 0.7). Figure 4.30b shows the air handling process if the processed air inlet state is 30 °C, 20 g/kg, and that of the supplied air is 58 °C, 12 g/kg, with a temperature increase of 28 °C.

Because the processed air is handled approaching the isenthalpic line, the supplied air temperature is usually very high, so a cooling system should be implemented to cool down the air. Heat recovery from the indoor exhaust air is a common way to cool the processed air. However, the cooling ability is limited since the supplied air temperature cannot be significantly lower than the indoor air temperature. Thus, other cooling sources such as high-temperature cooling water are preferable options to cool the processed air. Figure 4.31 illustrates a rotary wheel dehumidification system, which uses external cooling coils to precool the processed air and cool down the supplied air.

There are many air handling processes that use desiccant wheels, such as the ventilation cycle desiccant dehumidification process, the Dunkle cycle, and the recirculation cycle desiccant dehumidification process (La et al. 2010). Figure 4.32 shows the air handling process of the ventilation cycle. Seen from the psychrometric chart, there are four approximate isenthalpic processes, including the desiccant dehumidification process (f–g), the desiccant regeneration process (d–e), and the direct cooling processes (a–b and h–i).

As analyzed above, in the dehumidification process using the desiccant wheel, the processed air is handled approaching the isenthalpic line, which requires auxiliary heat recovery and direct cooling components. If the task is to dehumidify the processed air to a state with a humidity ratio lower than that of the indoor air, a regeneration temperature higher than 100 °C is usually required. To reduce the required regeneration temperature, it is advised to split one wheel into two (Ge et al. 2008), as depicted in Fig. 4.33.

Figure 4.34 compares the minimum regeneration temperature of the single-stage desiccant wheel system to that of the two-stage desiccant wheel system. The processed air (W) is 33.2 °C, 19.1 g/kg (Beijing design condition); the required supplied air humidity ratio is 8 g/kg; the dehumidification process is assumed to be along the

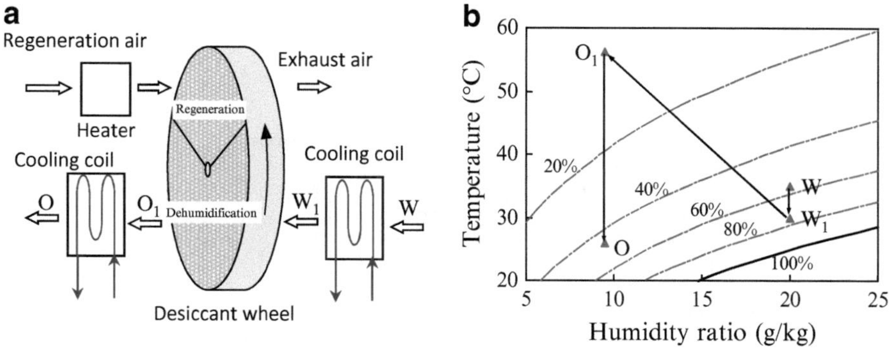

Fig. 4.31 Desiccant wheel with cooling devices: (**a**) schematic diagram and (**b**) air handling process in psychrometric chart

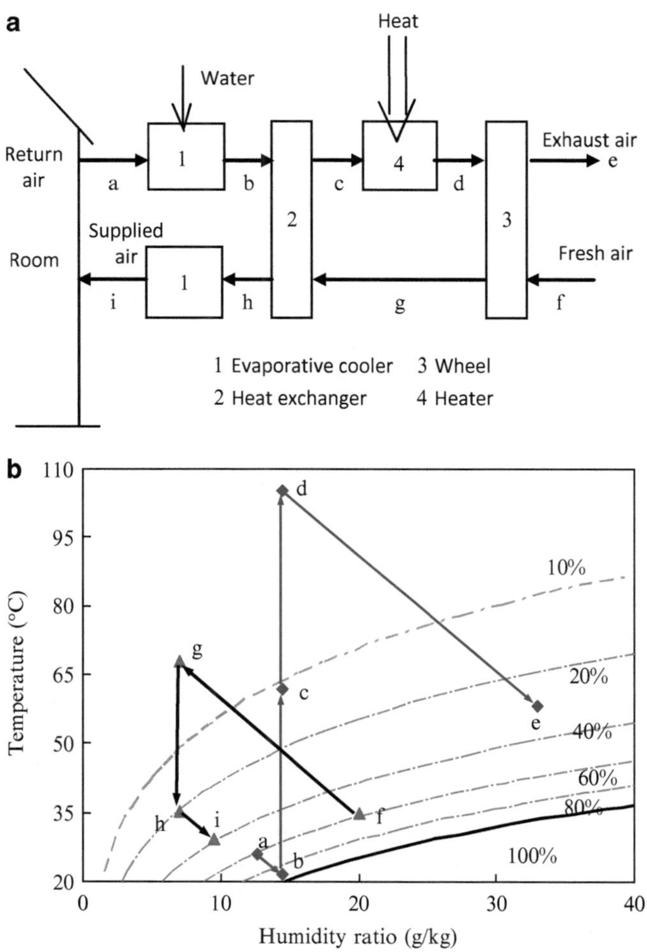

Fig. 4.32 Ventilation cycle solid desiccant dehumidification process: (**a**) schematic diagram and (**b**) air handling process in psychrometric chart

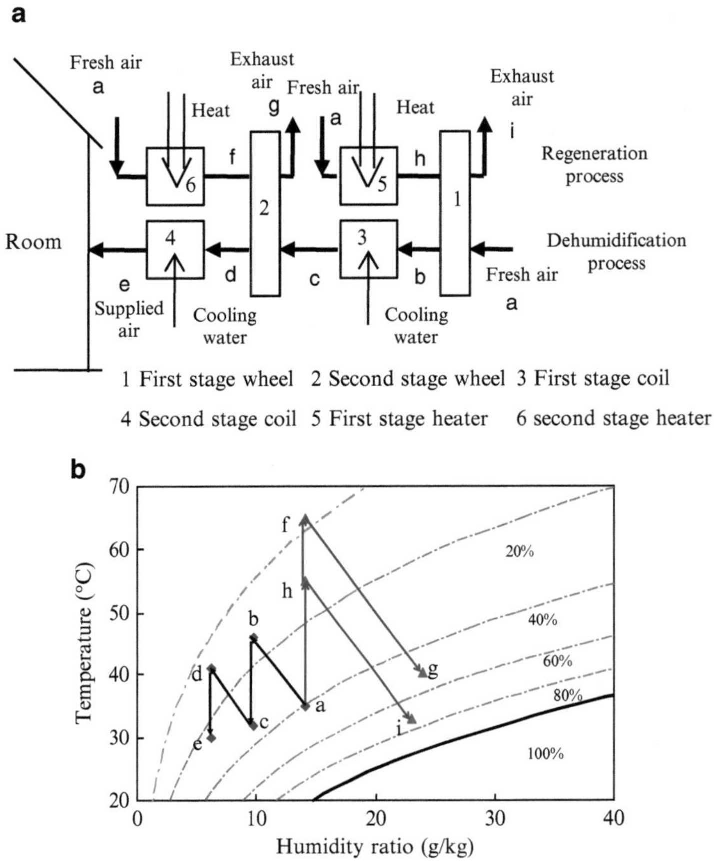

Fig. 4.33 Two-stage desiccant wheel system (La et al. 2011): (**a**) schematic diagram and (**b**) air handling process in psychrometric chart

isenthalpic line; and the outdoor air is used for regeneration. In the single-stage desiccant wheel system, the processed air is handled from point *W* to point *O* (61 °C, 8 g/kg, relative humidity of 6 %). The minimum regeneration temperature is the temperature at intersection point *M* obtained by the iso-relative humidity lines *O* and *W*. The minimum regeneration temperature is obtained under the following assumptions: (1) the temperature and humidity ratio differences between the air and the desiccant material are ignored; (2) the variance of the moisture content for the hygroscopic material inside the desiccant wheel is ignored, and a constant adsorbing capacity is maintained in the process, i.e., the state of the solid desiccant remains at the iso-relative humidity line corresponding to *O*; (3) the driving force of the mass transfer between the desiccant and the regeneration air required in the regeneration process is ignored, as an identical water vapor pressure for the desiccant and the regeneration air is sufficient to realize the desiccant regeneration process; and (4) the driving force of the heat transfer between the desiccant and the regeneration air required in the regeneration process is ignored,

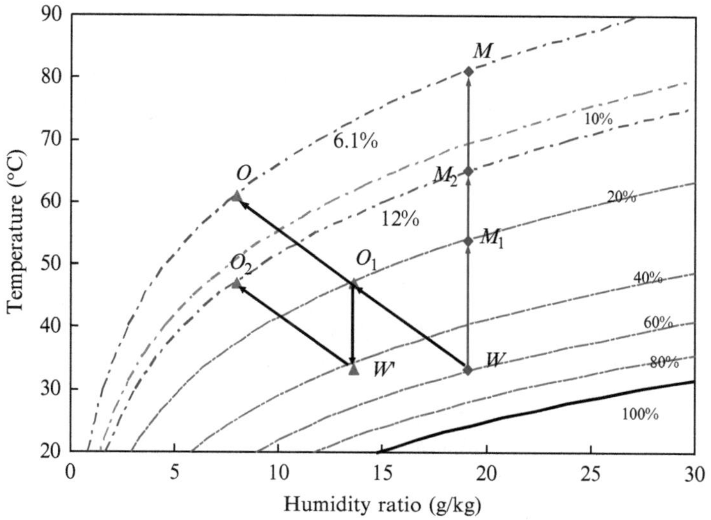

Fig. 4.34 Minimum regeneration temperatures of a single-stage desiccant wheel and a two-stage desiccant wheel

as the solid desiccant is heated to a state with the same temperature as the regeneration air (point M). Under these assumptions, the temperature at point M is the theoretical minimum regeneration temperature, which is significantly lower than the actual required regeneration temperature.

The same method can be applied to the two-stage desiccant wheel system to analyze its minimum regeneration temperature. Assuming that each wheel is responsible for half of the dehumidification task, the processed air state after the first wheel is O_1 (47 °C, 13.6 g/kg, relative humidity of 20 %); it is cooled to W' and is then handled by the second desiccant wheel to O_2 (47 °C, 8 g/kg, relative humidity of 12 %). The minimum regeneration temperature of the first wheel is 54 °C (relative humidity of 20 %), and the minimum regeneration temperature of the second is 65 °C (relative humidity of 12 %).

Table 4.8 compares these two systems, and it can be seen that when the wheel is divided into two wheels, the regeneration temperature can be reduced considerably. Under the same operating condition, the minimum regeneration temperature of the single-stage desiccant wheel system is 81 °C, while those of the two-stage desiccant wheel system are 65 and 54 °C. This can be explained by the fact that, in the two-stage desiccant wheel system, the desiccant maintains high relative humidity, which means a lower temperature can meet the regeneration requirement under the same humidity ratio.

4.3.2.3 Fixed Desiccant Bed

Daikin Industries, Ltd. developed an outdoor air handling processor called DESICA (Xi 2010), which combines solid desiccant materials with a vapor compression

Table 4.8 Minimum regeneration temperatures of a single-stage desiccant wheel system and a two-stage desiccant wheel system

	Air handling process	Typical parameters	Relative humidity of air leaving wheels	Minimum regeneration temperature, °C
Single stage	$W \rightarrow O$	W: 33.2 °C, 19.1 g/kg O: 61.0 °C, 8.0 g/kg M: 81.1 °C, 19.1 g/kg	6.1 %	81.1
Two stage	First stage $W \rightarrow O_1$ Second stage $W' \rightarrow O_2$	W: 33.2 °C, 19.1 g/kg W': 33.2 °C, 13.6 g/kg O_1: 47.0 °C, 13.6 g/kg O_2: 47.1 °C, 8.0 g/kg M_1: 53.8 °C, 19.1 g/kg M_2: 65.1 °C, 19.1 g/kg	First stage 20.3 % Second stage 12.0 %	First stage 53.8 Second stage 65.1

refrigeration system, as shown in Fig. 4.35. In contrast to the desiccant wheel, in which the air handling process approaches the isenthalpic line, the desiccant material is coated on the surface of both the evaporator and condenser in DESICA. As a result, the cooling and heating capacity can be transferred effectively to the desiccant material, and the air handling process is directly cooled (regeneration air is heated) during the dehumidification (regeneration) process.

- *Dehumidification process*: Outdoor air makes contact with the desiccant coated on the evaporator. Because the equilibrium humidity ratio on the surface of the desiccant is lower than that of the processed air, water vapor is transferred from the air to the desiccant. Adsorption heat of the dehumidification process, as well as sensible heat of the processed air, is taken away by the evaporator. The processed air is dried and cooled at the same time.
- *Regeneration process*: Indoor exhaust air makes contact with the desiccant coated on the condenser. As the equilibrium humidity ratio on the surface of the desiccant is higher than that of the regeneration air, water vapor is transferred from the desiccant to the air. The condenser provides desorption heat and heats the air. Desiccant regeneration is realized with a relatively low temperature compared to the desiccant wheel.

Air flow direction and refrigerant flow direction have to be changed at the same time to realize the shift between dehumidification and regeneration modes, as shown in Fig. 4.36. In the dehumidification process, when the desiccant fully adsorbs the water and has to be regenerated, the four-way valve of the refrigeration cycle changes direction to shift the functions of the two heat exchangers (i.e., the

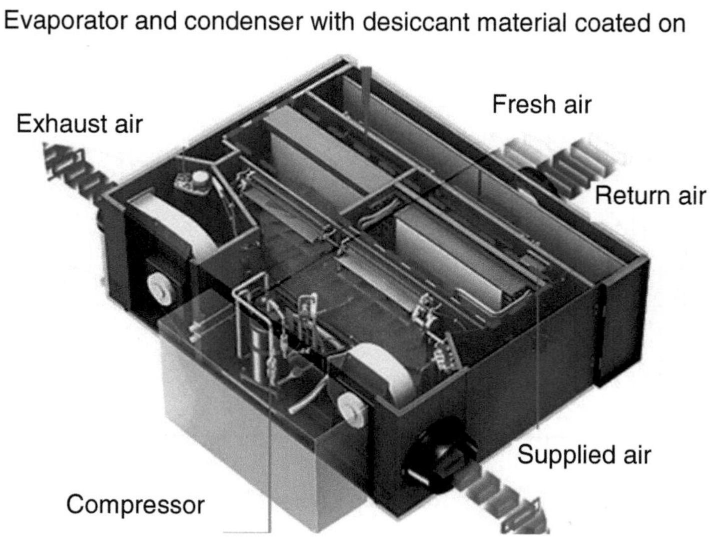

Fig. 4.35 Schematic diagram of DESICA

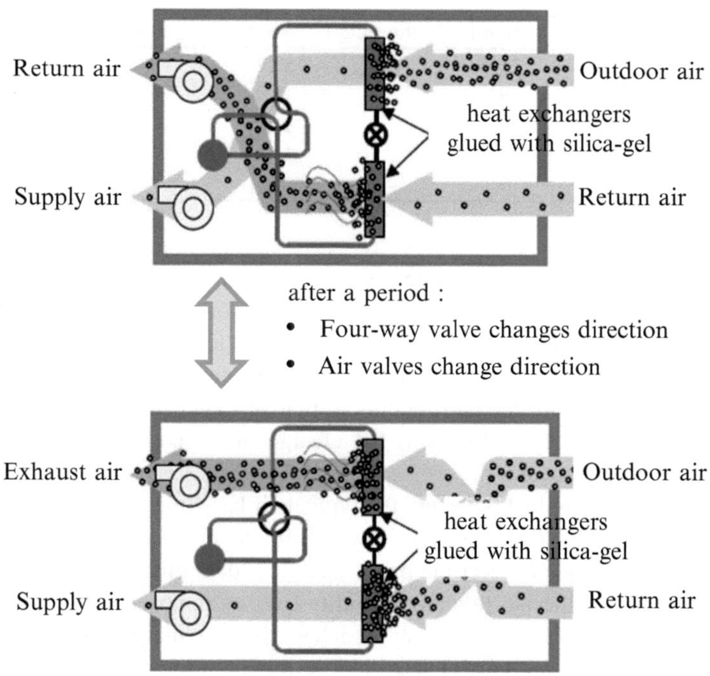

Fig. 4.36 Working principle of DESICA

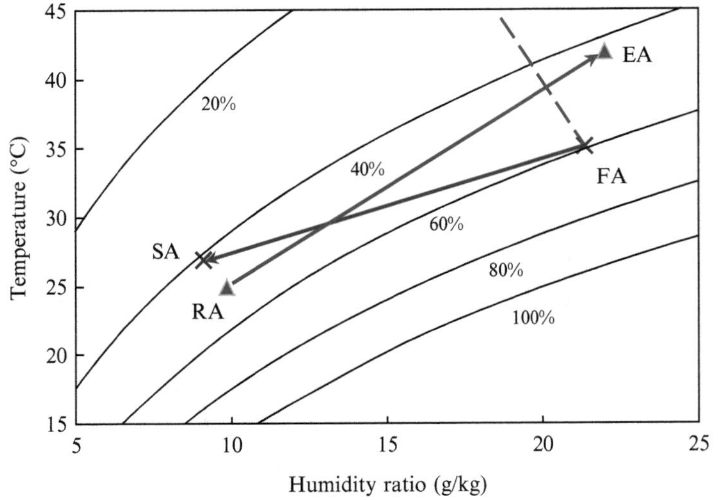

Fig. 4.37 Typical air handling process of DESICA

Table 4.9 Air states of DESICA

Point	Temperature (°C)	Humidity ratio (g/kg)	Relative humidity (%)	Enthalpy (kJ/kg)
FA	35.0	21.4	59.9	90.1
RA	25.0	9.9	49.9	50.2
SA	26.8	9.1	41.5	50.3
EA	41.9	22.0	42.4	98.9
SA_0	65.8	9.1	5.7	90.1

evaporator and condenser). At the same time, fan valves have to change the direction to send the indoor exhaust air to the new condenser, where the desiccant material is to be regenerated. If the regeneration process is complete, a four-way valve and fan valves change the direction back to continue the dehumidification process. Every 3–5 min, the shift occurs to realize the change between dehumidification and regeneration. However, the change of refrigerant direction leads to heating-cooling offset loss.

Figure 4.37 and Table 4.9 present the typical air handling process of DESICA. It can be seen that FA (relative humidity is around 60 %) reaches SA (relative humidity is around 40 %) after the processed air is dehumidified. The air handling process approaches the iso-relative humidity line, as opposed to the rotary wheel dehumidification process, which approaches the isenthalpic line. The indoor exhaust air changes from RA to EA to achieve regeneration, which also approaches the iso-relative humidity line. The exhaust air temperature is around 40 °C, which is much lower than the regeneration temperature in rotary wheel dehumidification. If the rotary wheel is used to handle FA, the state of the processed air changes along the dashed line in Fig. 4.37. The air parameters required to reach the same supplied humidity ratio, represented as SA_0, are listed in Table 4.9 (not shown in Fig. 4.37). Moreover, the processed outlet air temperature is higher than 65 °C, which is much higher than the 27 °C of DESICA.

Thus, this kind of solid desiccant outdoor air handling processor effectively combines desiccant material with an evaporator and condenser to realize internally cooled dehumidification and internally heated regeneration. It can be seen that the operating performance of DESICA is greatly improved compared to the desiccant wheel system.

References

China Institute of Building Standard Design & Research (2006) Selection and use of heat recovery devices for air-conditioning systems. China Planning Press, Beijing (in Chinese)

Ge TS, Dai YJ, Wang RZ, Li Y (2008) Experimental investigation on a one-rotor two-stage rotary desiccant cooling system. Energy 33:1807–1815

Huang X (2010) Theory and application of evaporative cooling method for air conditioning. China Architecture & Building Press, Beijing (in Chinese)

Jia CX (2006) Study on reinforcement dehumidification mechanism and application of composite desiccant based on silica gel. Doctoral dissertation, Shanghai Jiaotong University, Shanghai (in Chinese)

La D, Dai YJ, Li Y, Wang RZ, Ge TS (2010) Technical development of rotary desiccant dehumidification and air conditioning: a review. Renew Sustain Energy Rev 14:130–147

La D, Dai YJ, Li Y, Ge TS, Wang RZ (2011) Case study and theoretical analysis of a solar driven two-stage rotary desiccant cooling system assisted by vapor compression air-conditioning. Solar Energy 85:2997–3009

Lu YQ (2008) Practical design manual for heating and air-conditioning, 2nd edn. China Architecture & Building Press, Beijing (in Chinese)

MOHURD, China Institute of Building Standard Design & Research (2009) National technical measures for design of civil construction heating, ventilation and air conditioning. China Planning Press, Beijing (in Chinese)

Munters (2004) Samples of desiccant wheels. http://www.munters.us/en/us/

Pesaran AA, Mills AF (1987) Moisture transport in silica gel packed beds 1: theoretical study. Int J Heat Mass Transf 30:1037–1049

Suzuki K, Oya N (1983) Dehumidification design. China Architecture & Building Press, Beijing (in Chinese)

The 10th Design and Research Institute of Ministry of Electronics Industry (1995) Design handbook for air-conditioning, 2nd edn. China Architecture & Building Press, Beijing (in Chinese)

Xi GN (2010) Progress in technology research and development of Daikin. In: Proceeding of the annual conference of Chinese society of engineering thermophysics, Shanghai

Xie XY, Jiang Y (2010) Some views on design and thermal performance calculation methods of evaporative cooling air conditioning systems. Chin HV&AC 40(11):1–12 (in Chinese)

Zhang LZ (2005) Dehumidification technology. Chemical Industry Press, Beijing (in Chinese)

Zhang LZ, Niu JL (2002) Performance comparisons of desiccant wheels for air dehumidification and enthalpy recovery. Appl Therm Eng 22:1347–1367

Zhang YP, Zhang LZ, Liu XH, Mo JH (2006) Mass transfer in built environment. China Architecture & Building Press, Beijing (in Chinese)

Chapter 5
Key Components of the THIC System: Outdoor Air Processor Using Liquid Desiccant

Abstract Outdoor air handling method using liquid desiccant is analyzed in this chapter, which could also be adopted as the humidity control subsystem. Basic properties of liquid desiccant and the design principles for liquid desiccant outdoor air handling processors from the perspective of thermodynamics are presented. Then the performances of liquid desiccant outdoor air handling processors are investigated, including the processors with heat recovery module and with precooling module. At the end of this chapter, a comparison between the liquid desiccant dehumidification method and other dehumidification methods is discussed.

5.1 Basic Properties of Liquid Desiccant

5.1.1 Characteristics of Common Liquid Desiccants

The air dehumidification process using liquid desiccant depends on the low vapor pressure of the solution, and this subsection analyzes the liquid desiccant air-conditioning system based on the physical properties of liquid desiccants. Triethylene glycol was the first kind of liquid desiccant (Lof 1955) used for air dehumidification. Due to its high viscosity, triethylene glycol is prone to stagnation, and it adheres to device surfaces when circulating in desiccant air-conditioning systems, which is a significant disadvantage in terms of system stability. Furthermore, diethylene glycol, triethylene glycol, and other organic substances are volatile, so it is possible for them to be carried over by the air stream to the occupied space, which would be harmful to occupants. Because of these aforementioned drawbacks, the application of triethylene glycol has been restricted in air-conditioning systems in recent years, and it has gradually been replaced by metal halide solutions. The most common liquid desiccants used nowadays are lithium bromide (LiBr), lithium chloride (LiCl), and calcium chloride ($CaCl_2$) aqueous solutions.

X. Liu et al., *Temperature and Humidity Independent Control (THIC) of Air-conditioning System*, DOI 10.1007/978-3-642-42222-5_5,

LiBr is a stable substance that does not deteriorate, become volatile, or decompose in the atmosphere (Design Handbook for Air-conditioning 1995). It is a colorless crystal at room temperature, and it is nontoxic, odorless, and both salty and bitter in taste. Its molecular weight is 86.86, and its melting point and boiling point are 549 °C and 1,265 °C, respectively. LiBr is soluble in water, much more so than common table salt. The solubility of table salt at 20 °C is 35.9 g, while that of LiBr is about twice as high. The vapor pressure of LiBr solution is much lower than the saturated vapor pressure of water at the same temperature, indicating that LiBr solution has a strong ability to absorb moisture from air. LiBr aqueous solution is a strong corrosive medium for metallic materials, although the level of corrosion is lower than that of sodium chloride or calcium chloride solutions. Since LiBr aqueous solution with a mass concentration in the range of 60–70 % will crystallize at room temperature, the concentration of LiBr solution is generally not more than 70 % for common use.

LiCl is a white, cubic crystal salt, with a molecular weight of 42.4 (The 10th Design and Research Institute of Ministry of Electronics Industry 1995). Its melting point is 605 °C, and its boiling point is 1,350 °C. LiCl is also highly soluble in water. LiCl aqueous solution is colorless, transparent, nontoxic, and odorless; it has low viscosity, good heat transfer performance, and outstanding chemical stability. Under normal conditions, the LiCl solute does not decompose or become volatile. Since the vapor pressure of LiCl aqueous solution is low, it is an excellent desiccant for air dehumidification. The crystallization temperature of LiCl solution increases with its concentration. As a result, the mass concentration of the LiCl solution in dehumidification applications should not exceed 40 %, since it will crystallize at room temperature if its concentration is greater than 40 %. LiCl solution is corrosive to metal, while titanium and titanium alloys, molybdenum-containing stainless steel, nickel-copper alloys, synthetic polymers, and resins show resistance to corrosion from LiCl solution.

$CaCl_2$ is an inorganic salt with strong moisture absorption ability (Design Handbook for Air-conditioning 1995). Anhydrous calcium chloride is white, porous, diamond-shaped crystal block, with a slightly bitter and salty taste. Its melting point is 772 °C, and its boiling point is 1,600 °C. Anhydrous calcium chloride becomes a hydrated compound when it absorbs moisture from air. Calcium chloride aqueous solution still has an ability to absorb moisture, although this capability is significantly reduced compared to solid calcium chloride. $CaCl_2$ is also inexpensive and plentiful. Since calcium chloride solution is corrosive to metals, containers used to store it should be anticorrosive.

For these commonly used liquid desiccants, the boiling points of the solutes and water are rather different. Under atmospheric pressure, the boiling points of LiBr, LiCl, and $CaCl_2$ are all above 1,200 °C, while that of water is only 100 °C. As a result, the vapor pressure of the salt solutions is approximately equal to the partial pressure of the water vapor. As the liquid desiccant makes direct contact with the air and reaches equilibrium, the temperature and the water vapor pressure of the solution are equal to those of the air. The equivalent humidity ratio of the solution is defined as the humidity ratio of the air in equilibrium with the solution state. The states of the solution can be represented in a psychrometric chart according to the states of the air in equilibrium with the aqueous solution. Figure 5.1a–c shows the corresponding states of the LiBr,

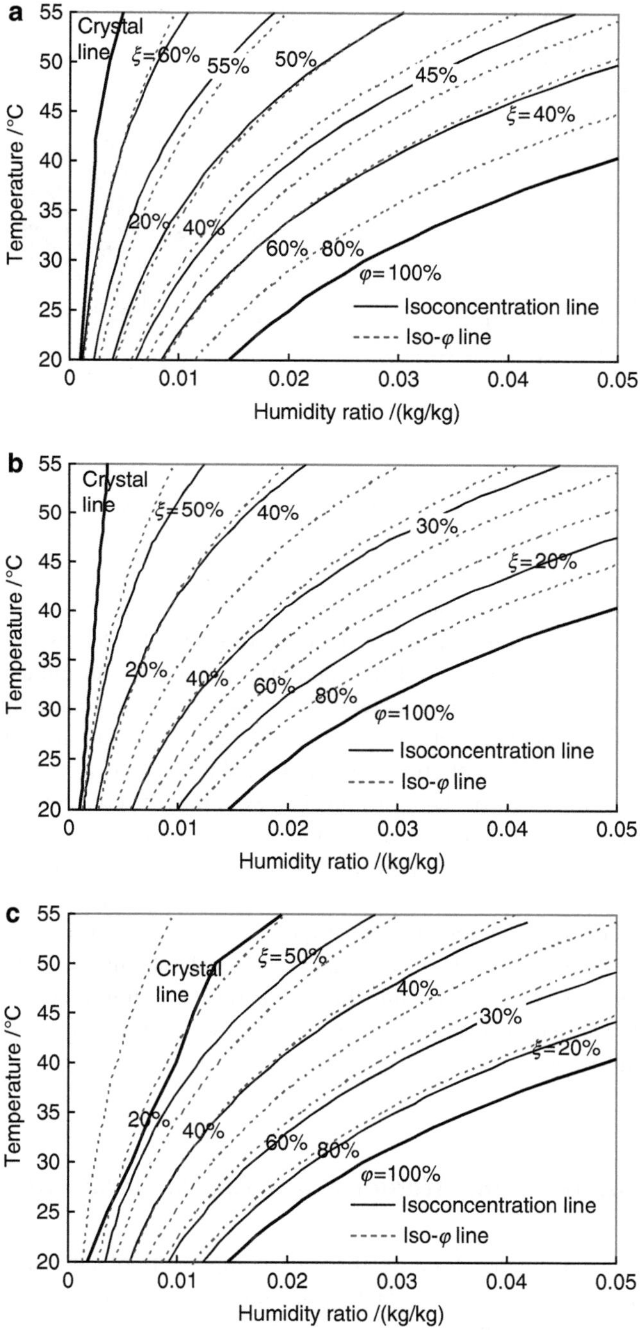

Fig. 5.1 States of commonly used liquid desiccants shown in psychrometric chart: (**a**) lithium bromide aqueous solution, (**b**) lithium chloride aqueous solution, and (**c**) calcium chloride solution

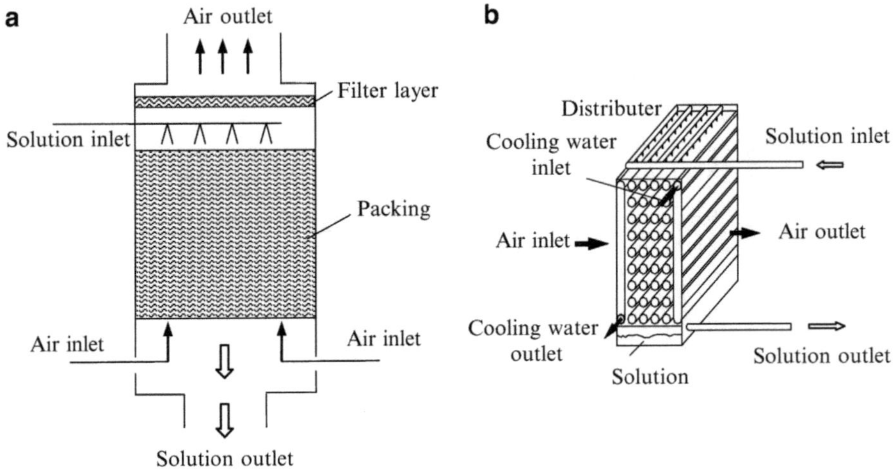

Fig. 5.2 Basic air-liquid desiccant contact devices: (**a**) adiabatic type and (**b**) internally cooled type

LiCl, and $CaCl_2$ aqueous solutions, respectively, where ξ represents the mass concentration of the solution (equal to the mass of solute divided by the mass of the solution) and φ represents the relative humidity of the air. As indicated by these figures, the lower the solution temperature, the lower its equivalent humidity ratio at the same desiccant concentration, and the stronger the dehumidification ability. Furthermore, the iso-concentration line of the solution almost coincides with the iso-relative humidity line of the moist air. For example, the concentrations with the iso-concentration lines of the LiBr solution, LiCl solution, and $CaCl_2$ solution corresponding to the 40 % iso-relative humidity line of the moist air are 46 %, 31 %, and 40 %, respectively. Due to the limitation of crystallization, the solution states cannot cover certain regions on the left side of the crystal line in the psychrometric chart. Among these three desiccants, the reachable air handling area of the $CaCl_2$ solution is the narrowest, while the LiBr and LiCl solutions can process the air to a similar lower humidity ratio. As a result, the dehumidification ability of the LiBr and LiCl solutions is better than that of the $CaCl_2$ solution.

5.1.2 Basic Handling Module Using Liquid Desiccant

Dehumidifiers and regenerators are the core components of air handling devices using liquid desiccant, the heat and mass transfer processes of which directly influence the performance of the entire system. Dehumidifiers/regenerators using liquid desiccant can be divided into two types: the adiabatic type and the internally cooled/heated type. As shown in Fig. 5.2, this distinction is based on whether there is extra cooling/heating medium introduced into the heat and mass transfer processes.

As the heat and mass transfer performance of dehumidifiers and regenerators is similar, the dehumidifier is taken as an example here and is described in detail.

The adiabatic dehumidifier usually has a spray tower structure or a packed tower structure. For the spray tower structure, the air pressure drop is small, but the desiccant carryover of the outlet air is more serious, which negatively affects indoor air quality and human health. The packed tower structure illustrated in Fig. 5.2a has a significantly higher dehumidification efficiency, and it has become more popular in recent years as a result. For the packed tower dehumidifier, the packing, which plays an important role in increasing the contact area between the air and the solution, can be categorized as either random packing or structured packing. Among the liquid desiccant experimental systems reported in the literature, random packing methods such as saddle packing, Berl saddle packing, and Pall ring packing were widely used in early studies. Structured packing has regular geometry, so the flow paths of the air and the solution are set. Compared to random packing, structured packing can provide a larger air-solution contact area and reduce flow resistance effectively at the same time.

In the dehumidifier, the latent heat of vaporization released during the absorption process increases the desiccant temperature and therefore increases the desiccant's vapor pressure. As a result, the driving force for the mass transfer between the air and the desiccant is greatly reduced, which adversely affects dehumidification. In light of this issue, a high solution flow rate is usually adopted in the adiabatic dehumidifier to suppress the solution's temperature rise, thereby maintaining a low water vapor pressure above the solution. In order to efficiently suppress the rise in temperature of the solution in the dehumidifier and improve dehumidification efficiency, a cooling medium can be introduced into the dehumidifier to cool the solution immediately, which forms the internally cooled dehumidifier, as illustrated in Fig. 5.2b. The internally cooled dehumidifier usually employs a falling film structure, and the cooling source could be cold air, water, refrigerant, etc.

Since the internally cooled device should strictly isolate the solution channel and the cooling medium channel, the manufacturing process is quite demanding. As a result, the packed adiabatic device is still the most popular type in the market. To combine the advantages of adiabatic and internally cooled devices, Jiang et al. (2003) proposed a new kind of spray module (shown in Fig. 5.3) that consists of an external cooling/heating medium and solutions flowing between modules and inside the module. In this spray module, the solution is first pumped from the bottom solution tank to the sensible heat exchanger to be cooled/heated by the external cooling/heating medium to improve the dehumidification/humidification ability; it is then sent into the solution distributor. The solution coming out of the distributor is sprayed uniformly on the surface of the packing to make direct contact with the air for heat and mass transfer, and afterward it flows back to the solution tank by gravity. The flow rate of the solution inside the module should meet the requirements of the heat and mass transfer processes, while the flow rate of the solution flowing between the modules should satisfy the match properties of heat and mass transfer (PhD Thesis by Li 2004). As a result, the solution flow rate

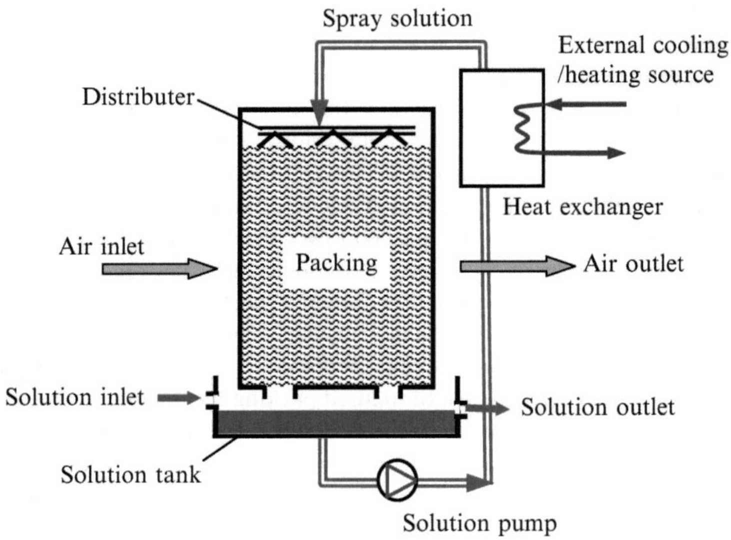

Fig. 5.3 Schematic of the basic air-liquid desiccant contact device with external cooling/heating

between the modules is only one-tenth of that inside the module. Based on the fundamental module given in Fig. 5.3, various dehumidification/regeneration devices can be constructed by combining a series of these modules.

5.2 Design Principles for Liquid Desiccant Outdoor Air Handling Processors

5.2.1 Match Properties of the Air Handling Process Using Liquid Desiccant

5.2.1.1 Sensible Heat Transfer Processes

Entransy is a thermological parameter for analyzing the heat transfer process (Guo et al. 2007; Cheng and Liang 2012), and the unmatched coefficient ξ based on entransy dissipation (ΔJ_{loss}) analysis is adopted to investigate the sensible heat transfer process. ΔJ_{loss} of a sensible heat transfer process, shown in Fig. 5.4a, is expressed as the shaded area in the *T*-*Q* (temperature-heat flux) chart in Fig. 5.4b, c. For the heat transfer process shown in Fig. 5.4b, the temperatures of both fluids change; the temperature of a single fluid changes (e.g., the evaporator) in the heat transfer process shown in Fig. 5.4c. In the figures, subscript h indicates the high-temperature fluid, c indicates the low-temperature fluid, in indicates the inlet fluid, and out indicates the outlet fluid.

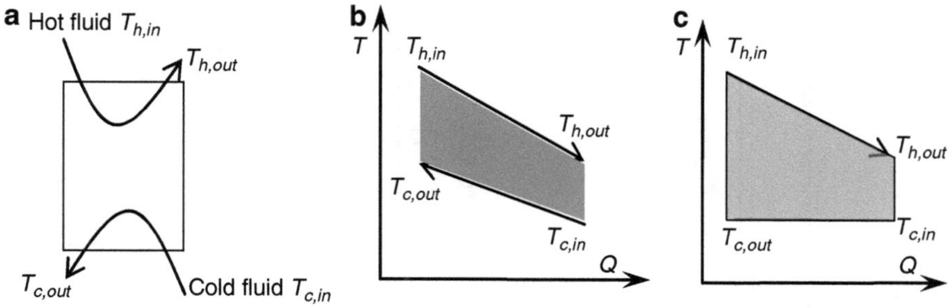

Fig. 5.4 Heat transfer process in a counterflow heat exchanger shown in *T-Q* chart: (**a**) hot and cold fluids, (**b**) temperatures of both fluids change, and (**c**) temperature of a single fluid changes

The thermal resistance (R_H) based on entransy dissipation is then calculated by Eq. (5.1), where ξ is the flow unmatched coefficient and is expressed as Eq. (5.2). The calculation of P in Eq. (5.2) is given as Eq. (5.3) for the different heat transfer processes shown in Fig. 5.4b, c:

$$R_{\mathrm{H}} = \frac{\Delta J_{\mathrm{loss}}}{Q^2} = \frac{\xi}{UA} \tag{5.1}$$

$$\xi = \frac{P}{2} \cdot \frac{e^P + 1}{e^P - 1} \tag{5.2}$$

$$\begin{aligned} P &= UA \cdot \left(\frac{1}{c_{p,\mathrm{h}} \dot{m}_{\mathrm{h}}} - \frac{1}{c_{p,\mathrm{c}} \dot{m}_{\mathrm{c}}} \right) \quad \text{for Fig. 5.4b,} \\ P &= UA \cdot \frac{1}{c_{p,\mathrm{h}} \dot{m}_{\mathrm{h}}} \quad \text{for Fig. 5.4c} \end{aligned} \tag{5.3}$$

where U is the heat transfer coefficient, kW/(m^2 · °C); A is the heat transfer area, m^2; c_p is the specific heat capacity, kJ/(kg · °C); and $\dot{m}$ is the mass flow rate, kg/s.

Figure 5.5 shows the heat transfer efficiency, $1/UA$, ξ, and thermal resistance varying with the heat capacity ratio and the number of transfer units NTU $= UA/(c_p\dot{m})_{\text{smaller side}}$. The heat capacity of the high-temperature fluid is 1 kW/°C, and the heat capacity ($c_p\dot{m}$) is treated as infinite for the process shown in Fig. 5.5c. The heat transfer efficiency for a heat exchanger is defined as the heat exchange in practice divided by the maximum heat exchange in theory. It can be seen that ξ is always greater than or equal to 1, but only when the calorific capacities of the two fluids are equal will ξ be equal to 1.

As indicated in Eq. (5.1) and Fig. 5.5b, $1/UA$ denotes the resistance caused by the limited heat transfer capacity; the thermal resistance will increase by a multiple of ξ from the base of $1/UA$ when the calorific capacities of the fluids are different. If the heat capacity ratio of the two fluids is equal to 1, the heat transfer process satisfies the flow matching condition. Then, unmatched coefficient ξ is equal to the thermal resistance (or entransy dissipation) of the process divided by that of a process with

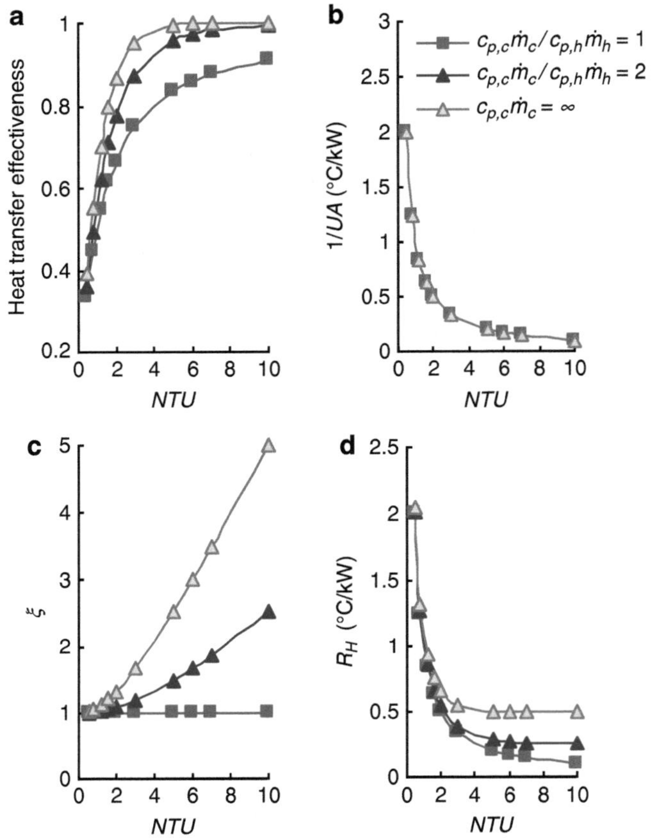

Fig. 5.5 Heat transfer performance of a counterflow heat exchanger: (**a**) heat transfer effectiveness, (**b**) 1/*UA*, (**c**) unmatched heat transfer coefficient, and (**d**) thermal resistance

the same heat exchange that satisfies the flow matching condition. Thus, limited heat transfer capability *UA* and flow mismatching are the main factors that result in elevated thermal resistance of the sensible heat transfer process. The thermal resistance can be reduced by improving the heat transfer coefficient or increasing the heat transfer area, while flow mismatching will lead to an increase of thermal resistance. Indeed, match property analysis helps to clarify the key issues for improving the heat transfer performance.

5.2.1.2 Dehumidification and Regeneration Processes Using Liquid Desiccant

The match properties and unmatched coefficient based on entransy dissipation analysis can also be adopted for the air handling process using liquid desiccant. The state of the LiBr aqueous solution is illustrated in a psychrometric chart in

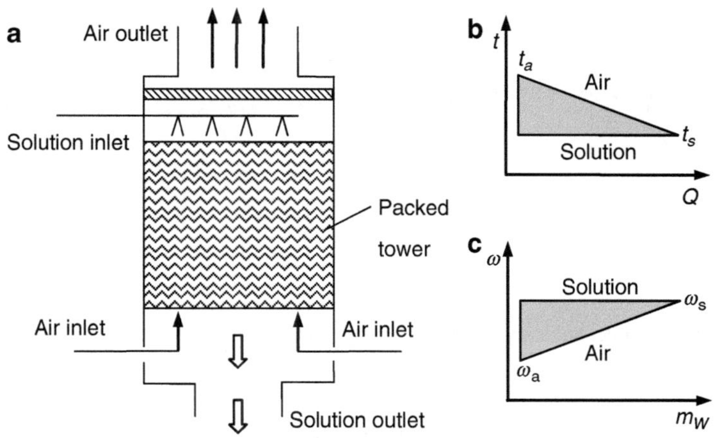

Fig. 5.6 Entransy dissipation in an air-liquid desiccant direct contact device: (**a**) operating schematic, (**b**) heat transfer process, and (**c**) moisture transfer process

Fig. 5.1, which indicates that the iso-concentration line of the liquid desiccant coincides with the iso-relative humidity line of the air. The solution state can also be expressed in the chart, and it can reach a larger region compared to water (which is located at the 100 % saturated state of air). The basic air-solution direct contact device is shown in Fig. 5.6a; as indicated by previous research (PhD Thesis by Liu 2007), the heat and mass transfer equations and analytical solutions of the air-desiccant process are similar to those of the air-water process. As a result, the match properties in the air-desiccant process can be investigated by way of analogy to the air-water process.

Based on the match properties in the coupled heat and mass transfer process between the air and water examined in previous research (Zhang et al. 2012), dissipations occur during the heat and mass transfer processes of the air-water packed tower due to limited transfer capability, flow mismatching, and parameter mismatching. Furthermore, dissipations occur during the air-water process for similar reasons (Liu et al. 2009; Zhang et al. 2013). For example, if the enthalpies of the inlet air and solution states are the same, the heat transfer dissipation can be illustrated by the shaded areas in the *T*-*Q* chart in Fig. 5.6b, and the mass transfer dissipation can be illustrated in the shaded areas of the ω-m_w (humidity ratio-moisture removal rate) chart in Fig. 5.6c. However, the variance of solution concentration is not taken into account.

For a certain air-water process, the inlet states of fluids are shown in Fig. 5.7a; similarly, Fig. 5.7b shows the inlet states of a certain air-desiccant process. The mass transfer resistance (R_M) based on entransy dissipation is then calculated by Eq. (5.4). ξ_M is the unmatched coefficient, expressed as Eq. (5.5), where $\Delta\omega_1$ and $\Delta\omega_2$ can be determined by Eq. (5.6) for the air-desiccant process shown in Fig. 5.7b. Similar to the sensible heat transfer process, $1/U_M A_M$ denotes the resistance caused by the limited mass transfer capacity, and the mass transfer resistance will increase by a multiple of

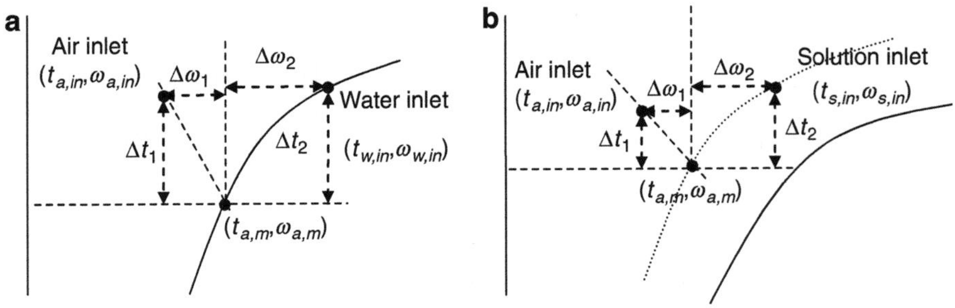

Fig. 5.7 States of inlet fluids shown in psychrometric chart: (**a**) air-water process and (**b**) air-liquid desiccant process

ξ_M from the base of $1/U_M A_M$ if the coupled heat and mass transfer process is unmatched:

$$R_M = \frac{\Delta M_{loss}}{m_w^2} = \frac{\xi_M}{U_M A_M} \tag{5.4}$$

$$\xi_M = \frac{a_{11}\cdot(\Delta\omega_1)^2 + a_{12}\cdot\Delta\omega_1\Delta\omega_2 + a_{22}\cdot(\Delta\omega_2)^2}{b_{11}\cdot(\Delta\omega_1)^2 + b_{12}\cdot\Delta\omega_1\Delta\omega_2 + b_{22}\cdot(\Delta\omega_2)^2} \tag{5.5}$$

$$\Delta\omega_1 = \omega_{a,m} - \omega_{a,in},\ \ \Delta\omega_2 = \omega_{w,in} - \omega_{a,m} \tag{5.6}$$

The expressions of $a_{11} \sim a_{22}$ and $b_{11} \sim b_{22}$ in Eq. (5.5) are calculated by

$$a_{11} = \frac{NTU_M}{2}\left(1 - e^{-2NTU_M}\right),\quad a_{12} = \frac{2p\cdot NTU_M}{2-m^*}\left(1 - e^{NTU_M(m^*-2)}\right),$$

$$a_{22} = \frac{1}{2}\frac{p^2\cdot NTU_M}{1-m^*}\left(1 - e^{2NTU_M(m^*-1)}\right),\quad b_{11} = \left(1 - e^{-NTU_M}\right)^2,$$

$$b_{12} = 2\left(1 - e^{-NTU_M}\right)\left(1 - p\cdot e^{NTU_M(m^*-1)}\right),\quad b_{22} = \left(1 - p\cdot e^{NTU_M(m^*-1)}\right)^2$$

$$p = \frac{m^*-1}{m^* e^{NTU_M(m^*-1)} - 1},\quad m^* = \frac{\dot{m}_a c_{p,e}}{\dot{m}_s c_{p,s}},\quad NTU_M = \frac{U_M A_M}{\dot{m}_a} \tag{5.7}$$

In contrast to the sensible heat transfer process, ξ_M of the coupled heat and mass transfer process is caused by both the unmatched flow rates and unmatched parameters. The condition for flow matching is that the heat capacity of the air is equal to that of the solution ($m^* = 1$). Taking a typical solution state as an example, match properties varying with the typical inlet air states are shown in Fig. 5.8, where $a_1 \sim a_{16}$ are the air inlet states around inlet solution state S (the temperature of the LiBr solution is 25 °C, and the concentration is 45 %). The mass transfer rates of $a_1 \sim a_{16}$ varying with NTU_M in this air-desiccant process are shown in Fig. 5.9a. An analytical method is used to calculate the values; the air flow rate is 1 kg/s, and the heat capacity ratio of two fluids is 1. The unmatched coefficient (ξ_M) of different air inlet states is illustrated in Fig. 5.9b, except for a_3 and a_{11}.

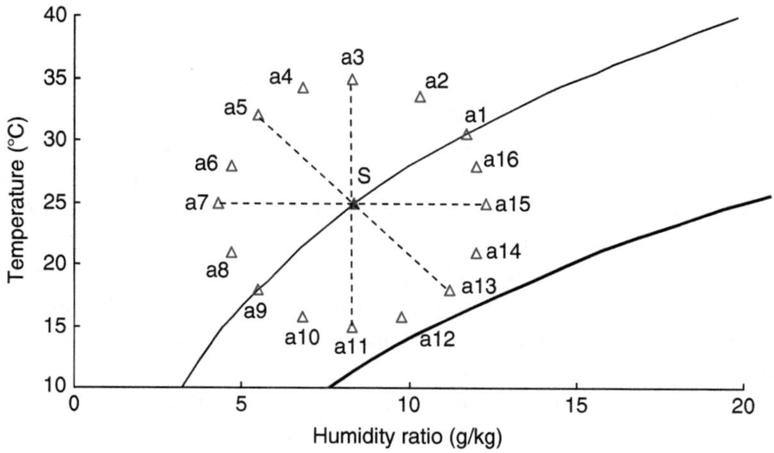

Fig. 5.8 Air and solution inlet states shown in psychrometric chart (*S* solution inlet state)

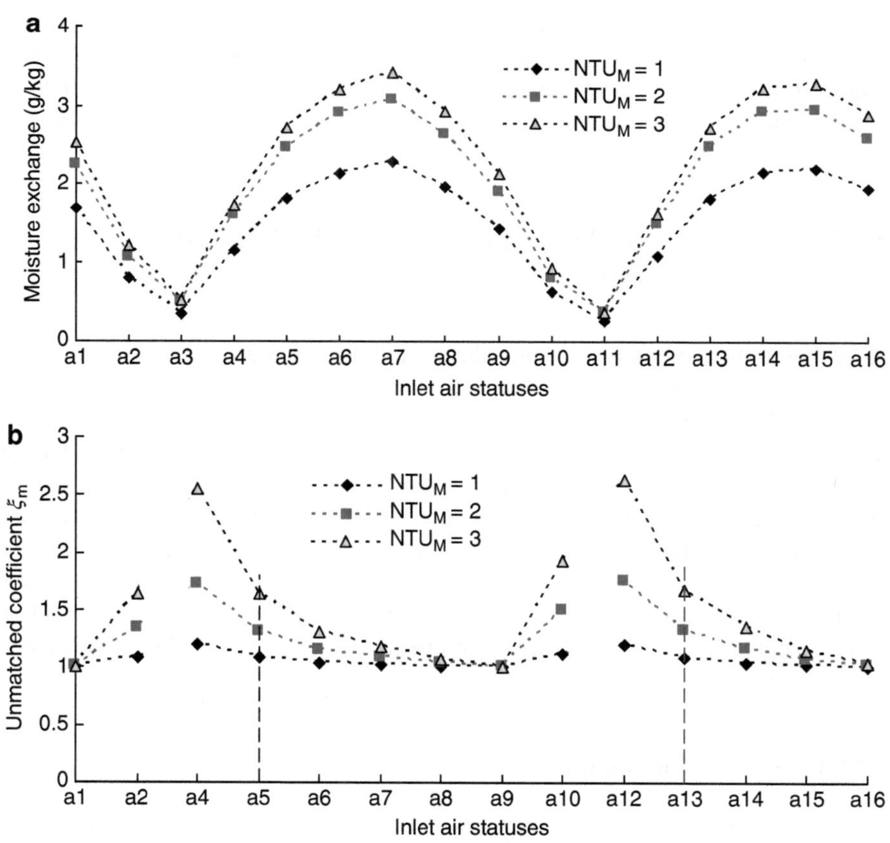

Fig. 5.9 Unmatched heat and mass transfer coefficients in an air-liquid desiccant direct contact device: (**a**) humidity ratio variance of inlet and outlet air through the device and (**b**) mass transfer unmatched coefficient

A comparison can be made between the heat and mass transfer processes that proceed along the isenthalpic line and those proceeding along the iso-concentration line. For the processes along the iso-concentration line (a_1-S and a_9-S), ξ_M is always 1; for the processes along the isenthalpic line (a_5-S and a_{13}-S), ξ_M is 1.08, 1.31, and 1.66 if NTU_M is 1, 2, and 3, respectively. This indicates that, for the processes along the isenthalpic line, 8, 31, and 66 % more mass transfer capacity is required due to the unmatched parameters when NTU_M is 1, 2, and 3, respectively. Therefore, the condition for parameter matching in the air-desiccant process is that the air handling process proceeds along the iso-concentration line of the liquid desiccant (or iso-relative humidity line of the humid air).

Thus, in the coupled heat and mass transfer process between the air and liquid desiccant, the greater the unmatched coefficient, the greater the required transfer ability for certain heat or moisture transfer devices. ξ_M evaluates the rate of increase for the entransy dissipation or transfer resistance of the process compared to that of the process satisfying the flow matching and parameter matching conditions. Similarly, the effects of the unmatched parameters on the process are magnified as input heat and mass transfer capacity increase. Match properties help to examine the performance of the heat and mass transfer process between the air and desiccant, and the following subsection demonstrates how the unmatched coefficient method is used to design an optimized outdoor air handling process using liquid desiccant.

5.2.2 *Performance Optimization of the Air Handling Processor Using Liquid Desiccant*

Air dehumidification methods that utilize liquid desiccant are becoming more and more popular (Khan 1998; Zhao 2002; Zhang et al. 2004; Li et al. 2005; Lazzarin and Castellotti 2007), and heat pump-driven liquid desiccant (HPLD) systems are developing very quickly in China (Chen 2005; Liu 2010). The cooling capacity from the evaporator and heat from the condenser are both utilized in HPLD systems, where the cooling capacity is used to cool the desiccant to enhance its dehumidification ability and the heat is used to regenerate the desiccant. For HPLD systems, there are different ways to utilize the condensation heat for desiccant regeneration. The issues of how to evaluate the performances of different HPLD systems and how to pursue a process with better energy efficiency are at the center of current research efforts.

5.2.2.1 System Operating Principles

Figure 5.10a, b shows the operating schematics of two basic HPLD processes. In these processes, an adiabatic dehumidifier/regenerator is adopted. A small part of the diluted solution after dehumidification is sent into the regenerator, and another small amount of concentrated solution after regeneration returns to the

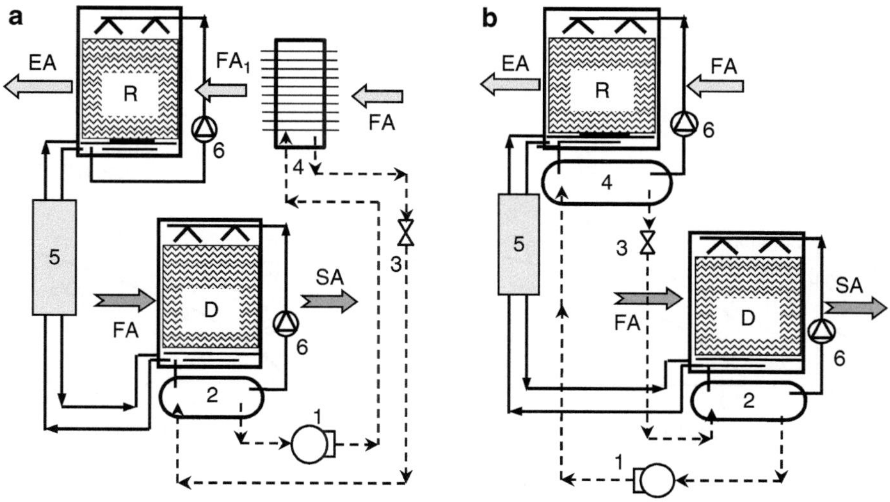

Fig. 5.10 Operating schematic of heat pump-driven liquid desiccant air handling processes with different regeneration modes: (**a**) Basic Type I, heating inlet air, and (**b**) Basic Type II, heating inlet solution (*FA* fresh air, *SA* supply air, FA_1 fresh air heated by the condenser, *EA* exhaust air, *D* dehumidifier, *R* regenerator, *1* compressor, *2* evaporator, *3* expansion valve, *4* condenser, *5* solution heat exchanger, *6* pump)

dehumidifier. As the variance of the desiccant concentration is usually lower than 0.5 %, the mass flow rates of these two parts of the solution circulating between the dehumidifier and the regenerator are almost identical for these HPLD systems (Liu et al. 2006, 2007). A solution heat recovery device is used for the heat recovery between the diluted solution sent to the regenerator and the concentrated solution sent to the dehumidifier. The concentrated solution sent to the dehumidifier is then mixed with the rest of the diluted solution after dehumidification and sent to the top of the dehumidifier with the help of a solution pump.

In contrast to the typical adiabatic process where all the solution circulates between the dehumidifier and the regenerator, in the adiabatic dehumidifier/regenerator examined here, most of the solution circulates only between the packed tower and the evaporator or condenser. The major difference between Basic Type I and Basic Type II systems is related to the different fluids that the condenser uses for heating. In Basic Type I, the condenser is an air-cooled type, and the regeneration air is heated by the condensation heat before flowing through the regenerator. A solution-cooled condenser is adopted in Basic Type II, and the solution is heated by the condensation heat instead of the regeneration air.

5.2.2.2 Performance Analysis and Design Principle

The detailed simulated results of the two basic systems under a typical condition are given in Appendix E of this book, including the system models, operating parameters, and unmatched coefficients of key components. The performance discrepancy

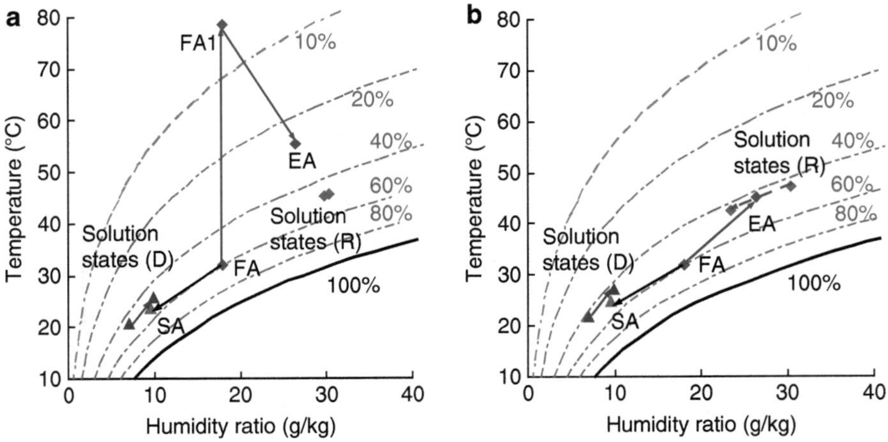

Fig. 5.11 Air and solution states of heat pump-driven liquid desiccant air handling processes shown in psychrometric chart: (**a**) Basic Type I and (**b**) Basic Type II

between these two systems and the primary explanations for this discrepancy are discussed here. The typical air handling processes shown are illustrated in Fig. 5.11 in a psychrometric chart. As indicated by the figure, the coupled heat and mass transfer process in the regenerator of Basic Type I proceeds close to the isenthalpic line, while the process of Basic Type II is near the iso-concentration line of the liquid desiccant. As indicated by the simulation results shown in Fig. 5.11 and E-1 in Appendix E, there are significant discrepancies in the operating performances of Basic Types I and II due to the difference between heating the regeneration air and heating the solution.

As indicated by the match properties of the components, the unmatched coefficients of the evaporator, the dehumidifier, and the solution heat exchanger are similar for Basic Types I and II. For the condenser and regenerator, there is a significant discrepancy of the unmatched coefficients between Basic Types I and II. In Basic Type I, ξ of the air-cooled condenser is more than twice that in Basic Type II under the typical condition. This means that, for the air-cooled condenser of Basic Type I, the thermal resistance caused by the unmatched flow rates increases significantly on the basis of $1/UA$ (UA is the input heat and mass transfer capacity). As indicated in Fig. 5.11a, the regeneration air in Basic Type I has to be heated to a relatively high temperature to satisfy the requirement for desiccant regeneration before flowing through the regenerator, which leads to a high ξ of the air-to-solution condenser. As a result, the unmatched coefficient of the regenerator is still significantly higher than that of Basic Type II. This implies that, in order to overcome the mass transfer resistance caused by unmatched flow rates and parameters, greater heat and mass transfer capacity or a higher heat source temperature than Basic Type II is required in Basic Type I. In summary, the match properties of Basic Type II are much better than those of Basic Type I.

The significant difference in unmatched coefficients of Basic Types I and II is similar to the performance discrepancy between the two HPLD systems. The values for humidity ratio of the supply air (9.5 g/kg), evaporating temperature (about 16 °C), and supply air temperature (about 23 °C) are similar for Basic Types I and II. However, the condensing temperature of Basic Type I (78.8 °C) is much higher than that of Basic Type II (51.1 °C), indicating that the condensing temperature increases due to the unmatched flow rates of the air-cooled condenser. The difference of the condensing temperature for these two systems results in the energy performance discrepancy: the coefficient of performance for the heat pump (COP_{hp}) of Basic Type I is only 1.89, lower than half of the COP_{hp} of Basic Type II (4.53).

Thus, the HPLD system, in which the solution is heated for regeneration, performs much better than the system in which the regeneration air is heated. For different inlet air and solution states, the unmatched coefficient of the process along the iso-concentration line is significantly lower than that along the isenthalpic line. The unmatched coefficient offers a principle for constructing an HPLD system: lowering the unmatched coefficients helps to improve system performance.

As indicated by the above analysis of different air handling systems using liquid desiccant, it is recommended that the heat source (rather than the air) be used to heat the solution for regeneration (Basic Type II) in order to achieve a more energy-efficient process. Also, the required temperature for desiccant regeneration is as low as about 50 °C for the solution-heated process (Basic Type II), which is much lower compared to the process in which air is heated for regeneration, providing the possibility to utilize waste heat or solar energy. If solar energy is adopted, the diluted solution could be heated by solar collectors. Then, the heated solution could be sprayed from the top of the regenerator and be regenerated. Moreover, the cooling water from the cooling tower could be used to cool the concentrated solution before it flows into the dehumidifier.

5.3 Performance of Liquid Desiccant Outdoor Air Handling Processors

For liquid desiccant outdoor air handling processors operating in summer, the outdoor air is dehumidified in the dehumidifier, and then the diluted solution is regenerated in the regenerator. According to whether there is indoor exhaust air that can be adopted as regeneration air, a distinction can be made between two different kinds of outdoor air handling processors that use liquid desiccant. If there is sufficient indoor exhaust air, the processor with enthalpy recovery from the indoor exhaust air and the process of utilizing the indoor exhaust air for desiccant regeneration could be adopted for outdoor air dehumidification. Alternatively, if there is not sufficient indoor exhaust air to be utilized directly, outdoor air could be adopted as the regeneration air, and a process using high-temperature chilled water to precool the outdoor air could be constructed to improve the performance of the outdoor air handling process in the THIC system. The following subsection examines the performance of these two kinds of liquid desiccant outdoor air handling processors, in which exhaust heat from the condenser is utilized to heat the desiccant coming into the regenerator, so the air handling processes are close to the iso-relative humidity line (rather than the isenthalpic line).

5.3.1 Outdoor Air Handling Processor with Enthalpy Recovery

Figure 5.12 demonstrates the operating principle of a two-stage heat pump-driven liquid desiccant outdoor air processor operating in summer, which is composed of a two-stage enthalpy recovery device (spray modules numbered I and II), a two-stage dehumidifier/regenerator, and a vapor compression refrigeration cycle. In this figure, the straight lines and dashed lines stand for liquid desiccant and refrigerant, respectively. The top channel is for the indoor exhaust air, and the bottom channel is for the outdoor air. The outdoor air first enters the two-stage enthalpy recovery device and then flows into the evaporator-cooled two-stage dehumidification modules (numbered III and IV) before being supplied into occupied spaces. In the same way, the indoor exhaust air first enters the enthalpy recovery device and then flows into the condenser-heated modules (numbered IV′ and III′) before being exhausted to the outdoor environment. The evaporator of the heat pump is adopted to further cool and dehumidify the outdoor air coming out of the enthalpy recovery device to the desired supplied temperature and humidity ratio.

The liquid desiccant in this outdoor air processor is divided into two parts. One part is stored in spray modules I and II (in Fig. 5.12) for the purpose of enthalpy recovery, the equivalent state of which is decided by the outdoor air and the indoor exhaust air simultaneously. The other part is stored in spray modules III, III′, IV, and IV′ to exchange heat with the evaporator and condenser, respectively. Taking the solution circulating between spray modules III and III′ as an example to illustrate the operating principle in summer, the solution is first cooled by the evaporator (labeled 5 in Fig. 5.12a) and then exchanges heat and mass with the outdoor air in spray module III, where the solution is diluted and heated. The solution heated by the condenser (labeled 3 in Fig. 5.12a) enters spray module III′ to be regenerated. The diluted solution and the regenerated solution are connected by solution pipes, and a plate heat exchanger (labeled 7 in Fig. 5.12a) is adopted for the internal heat recovery of the solution. The principle of spray modules IV and IV′ is similar to that of modules III and III′. It is obvious that the enthalpy recovery device can efficiently reduce the energy consumption of the outdoor air handling processor. The cooling capacity and condensation heat of the heat pump are both effectively utilized in this outdoor air processor. Two parallel compressors are utilized in the heat pump cycle to operate efficiently under the partial-load condition, so that the processor can have higher energy efficiency and control accuracy at partial load.

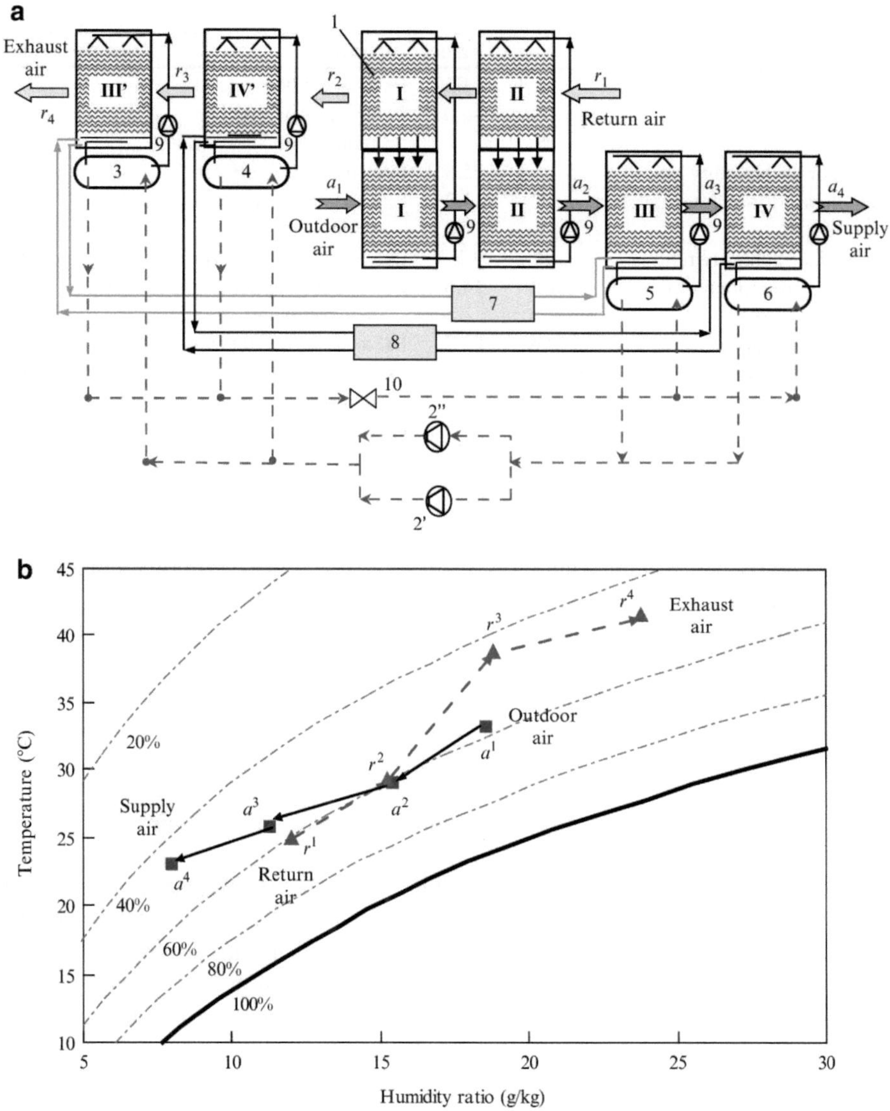

Fig. 5.12 Summer operation principle of the two-stage liquid desiccant outdoor air processor with enthalpy recovery: (**a**) operating schematic (*1* heat recovery module, *2* compressor, *3* condenser I, *4* condenser II, *5* evaporator I, *6* evaporator II, *7* plate heat exchanger I, *8* plate heat exchanger II, *9* solution pump, *10* expansion valve) and (**b**) air handling process shown in psychrometric chart

5.3.1.1 Tested Performance and Analysis at Full Load (Typical Hot and Humid Condition)

Because the heat pump system in the outdoor air processor utilizes two compressors working in parallel and the solution in each stage exchanges heat with an individual

Table 5.1 Test results of the outdoor air processor under the summer design condition

Parameter	Unit	Value	Parameter	Unit	Value
Outdoor air flow rate	m^3/h	4,058	Exhaust air flow rate	m^3/h	4,021
Outdoor air dry-bulb temperature	°C	36.0	Return air dry-bulb temperature	°C	26.0
Outdoor air humidity ratio	g/kg	25.8	Return air humidity ratio	g/kg	12.6
Supply air dry-bulb temperature	°C	17.3	Exhaust air dry-bulb temperature	°C	39.1
Supply air humidity ratio	g/kg	9.1	Exhaust air humidity ratio	g/kg	38.6
Power consumption of the compressors	kW	14.6	Power consumption of the solution pump	kW	1.92
Outdoor air cooling capacity	kW	82.7	Exhaust air heating capacity	kW	101.9
Outdoor air dehumidification rate	kg/h	80.1	Cooling capacity of the heat pump	kW	59.0
COP_{air}	W/W	5.0	COP_{hp} of the heat pump	W/W	4.0

evaporator (condenser), the dehumidification requirement at partial load can be achieved by adjusting the on-off time of these two compressors. As the required cooling temperature of the solution is about 15–20 °C, the evaporating temperature can be increased to about 10 °C. Meanwhile, the condensing temperature is only around 45 °C since the required heating temperature of the solution is about 40 °C. The test results of the outdoor air processor shown in Fig. 5.12 under the summer design condition are listed in Tables 5.1 and 5.2. The coefficient of performance COP_{air} is defined as

$$COP_{air} = \frac{\text{Cooling capacity of the outdoor air}}{\text{Power consumption of compressor and solution pumps}} \tag{5.8}$$

As shown in Tables 5.1 and 5.2, COP of the heat pump (COP_{hp}) in the outdoor air processor is 4.0, while COP of the processor (COP_{air}) is 5.0. It is worth noting that the superheating temperature is 8.3 °C, which is higher than the designed 5 °C, and the subcooling temperature is 4.9 °C, which is higher than the designed 3 °C. The high superheating temperature is due to the slightly smaller expansion valve. The superheating temperature can be lowered by regulating the electronic expansion valve, and then COP_{hp} can be improved further. Moreover, the refrigerant at the expansion valve outlet is 15.0 °C, which is 8 °C higher than the evaporating temperature. Similarly, the refrigerant at the evaporator outlet is 15.3 °C, while the suction temperatures of the two compressors are only 9.6 and 11.6 °C. However, the refrigerant temperature difference between the condenser outlet and the expansion valve inlet is less than 1 °C. These uncommon phenomena can probably be attributed to the unreasonable refrigerant piping layout, which results in a large pressure drop along the refrigerant pipes. Therefore, special attention should be paid to the layout of the refrigerant piping when designing the processors. The number of elbows and tees, as well as the length of copper pipe, should be minimized to reduce the pressure drop of the refrigerant along the way.

Table 5.2 Test results of the heat pump system

Temperature at the evaporating side			Temperature at the condensing side		
Evaporating temperature	°C	7.0	Condensing temperature	°C	45.0
Suction temperature of compressor I	°C	9.6	Exhaust temperature of compressor I	°C	67.2
Suction temperature of compressor II	°C	11.6	Exhaust temperature of compressorII	°C	66.8
Temperature at evaporator outlet	°C	15.3	Temperature at condenser outlet	°C	40.1
Temperature at expansion valve outlet	°C	15.0	Temperature at expansion valve inlet	°C	39.1

Table 5.3 Test results of the outdoor air processor at partial load under the summer condition

Parameter	Unit	Value	Parameter	Unit	Value
Outdoor air flow rate	m^3/h	4,058	Exhaust air flow rate	m^3/h	4,021
Outdoor air dry-bulb temperature	°C	30.0	Return air dry-bulb temperature	°C	26.0
Outdoor air humidity ratio	g/kg	17.4	Return air humidity ratio	g/kg	12.7
Supply air dry-bulb temperature	°C	17.3	Exhaust air dry-bulb temperature	°C	39.1
Supply air humidity ratio	g/kg	9.6	Exhaust air humidity ratio	g/kg	26.6
Power consumption of the compressors	kW	5.8	Power consumption of the solution pump	kW	1.43
Outdoor air cooling capacity	kW	42.7	Cooling capacity of the heat pump	kW	33.2
COP_{air}	W/W	5.9	COP_{hp} of the heat pump	W/W	5.7
Evaporating temperature	°C	11.0	Condensing temperature	°C	37.6
Superheating temperature	°C	5.6	Subcooling temperature	°C	2.7

5.3.1.2 Tested Performance and Analysis at Partial Load (Summer)

Based on the test results under the design condition, performance of the processor at partial load was then tested, and the main results are listed in Table 5.3. Since only one compressor is in operation at partial load, while the heat transfer area of the evaporator and that of the condenser are identical, the heat transfer between the refrigerant and the solution became more sufficient and the heat transfer temperature differences decrease accordingly. As a result, the evaporating temperature rises and the condensing temperature drops. As indicated by the test results listed in Table 5.3, the evaporating temperature increases to 11 °C, while the condensing temperature decreases by 7.4 °C compared to the full-load condition to 37.6 °C. Thus, COP_{hp} and COP_{air} increase to 5.7 and 5.9, respectively. As the outdoor air handling processor was running at partial load for more than 70 % of the time, the comprehensive energy efficiency of this liquid desiccant processor could be as high as 5.5. In addition, the expansion valve can meet the refrigerant flow rate requirement at partial load, so both the superheating temperature and the subcooling temperature will be very close to the design values.

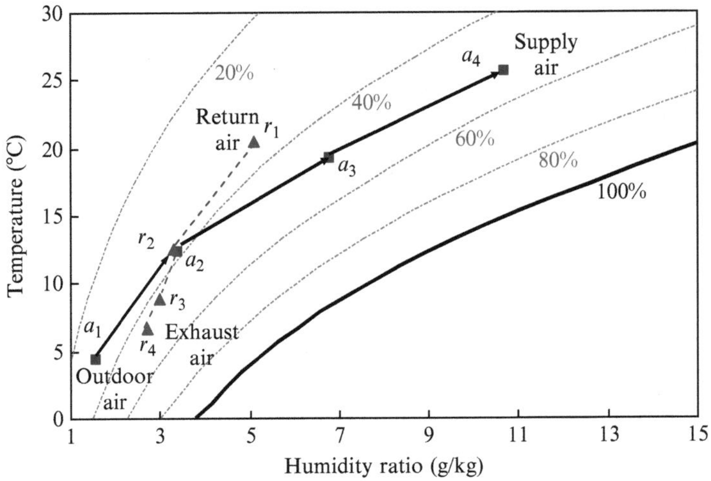

Fig. 5.13 Air handling process in winter of the two-stage liquid desiccant outdoor air processor with enthalpy recovery shown in psychrometric chart

Based on these test results, it can be seen that the heat pump-driven liquid desiccant outdoor air processor can effectively meet the dehumidification requirement in summer with high comprehensive energy efficiency. The humidity ratio of the dehumidified outdoor air can meet the demand for humidity control when the supply air temperature is higher than 17 °C, which could be supplied directly into the conditioned room without further reheating or cooling. Moreover, the enthalpy recovery modules are adopted in this process to recover energy from the return air. In summary, the outdoor air processor can meet the needs of running at full load and partial load with comprehensive energy efficiency up to 5.5.

5.3.1.3 Tested Performance of Air Humidification (Winter)

For the regions where the climate is cold and dry in winter, there are requirements for heating and humidification in the outdoor air handling process. If the condensation dehumidification method is adopted for outdoor air dehumidification in summer, the device cannot be adopted for humidification in winter, and an extra humidification device must be used to meet the requirement. Appropriately, the liquid desiccant outdoor air handling processor shown in Fig. 5.12 can be utilized for air humidification in winter.

In contrast to the summer operating mode illustrated in Fig. 5.12a, the four-way valve can be adopted to realize the interchange of the evaporator and the condenser in the heat pump cycle in winter. The goal of the liquid desiccant air handling processor in winter mode is to heat and humidify the outdoor air. Figure 5.13 shows the air handling process in winter of this processor using liquid desiccant.

Table 5.4 Annual performance of the liquid desiccant outdoor air handling processor

Performance index		Cooling and dehumidification		Enthalpy recovery	Heating and humidification	
Outdoor air	Temperature (°C)	36.0	30.0	35.9	6.4	4.3
	Humidity ratio (g/kg)	25.8	17.4	26.7	2.1	1.6
Return air	Temperature (°C)	26.0	26.0	26.1	20.5	22.1
	Humidity ratio (g/kg)	12.6	12.7	12.1	4.0	5.1
Supply air	Temperature (°C)	17.3	17.3	30.4	22.5	25.6
	Humidity ratio (g/kg)	9.1	9.6	19.5	7.2	10.7
Exhaust air	Temperature (°C)	39.1	39.1	32.6	7.0	6.6
	Humidity ratio (g/kg)	38.6	26.6	20.3	2.7	2.9
COP_{air}		5.0	5.9	62.5 %	6.2	4.6

The outdoor air (state a_1) first reaches state a_2 after the enthalpy recovery process, and then it flows through the packed tower to be heated and humidified by the sprayed solution to supply air state a_4, thereby meeting the indoor requirement. The indoor return air (state r_1) reaches state r_2 after the enthalpy recovery process and then flows through the two-stage packed tower to be cooled and dehumidified, finally reaching exhaust air state r_4.

Table 5.4 lists the annual tested performance of the outdoor air handling processor shown in Fig. 5.12. The processor can be adopted to satisfy the annual handling requirements of the outdoor air, including enthalpy recovery, cooling, and dehumidification in summer and heating and humidification in winter.

5.3.2 Outdoor Air Handling Processor with Precooling Module

In the THIC system, high-temperature chilled water (about 16–18 °C) is selected for indoor temperature control, and the energy efficiency of the high-temperature cooling source is much better than that of a conventional cooling source with a temperature of about 7 °C. In the outdoor air handling process, high-temperature chilled water can be adopted to precool the outdoor air, which helps to improve the energy performance of the cooling and dehumidification process. An HPLD outdoor air handling process combined with a precooling module that uses high-temperature chilled water to precool the outdoor air is introduced here.

Figure 5.14a, b shows the operating schematic of the processor and a psychrometric chart of the outdoor air handling process, respectively (Beijing SinoRefine Air Conditioning Technology Co.LTD). Outdoor air (OA) first flows through the precool coil and is precooled to OA_1 by the high-temperature chilled water. The air then flows into the dehumidifier using liquid desiccant and is processed to the supply air state with the required humidity ratio. In the liquid desiccant

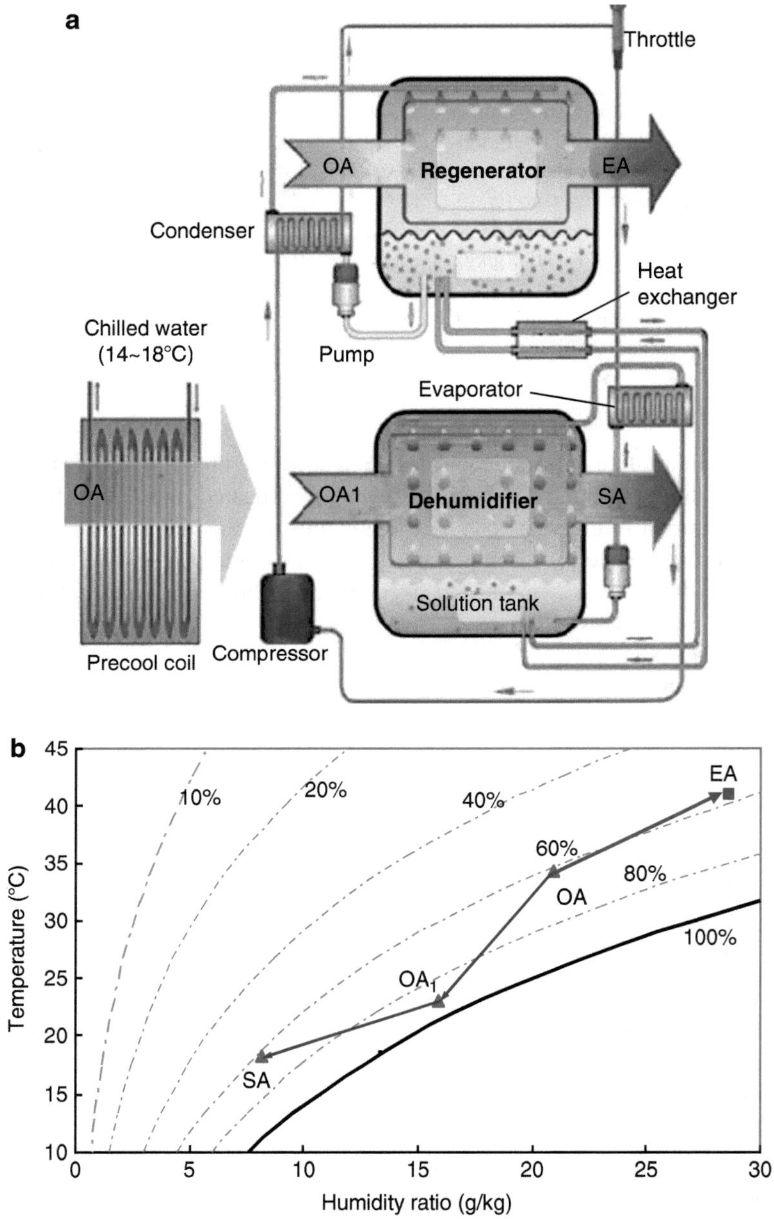

Fig. 5.14 Heat pump-driven liquid desiccant outdoor air handling processor with precooling module: (**a**) operating schematic and (**b**) air handling process in psychrometric chart

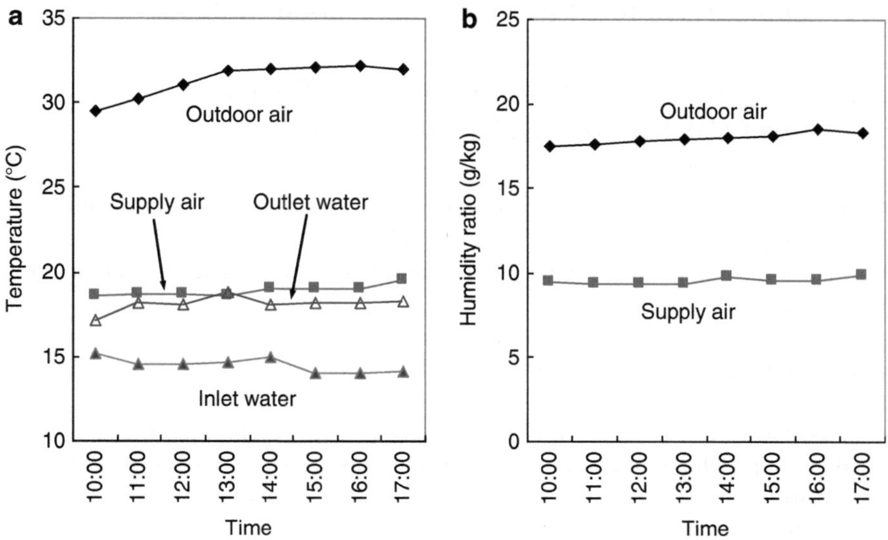

Fig. 5.15 Measured operating parameters of the liquid desiccant outdoor air processor: (**a**) temperature and (**b**) humidity ratio

dehumidification module, the concentrated solution is cooled by the evaporator before flowing into the dehumidifier. The diluted solution then leaves the dehumidifier, is heated by the condenser, and is finally concentrated in the regenerator. The outdoor air is used for desiccant regeneration, and the humid regenerated air is exhausted to the outdoor environment.

Figure 5.15 illustrates the measured hourly performance (including air temperature and humidity ratio) of an outdoor air processor in a departure hall. As indicated in Fig. 5.15a, the outdoor air temperature is around 30–32 °C, and the supply air temperature stabilizes at about 19 °C. The inlet and outlet water temperatures for precooling are about 14.5 °C and 18.0 °C, respectively. The outdoor air humidity ratio is about 18.0 g/kg, while the supply air humidity ratio for extracting indoor moisture is 9.6 g/kg.

This kind of outdoor air handling processor with a precooling module used in summer can also be adopted for air humidification in winter mode, in which the high-temperature chilled water in summer is replaced by hot water (about 35–40 °C) that preheats the outdoor air in the coils. The evaporator and condenser of the heat pump cycle also interchange with each other with the help of the four-way valve. After being preheated by the hot water, the outdoor air is further humidified by the sprayed solution in the packed power and is then supplied to the indoor conditioned space.

5.4 Comparison to Other Dehumidification Methods

5.4.1 Comparison of Desiccant Dehumidification and Condensation Dehumidification

Commonly used dehumidification methods include condensation dehumidification, liquid desiccant dehumidification, and solid desiccant dehumidification. The characteristics of these commonly utilized dehumidification methods are compared in this subsection.

Figure 5.16 illustrates the equilibrium states of water, liquid desiccant, and solid desiccant in a psychrometric chart. The water can only stay in the saturated line of the humid air with a relative humidity of 100 %, while the states of the liquid or solid desiccant can cover most of the air states in the psychrometric chart. In the condensation dehumidification method using cooling coils, the processed air is cooled to the dew point by the cooling medium, and vapor condensation occurs if the air temperature decreases further. As shown in the psychrometric chart, the water state can only stay on the saturation line, and, thus, the outlet state of the processed air using the condensation dehumidification method approaches the saturated state.

In the air handling process using liquid or solid desiccant, the iso-concentration line of the liquid desiccant (or the iso-water content line of the solid desiccant) almost coincides with the iso-relative humidity line of the humid air. If the desiccant dehumidification method is adopted, the handled air state could reach a much larger region in the psychrometric chart compared to the condensation dehumidification method. For the liquid desiccant dehumidification method, the air handling process approaches the process using chilled water when the liquid desiccant is extremely diluted, and the handled air cannot reach the region on the left side of the crystal line in the psychrometric chart due to the restriction of the crystallization of the solution. For the solid desiccant dehumidification method, the handled air cannot reach the region on the right side of the desiccant maximum adsorption capacity line due to the restriction of the maximum adsorption capacity of the solid desiccant. Moreover, there are significant discrepancies among the crystal lines for different liquid desiccants, as indicated in Fig. 5.1: the reachable region of the handled air using LiBr or LiCl solution is much wider than that of the handled air using $CaCl_2$ solution. Similarly, there are significant discrepancies among the maximum adsorption capacities for different kinds of solid desiccant; molecular sieves are appropriate for handling air with a lower relative humidity, while activated aluminum oxide is appropriate for handling air with a higher relative humidity.

In dehumidification, water changes from the vapor state to the liquid state, and a large amount of latent heat of vaporization is released during the dehumidification

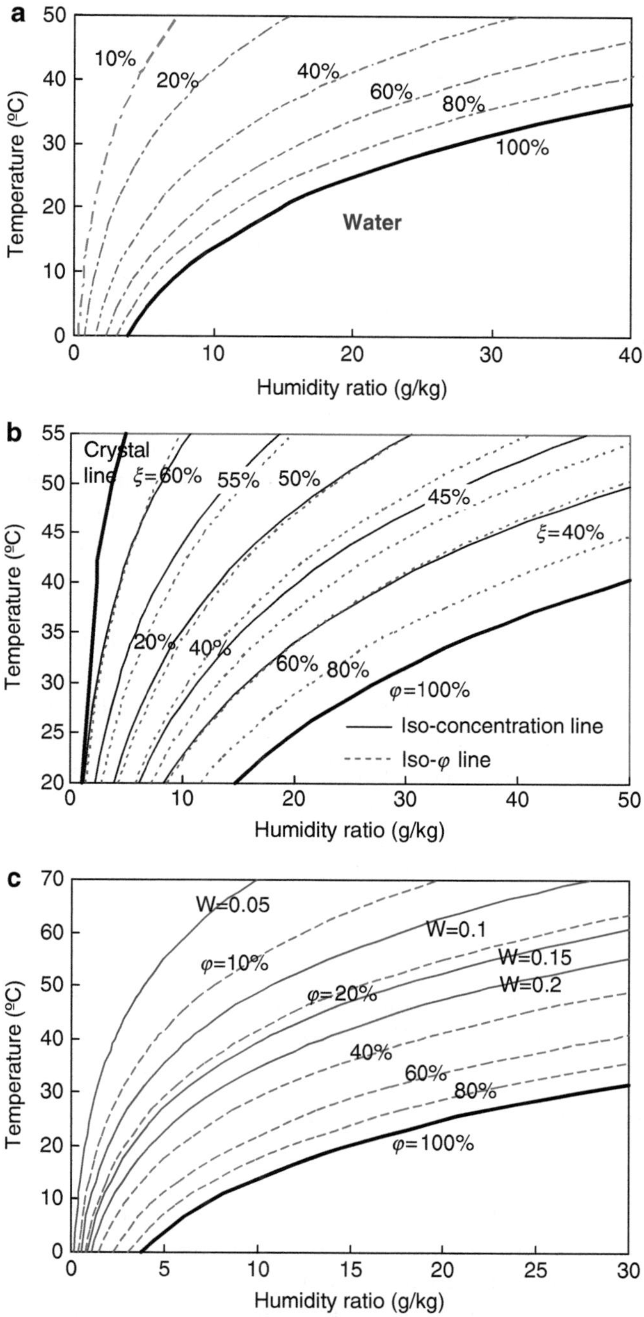

Fig. 5.16 States of water and desiccants shown in psychrometric chart: (**a**) water, (**b**) LiBr aqueous solution, and (**c**) silica gel

process. When the humidity ratio variance of the air is 1 g/kg, the relative latent heat variance is about 2.5 times greater than that of the air, with a temperature variance of 1 °C. Consequently, how to effectively extract the latent heat released during the dehumidification process is the key issue that influences the efficiency of the process. In the condensation dehumidification process, the cooling medium directly cools the air, and the released heat is removed directly and locally. For dehumidification processes using liquid or solid desiccant, it is difficult to realize local cooling or to extract the released heat directly (due to the complexity and difficulty in manufacturing internally cooled devices). Therefore, the air handling processes should be designed carefully. The cooling solutions commonly adopted in desiccant dehumidification systems are summarized as follows:

- Adiabatic liquid desiccant dehumidifier/regenerator + external cooling/heating device. This is the most commonly applied type in the market. The solution is cooled/heated by the external cooling/heating source before being supplied into the adiabatic dehumidifier/regenerator. For solutions with the same concentration, the lower the solution temperature, the better the dehumidification ability. Section 5.3 has introduced this type of outdoor air handling processor using liquid desiccant.
- Adiabatic solid desiccant dehumidifier (rotary wheel) + external cooling/heating device. The major difference between solid and liquid desiccants is that solid desiccants cannot be effectively cooled/heated by the external cooling/heating source directly due to the rotation of the wheel; therefore, the inlet air is cooled/heated instead. In contrast to the liquid desiccant process with cooling and heating solutions, the rotary wheel handling process has to cool the inlet air coming into the dehumidification zone of the wheel and heat the inlet regeneration air. Therefore, a higher regeneration temperature is required in the rotary wheel system to obtain the same supplied air humidity ratio.

5.4.2 Comparison of Liquid Dehumidification and Solid Desiccant Dehumidification

5.4.2.1 Similarities of Solid and Liquid Desiccant Dehumidification Methods

As shown in Fig. 5.16, the iso-concentration lines of the liquid desiccant (concentration = solute mass/solution mass) and the iso-water content lines of the solid desiccant (water content = water mass/dry solid desiccant mass) are essentially coincident with the iso-relative humidity lines of the air. In the dehumidification and regeneration processes, the expressions of the driving forces of the

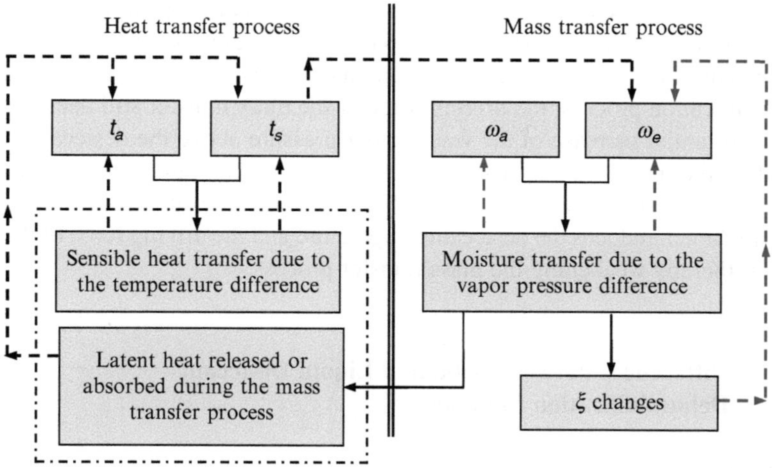

Fig. 5.17 Relationship between the heat transfer process and the mass transfer process in a liquid desiccant air contact device

heat and mass transfer between the solid desiccant and the air, and those between the liquid desiccant and the air, are similar. The driving force of the mass transfer in the dehumidification process is the water vapor pressure difference (humidity ratio difference) between the desiccant surface and the air. Therefore, lowering the temperature of the desiccant and reducing the water content of the solid desiccant (equivalent to improving the solution concentration) could improve the driving forces of the heat and mass transfer. The principle of the regeneration process is similar to the dehumidification process, with the only difference being the opposite mass transfer direction.

As shown in Fig. 5.17, the heat and mass transfer process between the air and liquid desiccant in the dehumidifier/regenerator is taken as an example to illustrate the interaction between the heat transfer process and the mass transfer process. The same characteristic applies to the heat and mass transfer process between the air and solid desiccant. On the one hand, the absorption/release of the latent heat of vaporization accompanying the mass transfer process influences the temperatures of the solution and the air, thus influencing the heat transfer process between them. On the other hand, the temperature variance of the solution (t_s) significantly affects the surface vapor pressure (or equivalent humidity ratio ω_e), thereby affecting the driving force of the mass transfer and the entire mass transfer process between the solution and the air. Due to the mutual influence of the heat transfer and mass transfer processes, the analysis of the heat and mass transfer should comprehensively integrate the interaction of the heat and mass transfer processes, rather than abandon either transfer process. In the dehumidification process, along with the moisture transfer from the humid air to the desiccant, a large quantity of latent heat of vaporization is released, which changes the temperatures of the air and desiccant. Although the concentration of the desiccant may not change significantly during the dehumidification process, the driving force of the mass transfer still decreases due to the substantial increase of the water vapor pressure above the desiccant surface caused by the increased desiccant temperature. Similarly, during the regeneration process, a large amount of heat is needed to desorb the moisture from the desiccant to the air, which reduces the desiccant temperature and the driving force of the mass transfer, thereby weakening the mass transfer process.

5.4.2.2 Differences Between Solid and Liquid Desiccant Dehumidification Methods

The most fundamental difference between solid desiccant and liquid desiccant dehumidification methods lies in the movability of the liquid desiccant. This seemingly insignificant distinction leads to considerable differences in the air handling processes and energy efficiency. Compared to the fixed solid desiccant bed method, the liquid desiccant dehumidification method can realize continuous dehumidification and regeneration processes. Therefore, stable outlet air parameters could also be achieved by circulating the solution between the dehumidifier and the regenerator. For the rotary wheel dehumidification method, the desiccant bonds with the base material and rotates with the wheel to achieve continuous dehumidification and regeneration processes, leading to a stable state of the outlet air. The main differences between the solid desiccant wheel and the liquid desiccant device are summarized in the following two points:

1. The cooled dehumidification process and heated regeneration process can be achieved easily in the liquid desiccant air processor by directly cooling or heating the solution close to the iso-relative humidity handling process. For the solid desiccant wheel, the air-conditioning process can only be realized by cooling the air (rather than the desiccant) close to the isenthalpic handling process, and the regeneration process can only be achieved by heating the air (rather than the solid desiccant) close to the isenthalpic handling process. Previous research results have shown that cooling the desiccant results in much better dehumidification performance than cooling the air and heating the desiccant is also superior to heating the air. This can explain the lower regeneration temperature of the liquid desiccant compared to the desiccant wheel (with the same handled humidity ratio of air).
2. A solution-to-solution heat exchanger can be adopted in the liquid desiccant system to precool the concentrated solution entering the dehumidifier or preheat the diluted solution entering the regenerator, thereby reducing the heating/cooling offset caused by the solution circulating between the dehumidifier and

the regenerator. Taking the desiccant wheel as an example, in the solid desiccant process, the solid desiccant in the dehumidification zone directly enters the regeneration zone, while the solid desiccant in the regeneration zone also directly enters the dehumidification zone. This means that heating/cooling offset of the solid desiccant does exist. Since the solid desiccant cannot flow like a solution can, it must adhere to the base material in order to achieve the whole dehumidification and regeneration process. The thermal mass of the base wheel material is not negligible. Thus, the hot air should heat the desiccant and the base wheel material at the same time, while the cold air should simultaneously cool the desiccant and the base wheel material. This will lead to extra heat during the regeneration process and extra cooling consumption during the dehumidification process.

The movability of liquid accounts for the significant differences between liquid desiccant systems and solid desiccant systems. For the liquid desiccant system, where the solution is cooled before flowing into the dehumidifier or heated before flowing into the regenerator, the whole handling process is close to the iso-relative humidity line, leading to a lower required regeneration temperature. For the heat pump-driven liquid desiccant system examined in Sect. 5.3, the regeneration temperature (i.e., condensing temperature) can be 40–50 °C to meet the demand for regeneration.

For the desiccant wheel, the regeneration of the solid desiccant can only be achieved by heating the air, and the cooling in the dehumidification process is only made possible by cooling the processed air. As a result, the solid desiccant process is close to the isenthalpic line, with a large heat and mass transfer unmatched coefficient (Zhang et al. 2012). To achieve the same dehumidification effect, the temperature of the air dehumidified by the solid desiccant rises significantly (approximate isenthalpic temperature rise). In general, the regeneration air temperature could be above 100 °C, which means that a very-high-temperature heat source is needed. By dividing a single desiccant wheel into two desiccant wheels, the water content of the solid desiccant can be reduced, and the regeneration temperature can be lowered to 60–80 °C. In contrast, the dehumidification and regeneration processes in DESICA, the heat pump-driven solid desiccant air-conditioning apparatus developed by Daikin Industries, Ltd., are close to the iso-relative humidity line. Since the evaporator is the cooling source, the regeneration temperature is only 40–50 °C. Table 5.5 summarizes the performance comparison of the various liquid and solid desiccant air-conditioning processes described in this book, with the task of dehumidifying the outdoor air (e.g., a humidity ratio of 20 g/kg) to the desired level (e.g., a humidity ratio of 8–10 g/kg).

Table 5.5 Comparison of liquid and solid desiccant processes

Process	Driving force	Section in this book	Dehumidification and regeneration processes	Cooling source	Regeneration temperature	Remarks
Liquid desiccant	Heat pump	5.3	Close to iso-relative humidity line	Evaporator in the heat pump	40–50 °C	Continuous dehumidification
Solid desiccant	Waste heat (single wheel)	4.3.2.1	Close to isenthalpic line	Cooling water, chilled water	Above 100 °C	Continuous dehumidification
	Waste heat (double wheel)	4.3.2.1	Close to isenthalpic line	Cooling water, chilled water	60–80 °C	Continuous dehumidification
	Heat pump (DESICA)	4.3.2.2	Close to iso-relative humidity line	Evaporator in the heat pump	40–50 °C	Intermittent dehumidification

References

Beijing SinoRefine Air Conditioning Technology Co.LTD. http://www.sinorefine.com.cn/mainFrm.aspx
Chen XY (2005) Application study on liquid desiccant air conditioning system. Master thesis, Tsinghua University, Beijing (in Chinese)
Cheng XT, Liang XG (2012) Computation of effectiveness of two-stream heat exchanger networks based on concepts of entropy generation, entransy dissipation and entransy-dissipation-based thermal resistance. Energy Convers Manage 58:163–170
The 10th Design and Research Institute of Ministry of Electronics Industry (1995) Design handbook for air-conditioning, 2nd edn. China Architecture & Building Press, Beijing (in Chinese)
Guo ZY, Zhu HY, Liang XG (2007) Entransy – a physical quantity describing heat transfer ability. Int J Heat Mass Transf 50:2545–2556
Jiang Y, Li Z, Liu XH, Chen XY (2003) An air-liquid desiccant direct contact device with external cooling/heating, ZL03249068.2 (Chinese Patent)
Khan AY (1998) Cooling and dehumidification performance analysis of internally-cooled liquid desiccant absorbers. Appl Therm Eng 18:265–281
Lazzarin RM, Castellotti F (2007) A new heat pump desiccant dehumidifier for supermarket application. Energy Build 39:59–65
Li Z (2004) Principles of thermodynamic analysis of humid air process and its application in liquid desiccant air conditioning systems. Doctoral dissertation, Tsinghua University, Beijing (in Chinese)
Li Z, Liu XH, Jiang Y, Chen XY (2005) New type of fresh air processor with liquid desiccant total heat recovery. Energy Build 37:587–593
Liu XH (2007) Combined heat and mass transfer characteristic in air handling process using liquid desiccant. Doctoral dissertation, Tsinghua University, Beijing (in Chinese)
Liu SQ (2010) Research on liquid desiccant air handling device driven by heat pump. Doctoral dissertation, Tsinghua University, Beijing (in Chinese)
Liu XH, Zhang Y, Qu KY, Jiang Y (2006) Experimental study on mass transfer performances of cross flow dehumidifier using liquid desiccant. Energy Convers Manage 47(15–16):2682–2692
Liu XH, Jiang Y, Chang XM, Yi XQ (2007) Experimental investigation of the heat and mass transfer between air and liquid desiccant in a cross-flow regenerator. Renew Energy 32(10):1623–1636
Liu XH, Li Z, Jiang Y (2009) Similarity of coupled heat and mass transfer process between air-water and air-liquid desiccant contact system. Build Environ 44:2501–2509
Lof GOG (1955) Cooling with solar energy. Congress on Solar Energy, Tucson, AZ, pp 171–189
Zhang XS, Yin YG, Cao YR (2004) Experimental study on the performance of liquid desiccant cooling air conditioning with energy storage. J Eng Thermophys 25(4):546–549 (in Chinese)
Zhang T, Liu XH, Zhang L, Jiang Y (2012) Match properties of heat transfer and coupled heat and mass transfer processes in air-conditioning system. Energy Convers Manage 59:103–113
Zhang T, Liu XH, Jiang Y (2013) Performance comparison of liquid desiccant air handling processes from the perspective of match properties. Energy Convers Manage 75:51–60
Zhao Y (2002) Investigation on solar liquid desiccant air-conditioner. Doctoral dissertation, Southeast University, Nanjing (in Chinese)

Chapter 6
Key Components of the THIC System: High-Temperature Cooling Sources

Abstract In THIC systems, the required temperature of the high-temperature cooling source during summer is significantly higher than that of conventional systems. As there is no dehumidification requirement, the chilled water temperature could be increased from 5 to 7 °C in conventional systems to 16–18 °C in THIC systems. This offers the opportunity to utilize natural cooling sources, e.g., deep phreatic water, ground-source heat exchangers, and direct or indirect evaporative cooling methods in dry regions.l If no natural cooling sources are available, a vapor compression system can be applied instead. Due to the higher evaporating temperature of the vapor compression cycle, the operating compression ratio of chillers in THIC systems is significantly different from that in conventional systems. Therefore, new requirements are proposed in terms of the chiller design and device development of the vapor compression cycle for THIC systems. In this chapter, typical high-temperature cooling sources are introduced.

In THIC systems, the required temperature of the high-temperature cooling source in summer is significantly higher than that of conventional systems. As there is no dehumidification requirement, the chilled water temperature could be increased from about 5–7 °C in conventional systems to about 16–18 °C in THIC systems. This offers the possibility to utilize natural cooling sources, e.g., deep phreatic water, ground heat exchangers, and direct or indirect evaporative cooling methods in certain dry regions. If no natural cooling sources can be adopted directly for temperature control in the THIC system, a vapor compression refrigeration system can be utilized instead. Owing to the increased evaporating temperature of the vapor compression refrigeration cycle, the operating compression ratio of chillers in THIC systems is significantly different from that in conventional systems. Thus, new requirements are proposed for system design and device development of the vapor compression refrigeration cycle for THIC systems. In this chapter, common types of high-temperature cooling sources are examined.

X. Liu et al., *Temperature and Humidity Independent Control (THIC) of Air-conditioning System*, DOI 10.1007/978-3-642-42222-5_6,

6.1 Underground Embedded Pipe Cooling

6.1.1 Operating Principle

Soil provides excellent energy storage, and with the help of a ground heat exchanger, soil could be adopted as the heat sink/source for THIC air-conditioning systems. In THIC systems, the required cooling source temperature in summer is usually about 16–18 °C for temperature control, and this makes it possible to utilize the ground heat exchanger as the high-temperature cooling source directly.

The operating schematics of a ground heat exchanger in summer and winter are illustrated in Figs. 6.1 and 6.2, respectively. During summer, the soil is adopted as the cooling source, and heat is released to the soil. The high-temperature cooling source for temperature control is obtained by the ground heat exchanger. During winter, the soil is adopted as the heating source, and heat is removed from the soil.

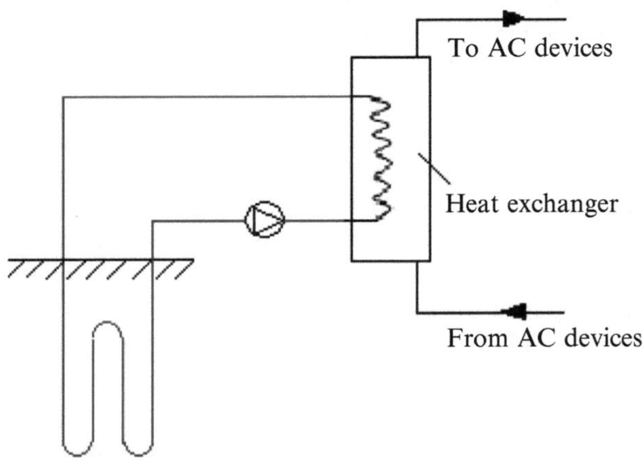

Fig. 6.1 Ground heat exchanger (summer)

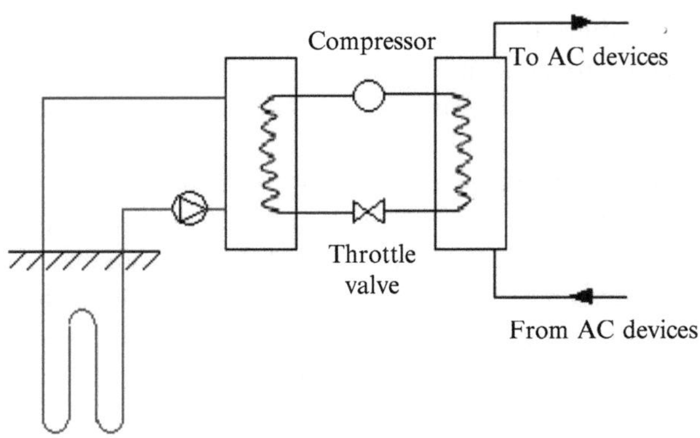

Fig. 6.2 Ground heat exchanger (winter)

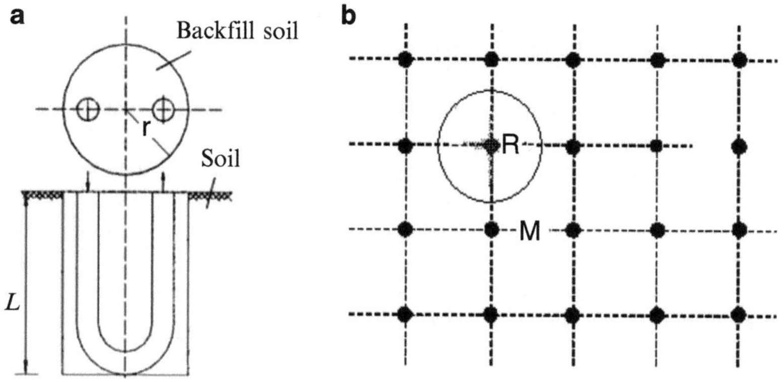

Fig. 6.3 Schematic of a vertical buried ground heat exchanger: (**a**) U-type vertical buried pipe and (**b**) arrangement of multiple U-type pipes

The heat transfer process proceeds between the evaporator of the heat pump cycle and the ground heat exchanger. Then, the condenser of the heat pump produces hot water with an appropriate temperature for heating.

The ground heat exchanger could take the form of horizontal pipes embedded in the soil at a depth of about 2–5 m or vertical pipes buried at a depth of about 50–100 m. Horizontal pipes cover a larger area than vertical pipes, which is the main reason leading to that vertical buried pipes are the most popular type in China. Figure 6.3a illustrates the U-type buried pipe commonly adopted in the vertical ground heat exchanger, where the U-type pipe is deposited in a drilled hole with a diameter ($2r$) of about 100 mm and backfill material is filled in around it. The diameter of the buried pipe is about 20 mm. An arrangement of U-type buried pipes aligned in a grid is shown in Fig. 6.3b, where the black points represent the drilled holes, with an interval of about 3–6 m between neighboring holes.

For ground heat exchangers with vertical buried pipes, there are two extreme cases according to whether there is sufficient groundwater flow. The most practical heat exchange process lies in the middle of these two extreme cases:

- *Extreme case I*: There is limited or no groundwater flow. Under this circumstance, the influence of groundwater flow on the ground heat exchanger could be negligible. If the ground heat exchanger is adopted for cooling in summer, heat should be drawn from the soil in winter to ensure the annual heat balance of the soil. Otherwise, the ground heat exchanger operating only in cooling mode will cause the underground soil to overheat, and the ground heat exchanger will be useless in the end. If the vertical buried pipes are adopted for both heating and cooling, the ground heat exchanger could be regarded as a heat exchanger with heat storage that transfers heat between summer and winter. Therefore, the temperature for heat release (obtaining cooling) in summer is not only related to the local annual mean temperature; it is also determined by the heat removal temperature in winter and the annual heat balance between summer and winter.

Table 6.1 Annual average temperatures of typical cities in China

City	Harbin	Changchun	Xining	Urumchi	Hohhot	Lhasa	Shenyang
$T_{Average}$ (°C)	3.6	4.9	5.7	5.7	5.8	7.5	7.8
City	Yinchuan	Lanzhou	Taiyuan	Beijing	Tianjin	Haikou	Xi'an
$T_{Average}$ (°C)	8.5	9.1	9.5	11.4	12.2	23.8	13.3
City	Zhengzhou	Jinan	Luoyang	Kunming	Nanjing	Guiyang	Shanghai
$T_{Average}$ (°C)	14.2	14.2	14.6	14.7	15.3	15.3	15.7
City	Hefei	Chengdu	Hangzhou	Wuhan	Changsha	Nanchang	Chongqing
$T_{Average}$ (°C)	15.7	16.2	16.2	16.3	17.2	17.5	18.3
City	Fuzhou	Nanning	Guangzhou	Taipei	Shijiazhuang		
$T_{Average}$ (°C)	19.6	21.6	21.8	22.1	12.9		

- *Extreme case II*: The groundwater flow is sufficient. Under this circumstance, there is sufficient heat exchange between the groundwater flow and the soil so that the ground heat exchanger could be regarded as an equivalent heat exchanger with the groundwater flow. Thus, there is no requirement for equal heat exchange between summer and winter. Previous research has shown that the temperature of soil at a depth of more than 10 m seldom varies as the outdoor climate or seasons change, and it is nearly equal to the local annual average temperature. Table 6.1 lists the annual average temperatures ($T_{Average}$) of typical cities in China; it can be seen that the annual average temperatures of some cities are lower than 15 °C. For example, the annual average temperature of Beijing is 11.4 °C. Thus, if the soil temperature permits, the soil could be adopted as the natural cooling source for indoor temperature control through ground heat exchangers.

6.1.2 Analysis of the Characteristics of the Heat Transfer Process

6.1.2.1 Method of Analysis of the Heat Transfer Process

If the heat transfer between the groundwater flow and the soil is omitted (extreme case I), there should be equal heat exchange between summer and winter in order for this ground heat exchanger to satisfy a cyclic steady-state soil temperature. The ground heat exchanger with vertical buried pipes could then be regarded as a heat exchanger with thermal storage between the equivalent heating source ($t_{w,1}$) in summer and the equivalent cooling source ($t_{w,2}$) in winter, as shown in Fig. 6.4. The left part of the figure illustrates the heat transfer process when the soil absorbs heat, while the right part shows the process when the soil releases heat.

As the soil absorbs heat, heating source $t_{w,1}$ transfers heat to the soil through medium h_1A (where h_1 is the heat transfer coefficient and A is the heat transfer area). The temperature increase of the soil after absorbing heat is ΔT_2. During the heat release stage, cooling source $t_{w,2}$ absorbs heat from the soil through medium h_2A and absorbs the heat stored by the soil. The temperature decrease of the soil after releasing heat is ΔT_2.

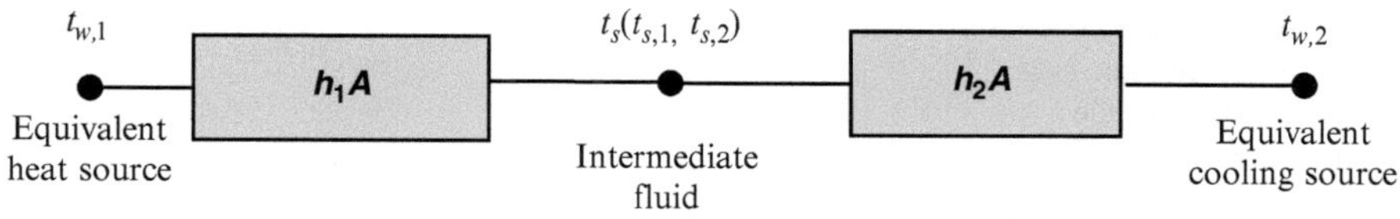

Fig. 6.4 Equivalent heat transfer between the intermediate fluid and the heat/cooling sources

Fig. 6.5 Temperature levels of the periodic heat transfer process of the ground heat exchanger

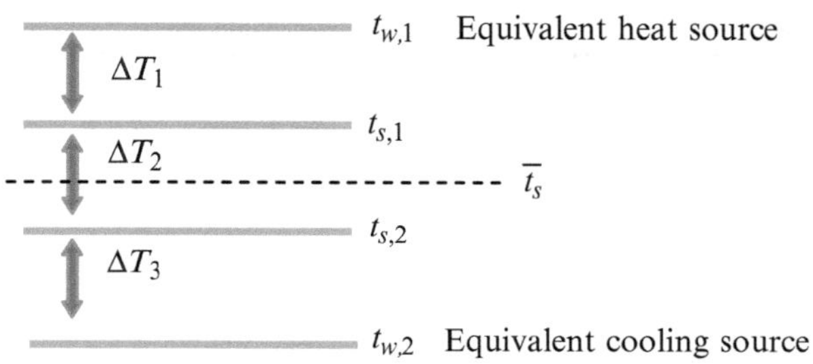

The entire heat exchange process could be regarded as an equivalent heat transfer process between the heating and cooling sources during different times, and the soil affects thermal storage and the heat transfer medium. The larger the temperature difference between the heating and cooling sources, the greater the heat exchange; in other words, the larger the heat transfer ability inputs during the heat absorption and heat release stages (h_1A and h_2A, respectively), the greater the heat exchange. Figure 6.5 shows the temperature levels of the heating source, the cooling source, and the intermediate medium (soil) of this heat transfer process. The following results can be obtained:

- As $\frac{\Delta T_2}{\Delta T_1+\Delta T_2+\Delta T_3} \to 0$, the thermal resistance of the process is mostly composed of the thermal resistance between the fluids and the intermediate. The key issue for improving the heat transfer performance of this ground heat exchanger is to improve h_1A and h_2A as much as possible.
- As $\frac{\Delta T_1+\Delta T_3}{\Delta T_1+\Delta T_2+\Delta T_3} \to 0$, the heat exchange process mostly relies on the temperature variance of the intermediate (soil), so improving h_1A and h_2A has less influence on improving the heat exchange performance of the entire process.

If there is sufficient groundwater flow (extreme case II), there should be a sufficient heat exchange process between the soil and the groundwater flow. As a result, there is no requirement to ensure equal heat exchange between summer and winter. In this scenario, the ground heat exchanger can be regarded as the heat transfer process between the fluids inside the buried pipes and the groundwater flow (using the soil as the intermediate), as shown in Fig. 6.6.

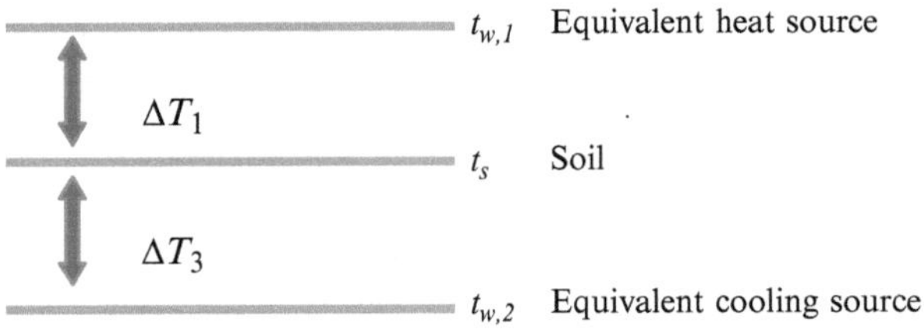

Fig. 6.6 Temperature levels of the periodic heat transfer process of the ground heat exchanger with sufficient groundwater flow rate

Table 6.2 Soil parameters

	Thermal conductivity [W/(m · °C)]		Specific heat	Density
Type	Arid soil	Saturated soil	[J/(kg · °C)]	(kg/m^3)
Grit	0.197	0.60	930	837
Fine dinas	0.193	0.60	930	837
Sub-dinas	0.188	0.60	600	2,135
Sub-clay	0.256	0.60	1,260	1,005
Dense rock	1.068	–	2,000	921
Rock	0.930	–	1,700	921
Clay	1.407	–	1,850	1,842
Wet sand	0.593	–	1,420	1,507

6.1.2.2 Heat Gain Analysis of a Single Pipe (Extreme Case I)

When there is limited or no groundwater flow, the ground heat exchanger can be regarded as a periodic heat exchange process with thermal storage. Then, the thermal resistance of this heat exchange process can be expressed as follows:

$$R_{\mathrm{t}} = \frac{t_{\mathrm{w},1} - t_{\mathrm{w},2}}{Q} \tag{6.1}$$

where Q is the total amount of heat charged/discharged to the soil during the heat release/removal stages of the ground heat exchanger and $t_{\mathrm{w},1}$ and $t_{\mathrm{w},2}$ are the temperatures of the equivalent heating and cooling sources, respectively (i.e., the water temperatures inside the buried pipes in summer and winter, respectively), as shown in Fig. 6.5.

The key issues influencing the performance of the ground heat exchanger include the type, thermal conductivity, density, and water concentration of the soil. The key parameters of different soils are listed in Table 6.2 (Song et al. 1998). Although the thermal conductivity of scree-type soils is high, the construction cost is prohibitively expensive. Thus, sand and clay soil types are the most appropriate for ground heat exchangers. Moreover, as the thermal conductivity of water is 0.60 W/(m · °C), which is higher than common soils, wet soils have higher thermal conductivity. And if the soils are wet enough or if the groundwater level is high enough, the buried pipes will be at or lower than the groundwater level. This means that the thermal conductivity of the soil could be regarded as equal to that of the water.

Table 6.3 Analysis of the heat transfer process for different soils (extreme case I)

Type	$R_t \times 10^8$ (°C/J)	Proportion of thermal resistance			$\Delta T = 20$ °C		$\Delta T = 30$ °C	
		Stage A	Stage B	Stage C	Q_0 (MJ/m)	q_0 (W/m)	Q_0 (MJ/m)	q_0 (W/m)
Grit	17.2	11.2 %	77.6 %	11.2 %	58.0	5.6	87.0	8.4
Fine dinas	17.2	11.2 %	77.6 %	11.2 %	58.0	5.6	87.0	8.4
Sub-dinas	14.6	13.2 %	73.5 %	13.2 %	68.6	6.6	102.9	9.9
Sub-clay	14.6	13.2 %	73.6 %	13.2 %	68.3	6.6	102.5	9.9
Wet sand	12.4	15.6 %	68.8 %	15.6 %	80.7	7.8	121.1	11.7
Rock	10.8	17.9 %	64.3 %	17.9 %	92.7	8.9	139.1	13.4
Dense rock	9.54	20.2 %	59.6 %	20.2 %	104.8	10.1	157.2	15.2
Clay	6.94	27.8 %	44.4 %	27.8 %	144.2	13.9	216.3	20.9

Notes:
1. For grit, fine dinas, sub-dinas, and sub-clay soils, the results are calculated based on the parameters of saturated soil, $\lambda = 0.60$ W/(m · °C)
2. Stage A, heat charging process; Stage B, soil conduction process; Stage C, heat discharging process
3. ΔT, temperature difference between the high-temperature chilled water in summer and the low-temperature heating water in winter
4. Q_0, heat transfer rate per unit length of buried pipe; q_0, average unit length heat flux of the buried pipe

Table 6.3 lists soils with different characteristics, including the total amount of charged/discharged heat Q, total thermal resistance R_t, and the proportions of the total thermal resistance of the heat release stage, the soil, and the heat removal stage per unit length (Zhang et al. 2011a). The characteristics of the soils are listed in Table 6.2; the equivalent heat transfer coefficient (h) between the drilled holes' surfaces is 5 W/(m^2 · °C). This value is calculated based on the heat-conducting properties of the backfill material and the shell of the pipes, as well as the thermal conductivity of the inner shell of the pipes. The duration of the heat release period and that of the heat removal period are 4 months, with the remaining 4 months being the adiabatic period. The diameter of the drilled hole is 0.2 m (i.e., $r = 0.1$ m, as shown in Fig. 6.3), and the interval between neighboring pipes is 5 m (i.e., $M = 5$ m, and equivalent radius R is 3.2 m). As indicated by the calculated results listed in Table 6.3, the following conclusions are obtained:

- Under identical conditions, the total amount of heat charged/discharged by soil type is listed in descending order as follows: clay, dense rock, rock, wet sand, sub-clay, sub-dinas, grit, and fine dinas. The total amount of heat charged/discharged by clay is about 2.5 times greater than that of either grit or fine dinas.
- The thermal resistance of the soil itself accounts for a considerable portion of the total thermal resistance; even clay accounts for about 44 % of the total thermal resistance. And the proportions of total thermal resistance of the other types of soils listed in Table 6.3 are 60–80 %. Thus, the performance of the heat exchange process using vertical buried pipes is mainly restricted by the thermal resistance of the soil itself. If the soil characteristics can be determined,

Table 6.4 Analysis of the heat transfer process of different soils (extreme case II)

$\Delta T'$ (°C)	5	10	15	20
Mean heat transfer rate per length of buried pipe (W/m)	15.7	31.4	47.1	62.8

increasing the interval between neighboring buried pipes can help to reduce the thermal resistance of the soil to a significant degree.

- The heat flux of the buried pipe (unit: W/m) and the total charged heat for the cold or heat storage stage (unit: J/m) are related to the total thermal resistance of the process (influenced by the soil characteristics and the interval between neighboring pipes). However, they are also affected by the temperature difference between the fluids releasing heat in summer and obtaining heat in winter.

6.1.2.3 Heat Gain Analysis of a Single Pipe (Extreme Case II)

In this case, there is sufficient groundwater flow, and the heat exchange process of the buried pipe can be regarded as the heat transfer process between the fluid inside the pipe and the soil (with mean annual temperature). As the groundwater flow is sufficient, the soil temperature can be regarded as equal to the local average annual temperature. Moreover, there is no requirement for equal heat exchange between summer and winter in extreme case II. Table 6.4 lists the mean heat transfer rates per length of pipe, where $\Delta T'$ is the temperature difference between the equivalent heating source in summer (or the equivalent cooling source in winter) and the soil. The equivalent heat transfer coefficient between the drilling surfaces of the vertical buried pipes is 5 W/(m^2 · °C), with a drilled diameter of 0.2 m.

In comparing extreme cases I and II, it can be seen that once the equivalent heating sources in summer and winter are determined, the mean heat transfer rate per pipe length in extreme case II is significantly higher than that in extreme case I. The temperature difference between the equivalent heating source in summer and the local mean annual temperature and that between the equivalent cooling source in winter and the local mean annual temperature are both 10 °C (i.e., the temperature difference between the equivalent cooling source and heating source is 20 °C). As shown in Table 6.3, the mean heat transfer rate per length of pipe in extreme case II is 31.4 W/m, while that in extreme case I is only 5.6–13.9 W/m.

The actual performance of the ground heat exchanger lies in the middle of these two extreme cases. In extreme case I, the ground heat exchanger with buried pipes can be regarded as the heat transfer process between the equivalent cooling and heating sources; the temperature difference between the sources (ΔT) is the driving force of this process. The heat exchange of the ground heat exchanger (Q) is influenced by ΔT. In extreme case II, the heat absorption/release stages of the ground heat exchanger can be regarded as the heat transfer process with the soil. The driving force for this process is the temperature difference between the fluids inside the buried pipes and the soil ($\Delta T'$). The heat exchange is also significantly influenced by this temperature difference. In analyzing this type of ground heat exchanger, the temperature for absorbing or releasing heat is as important as the heat exchange.

Furthermore, the heat pump cycle significantly influences the heat exchange process for soil in winter. If the wet soil around the ground heat exchanger is frozen in winter, it will expand to a certain extent, increasing the heat transfer coefficient between the pipes and the soil. If the heat pump stops, the frozen soil will melt, leading to shifting of the soil and a possible increase in space between the soil and the pipes. As a result, the heat transfer coefficient will be reduced substantially due to the presence of air. To avoid these situations, it is recommended that the drilled holes be backfilled with sandy soil. In general, backfilling with sandy soil assists the operation of the ground heat exchanger in winter, while backfilling with clay is beneficial in summer.

As the only device in the heat exchange process between the air-conditioning system and the soil, the performance of the ground heat exchanger influences the energy efficiency of the entire system. The operating modes (continuous operation and intermittent operation) of the ground heat exchanger also affect the performance of this heat exchange process. For the same charged/discharged heat, one solution is to choose the intermittent operation mode with high heat flux and short operation duration, while another is to select the continuous operation mode with low heat flux. The heat transfer rate of the ground heat exchanger affects the temperature distribution of the surrounding soil and then influences the outlet water temperature of the buried pipes. Research has shown that the variance of soil temperature is significant if the intermittent operation mode is adopted due to low heat conductivity of the soil. This may have a negative influence on the heat transfer process. Thus, the intermittent operation mode with high heat flux should be avoided, and it is better to choose the continuous operation mode with low heat flux for the ground heat exchanger.

6.2 Producing Chilled Water Using the Evaporative Cooling Method

6.2.1 Elemental Ways to Produce Chilled Water Using the Evaporative Cooling Method

In areas in northwestern China such as Xinjiang province, the outdoor climate is dry in summer. Chilled water can be produced through evaporative cooling between the air and water and supplied as the high-temperature cooling source in THIC systems.

6.2.1.1 Direct Evaporative Cooler

Chilled water is produced through the heat and mass transfer processes between air and water in the direct evaporative cooler. Figure 6.7a, b shows the basic module of the direct evaporative cooler and the air handling process, respectively, in a psychrometric chart. The maximum temperature of the chilled water (in theory) is the wet-bulb temperature of the inlet air.

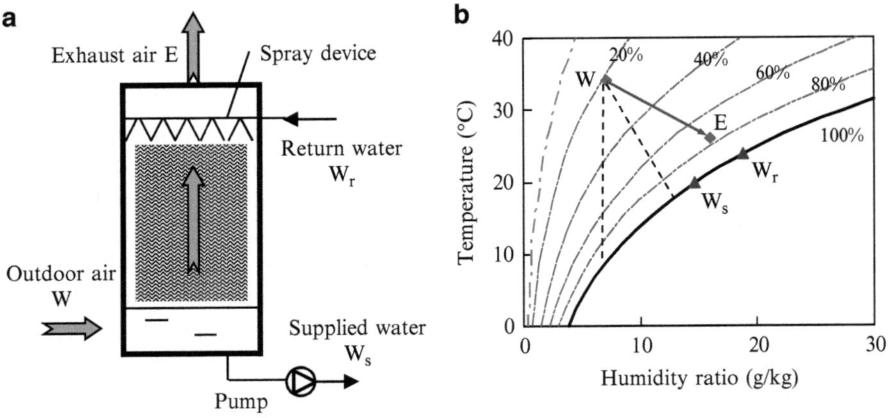

Fig. 6.7 Direct evaporative cooler: (**a**) schematic diagram and (**b**) air handling process in psychrometric chart

6.2.1.2 Indirect Evaporative Cooler

Figure 6.8a shows the schematic of an indirect evaporative cooler (Chinese patent by Jiang et al. 2002). For this innovative process, outdoor air is cooled down in a quasi-counter-flow air-water heat exchanger to a state approaching the saturation line, as shown in Fig. 6.8b; it then makes contact with the water to be cooled down through evaporative cooling. This flow pattern guarantees that the air inlet temperature for evaporative cooling is sufficiently low, helping to produce chilled water with a temperature as low as the outdoor air dew point temperature (in theory).

Figure 6.8b shows the air handling process of the indirect cooler in a psychrometric chart; W indicates the outdoor air, and E is the exhaust air. Outdoor air makes contact with chilled water at W_s in the quasi-counter-flow air-water heat exchanger, to be cooled from W to W'. Air at W' makes contact with water at W_r to realize direct evaporative cooling and reaches state E. Water at W_s is divided into two parts: one for regulating the indoor thermal environment of the THIC system and the other to cool down the outdoor air in the quasi-counter-flow air-water heat exchanger. For the air-water heat exchanger, the water outlet temperature approaches the dry-bulb temperature of outdoor air W due to a relatively large heat transfer area input. After mixing with water returning from the occupied zone, water with a state of W_r makes contact with air in the quasi-counter-flow tower to realize direct evaporative cooling. The core technique of this indirect evaporative cooler is the quasi-counter-flow heat and mass transfer process between the air and water, which can reduce irreversible loss and produce chilled water with a lower temperature than the direct evaporative cooler. In ideal conditions, the outlet chilled water temperature can be as low as the dew point temperature of the outdoor air.

The design of the indirect evaporative cooler has been presented in previous research (Jiang and Xie 2010). The core devices are the quasi-counter-flow air-water heat exchanger and the counter-flow air-water tower; the key technique is to make sure that the counter-flow transfer process takes place. Taking the air-water heat

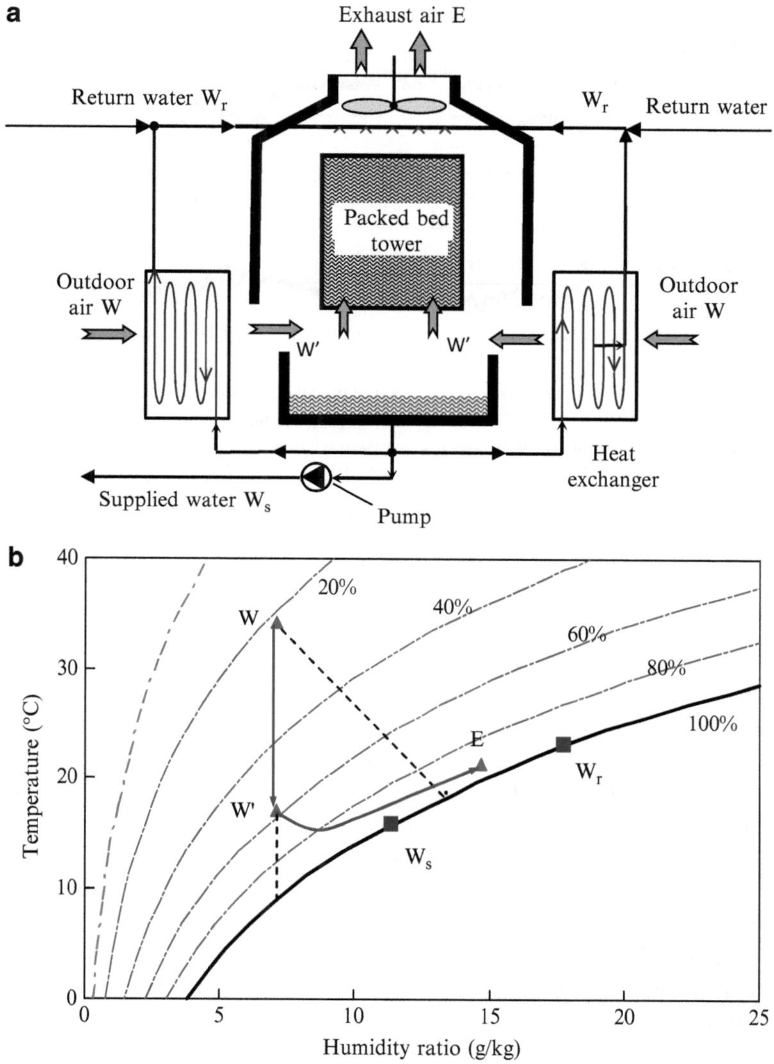

Fig. 6.8 Indirect evaporative cooler: (**a**) schematic diagram and (**b**) air handling process in psychrometric chart

exchanger as an example, in order to achieve the highest cooling capacity, the temperature difference between the air and the water in the cooling process has to be uniform. It is important to control the temperature rise of the water effectively after one row of cooling coils; the more the temperature rises, the less uniform it will be. If the designed outlet water temperature is 18 °C, the inlet air temperature is 33 °C, and the outlet air temperature is 21 °C, then the temperature difference between the inlet water and the outlet air is only 3 °C. Therefore, the temperature increase after one row of cooling coils cannot exceed 1 °C. In order to heat water from 18 to 30 °C, at least 12 rows of coils are needed.

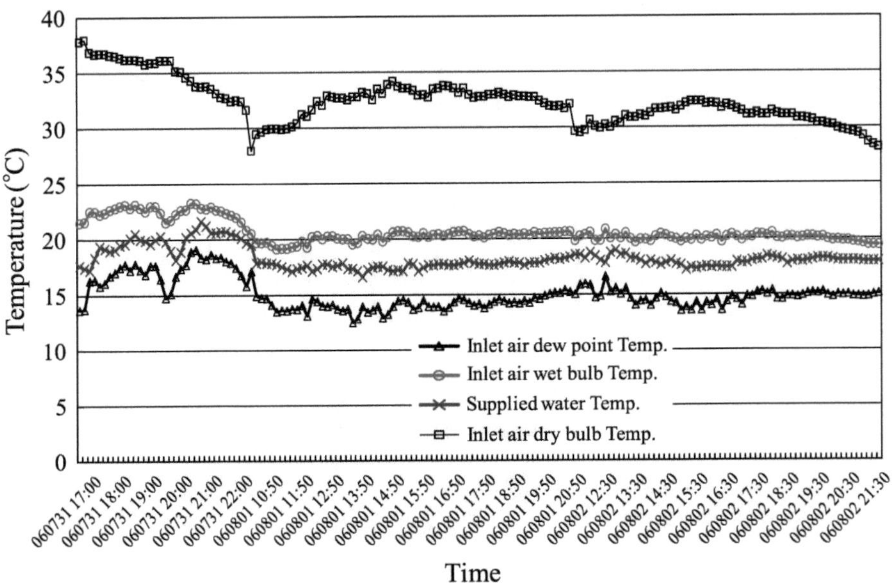

Fig. 6.9 Test results for the indirect evaporative cooler

The test results for this indirect evaporative cooler indicate that the outlet chilled water temperature is approximately equal to the average of the outdoor air dew point temperature and the wet-bulb temperature, as shown in Fig. 6.9. In the process of producing chilled water, the costs include the energy consumption of fans and pumps for air and water, respectively. In contrast to the conventional water chiller adopting a vapor compression refrigeration cycle, a compressor is no longer needed in the indirect evaporative cooler, and the *COP* (defined as the cooling capacity divided by the power consumption of fans and pumps) of the indirect evaporative cooler is much higher than that of the conventional water chiller. The test results show that, under the climate condition in Urumchi, the *COP* can be as high as 12–13; the drier the outdoor air, the colder the supplied chilled water, and the higher the *COP*.

6.2.2 Performance Analysis of the Evaporative Water Cooler

A direct evaporative cooler can produce water approaching the wet-bulb temperature of the inlet air, and an indirect evaporative cooler can produce water approaching the dew point temperature of the inlet air, as analyzed. For the different kinds of chilled water production methods using evaporative cooling, the outlet water temperature can be written as (Xie and Jiang 2010)

Table 6.5 Supplied water temperatures of direct and indirect evaporative coolers

	Outdoor design temp. in summer					
Location	Dry-bulb temp. (°C)	Wet-bulb temp. (°C)	Dew point temp. (°C)	Humidity ratio (g/kg)	Water temp. of direct evaporative cooler (°C)	Water temp. of indirect evaporative cooler (°C)
Altay	30.6	18.7	12.6	9.9	19.9	15.8
Karamay	34.9	19.1	9.4	8.2	20.7	14.1
Ili	32.2	21.4	15.7	12.9	22.5	18.6
Urumchi	34.1	18.5	7.5	8.5	20.1	12.6
Turpan	40.7	23.8	12.3	11.8	25.5	17.7
Kumul	35.8	20.2	11.3	9.9	21.8	15.8
Kashgar	33.7	19.9	13.4	11.4	21.3	16.9
Khotan	34.3	20.4	13.6	12.2	21.8	17.2

$$t_{\mathrm{w}} = t_{\mathrm{o}} - \eta_{\mathrm{tower}} \cdot \left\{t_{\mathrm{o}} - \left[t_{\mathrm{wb,o}} - \eta_{\mathrm{l}} \cdot \left(t_{\mathrm{wb,o}} - t_{\mathrm{dp,o}}\right)\right]\right\} \tag{6.2}$$

where t_{o}, $t_{\mathrm{wb,o}}$, and $t_{\mathrm{dp,o}}$ are the outdoor dry-bulb temperature, wet-bulb temperature, and dew point temperature, respectively; η_{tower} is the efficiency of the spray tower used for evaporative cooling; and η_{l} is the efficiency of the heat exchanger used for air precooling in the indirect evaporative cooling process. η_{tower} and η_{l} can be written as

$$\eta_{\mathrm{tower}} = (t_{\mathrm{o}} - t_{\mathrm{w}})/(t_{\mathrm{o}} - t_{\mathrm{wb,ain}}) \tag{6.3}$$

$$\eta_{\mathrm{l}} = (t_{\mathrm{o}} - t_{\mathrm{A}})/\left(t_{\mathrm{o}} - t_{\mathrm{dp,o}}\right) \tag{6.4}$$

where $t_{\mathrm{wb,ain}}$ is the inlet air wet-bulb temperature for the direct evaporative cooler and t_{A} is the outlet air temperature of the air-water heat exchanger used in the indirect evaporative cooler.

For the direct and indirect evaporative coolers, the efficiency values in Eq. (6.2) are as follows:

- Direct evaporative cooler: $\eta_{\mathrm{l}} = 0, 0 < \eta_{\mathrm{tower}} < 1$
- Indirect evaporative cooler: $0 < \eta_{\mathrm{l}} < 1, 0 < \eta_{\mathrm{tower}} < 1$

For the indirect evaporative cooler shown in Fig. 6.8, η_{l} is 70–80 %, and η_{tower} can be as high as 90 %. Table 6.5 shows the outlet temperature of the cold water produced in different cities in northwestern China, in which η_{l} is 75 % and η_{tower} is 90 %.

The temperature of the chilled water produced with different evaporative cooling methods can be calculated using Eq. (6.2). Taking outdoor climate parameters into account, the appropriate zones for these two evaporative cooling methods are shown in Fig. 6.10, where an outlet water temperature of 18 °C is chosen as the dividing line. It can be seen that the indirect evaporative cooling method is applicable in more areas compared to the direct evaporative cooling method. For the locations where both methods can both be used, the main focus is not the temperature of the outlet chilled water, but the economics of the whole system (including equipment investment and running cost) based on satisfying indoor temperature control.

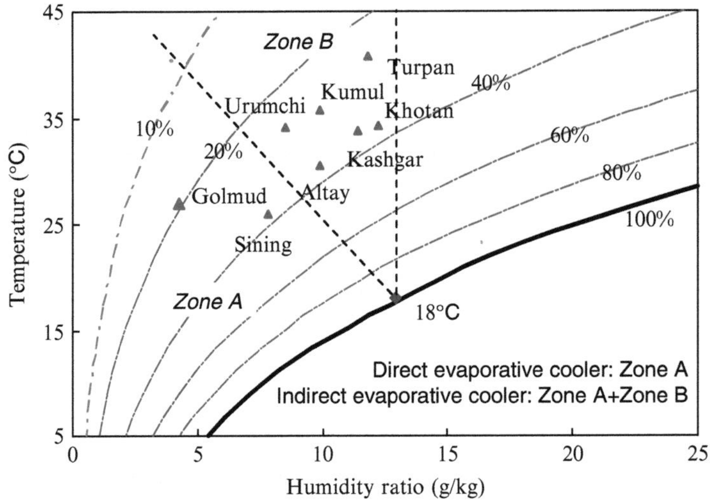

Fig. 6.10 Climate zones suitable for indirect and direct evaporative cooling methods (Xie and Jiang 2010)

For both indirect and direct evaporative cooling methods used to produce chilled water, the components that consume power are the water pumps and fans. Water pumps are used for water circulation, and if the cooling capacity and water temperature difference are identical for both methods, then the power consumed by the pumps could be regarded as the same for the two methods. For the indirect evaporative cooler, due to the quasi-counter-flow air-water heat exchanger, the fan power consumption is higher than that of the direct evaporative cooler. Therefore, when the outdoor air is sufficiently dry and the water produced by the direct evaporative cooler can meet the temperature control demand, the direct evaporative cooler is preferable.

6.3 Vapor Compression Cooling Sources

In the temperature control subsystem, since the required temperature of the cooling source in the THIC system is about 16–18 °C (compared to 7 °C in conventional systems), natural cooling sources can be utilized directly as high-temperature cooling sources. If no natural cooling sources are available, mechanical chillers are needed to satisfy the requirement for the high-temperature chilled water or refrigerant to buildings. The cooling sources that produce chilled water or refrigerant with a temperature of 16–18 °C in THIC systems are referred to as high-temperature chillers in the following analysis in order to distinguish them from the chillers adopted in conventional systems. There is a significant increase in the evaporating temperature of high-temperature chillers in the THIC system compared to conventional chillers, and theoretically, the *COP* of these high-temperature chillers is much higher than that of conventional chillers.

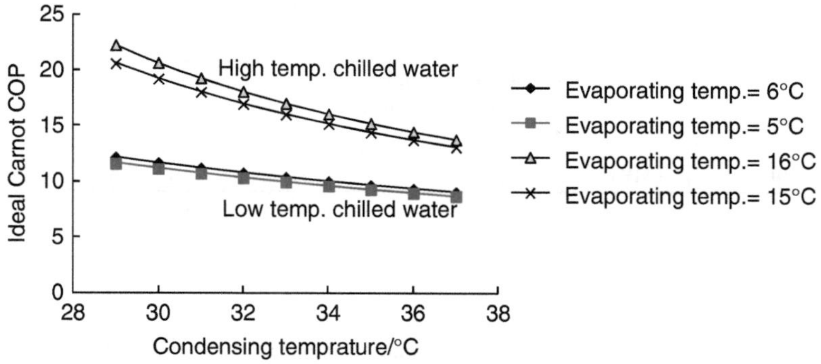

Fig. 6.11 *COP* of the reverse Carnot cycle with different evaporating and condensing temperatures

6.3.1 Main Features of High-Temperature Water Chillers

6.3.1.1 Analysis of High-Temperature Water Chillers from the View of the Ideal Carnot Cycle

For refrigeration cycles with the same evaporating temperature and condensing temperature, the reverse Carnot cycle has the highest energy efficiency, which is the theoretical limitation of actual refrigeration equipment (Yan et al. 2010). Figure 6.11 shows the *COP* of the reverse Carnot cycle with different evaporating temperatures and condensing temperatures based on the second law of thermodynamics. It is obvious that under the same condensing temperature, the *COP* of the reverse Carnot cycle increases significantly with the increase of evaporating temperature. In the THIC system, chillers are required to produce high-temperature chilled water (about 16–18 °C) to satisfy the temperature control demand, which is much higher than that of chilled water in conventional chillers (usually 7 °C). Thus, the high-temperature water chiller could have a higher evaporating temperature and a higher *COP* in theory. For example, if the condensing temperature is 37 °C, the *COP* of the reverse Carnot cycle for the conventional low-temperature water chiller is 9, while the *COP* of the high-temperature water chiller is 13, meaning that the high-temperature water chiller is much more energy efficient than the conventional chiller.

6.3.1.2 Compression Ratio of the High-Temperature Water Chiller

Compared to the conventional water chiller producing chilled water with a temperature of about 7 °C, the high-temperature water chiller in the THIC system operates at a relatively small compression ratio. Table 6.6 lists the compression ratios (ratio of condensing pressure p_k to evaporating pressure p_0) of the conventional low-temperature water chiller and the high-temperature water chiller. For conventional

Table 6.6 Compression ratios of conventional low-temperature water chillers and high-temperature water chillers under design conditions

		Conventional low-temperature water chiller		High-temperature water chiller	
Refrigerant	p_k (MPa)	p_0 (MPa)	p_k/p_0	p_0 (MPa)	p_k/p_0
R22	1.39–1.53	0.55–0.58	2.3–2.8	0.77–0.81	1.7–2.0
R134a	0.91–1.02	0.32–0.35	2.6–3.1	0.47–0.51	1.8–2.2

Table 6.7 Compression ratios of conventional water chillers and high-temperature chillers under different working conditions

		Conventional low-temperature water chiller		High-temperature water chiller	
Refrigerant	p_k (MPa)	p_k/p_0	Compression ratio variation	p_k/p_0	Compression ratio variation
R22	1.19–1.53	2.0–2.8	27 %	1.5–2.0	27 %
R134a	0.77–1.02	2.2–3.1	29 %	1.5–2.2	29 %

water chillers, the compression ratio is 2.3–2.8 using R22 as the refrigerant and 2.6–3.1 using R134a; evaporating temperature t_0 is 3–5 °C, and condensing temperature t_k is 36–40 °C. Taking t_0 as 14–16 °C and t_k as 36–40 °C in the design of the high-temperature water chillers, the compression ratios of R22 and R134a chillers are reduced to about 1.7–2.0 and 1.8–2.2, respectively.

When the chiller operates in the partial-load condition, there is little change in the evaporating temperature of the vapor compression cycle, but the condensing temperature varies significantly with outdoor climatic parameters. A lower ambient temperature (or wet-bulb temperature using the cooling tower) leads to a lower condensing temperature. As a result, the compression ratio deviates from the rated design condition. For conventional water chillers, if evaporating temperature t_0 is 3–5 °C and condensing temperature t_k is 30–40 °C, the compression ratios of R22 and R134a vapor compression water chillers are 2.0–2.8 and 2.2–3.1, respectively. For high-temperature water chillers, if evaporating temperature t_0 is 14–16 °C and condensing temperature t_k is 30–40 °C, the compression ratios of R22 and R134a vapor compression water chillers are 1.5–2.0 and 1.5–2.2, respectively. Table 6.7 shows the compression ratios of both conventional and high-temperature chillers. It can be seen that the compression ratio of the high-temperature chillers decreases by about 30 % compared to the conventional chillers when operating at partial load.

From the above analysis, it is clear that the high-temperature water chiller has a lower compression ratio compared to the conventional chiller, which is beneficial for improving the chiller's *COP*. As the decrease of the compression ratio introduces new requirements for equipment manufacturers and the system operation of chillers, relevant components need to be redesigned and re-optimized.

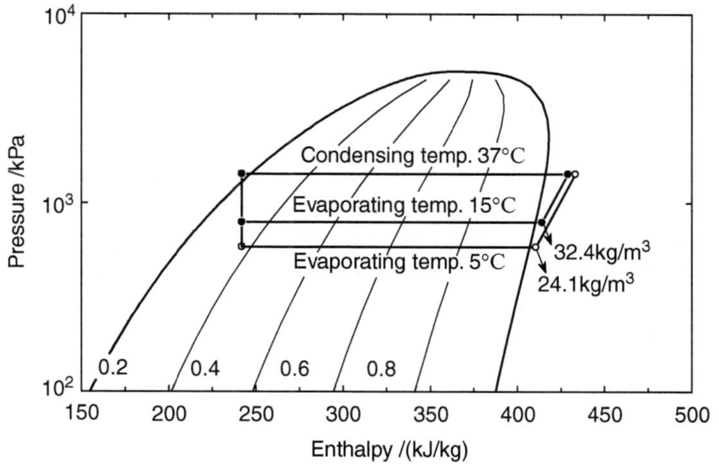

Fig. 6.12 Refrigeration cycle (R22) in pressure-enthalpy chart

6.3.1.3 Can Conventional Chillers Be Used as High-Temperature Chillers?

From the refrigeration efficiency analysis of the reverse Carnot cycle, we can see that high-temperature chilled water can raise the evaporating temperature, which can substantially improve the efficiency of the refrigeration system. But can conventional chillers be used directly as high-temperature chillers? In order to answer this question, the influence of improving evaporating temperature on the entire refrigeration system is investigated in this subsection. Figure 6.12 presents the refrigeration cycles of producing high-temperature water and conventional chilled water in a pressure-enthalpy diagram, in which R22 is taken as an example.

Compared with the conventional chiller, the high-temperature water chiller has a smaller compression ratio. When the condensing temperature is 37 °C, the compression ratio of the chiller (using R22 as refrigerant) is 2.4 and 1.8 when the evaporating temperature is 5 °C and 15 °C, respectively. For screw compressors or scroll compressors that have fixed volume ratios, there will be significant overcompression loss if a chiller designed for producing conventional chilled water is used directly to produce high-temperature chilled water.

In the case of the conventional low-temperature refrigeration cycle, as the evaporating temperature is 5 °C (5 °C superheat), the suction gas density is 24.1 kg/m^3. In the case of the high-temperature refrigeration cycle, as the evaporating temperature is 15 °C (5 °C superheat), the suction gas density is 32.4 kg/m^3, which is 34 % higher than that of the low-temperature refrigeration condition. Consequently, the refrigerant flow of the fixed suction volume compressor increases by about 34 %, and the system capacity increases proportionally as well.

A higher system cooling capacity requires a greater heat transfer ability of the evaporator and condenser. However, an overly high system capacity leads to overload of the compressor motor, which affects the chiller's normal operation. Decreasing the system pressure difference and increasing refrigerant flow of the compressor require a considerably larger expansion valve opening. Under these circumstances, if a conventional low-temperature cooling system is used to produce high-temperature chilled water directly, an inappropriate expansion device will result in more superheat, and system performance will not be fully achieved. Furthermore, the loss of pressure ratio between the condenser and the evaporator will affect the oil return of the cycle. In summary, for the sake of safe operation, the outlet chilled water temperature of conventional low-temperature water chillers is generally limited to a maximum of 12–14 °C. In other words, conventional low-temperature chillers can rarely be used directly as high-temperature chillers.

6.3.1.4 Key Issues for the Design and Development of High-Temperature Water Chillers

Because of the increase of evaporating pressure, the compression ratio of high-temperature water chillers is much lower than that of conventional low-temperature water chillers, which results in new requirements for compressors and other components. Analyses of piston, scroll, screw, and centrifugal compressors for high-temperature chillers are summarized below.

- *Fixed Volume Ratio Compressors (Scroll, Screw, etc.)*

  For scroll, screw, and other fixed volume ratio compressors, overcompression or insufficient compression processes exist when the external compression ratio is not equal to the internal compression ratio. For such compressors, increasing the evaporating temperature inevitably leads to an increase in overcompression loss and a reduction of compressor efficiency. If these kinds of compressors are used under high-temperature conditions, the actual *COP* values will be far lower than the theoretical values.

- *"Self-Adaptive" Compressors (Piston, Centrifugal, etc.)*

  The suction and discharge pressures of the piston compressor are equal to the evaporating and condensing pressures, respectively. As a consequence, there is no overcompression loss. The centrifugal compressor is another type of "self-adaptive" compressor since there is no internal compression process in the compressor. Among these "self-adaptive" compressors, as the internal efficiency of piston compressors is relatively lower, centrifugal compressors are more suitable for high-temperature water chillers. To achieve efficient operation with a small compression ratio, the centrifugal compressor should adjust the inlet guide vanes and the rotation speed.

Since the refrigerant pressure difference between the inlet and the outlet of the throttle device is remarkably reduced, the throttle device of the conventional water chiller requires some regulations to satisfy the small compression ratio condition in the high-temperature water chiller.

According to the above analyses, the basic design principles of high-temperature water chillers compared to conventional water chillers can be summarized as follows:

- *Compressor*: a smaller compression ratio (centrifugal and piston compressors or specially designed compressors with fixed volume ratios) and a higher rated motor power
- *Throttle device*: greater capacity and superior regulation performance under a lower compression ratio and a lower working pressure difference
- *Evaporator and condenser*: a greater capacity and higher heat transfer efficiency (by improving the heat transfer coefficient or area to strengthen the heat transfer ability)
- *Oil return system*: working properly under the small compression ratio condition

6.3.2 Development Cases of the High-Temperature Water Chiller

6.3.2.1 Development Case I: Centrifugal Refrigerant Cycle (GREE)

Optimal Design of the Chiller

The high-temperature centrifugal water chiller developed by GREE (Gree Electric Appliances Inc.), shown in Fig. 6.13, is adopted as an example in this subsection to illustrate the development process of the high-temperature centrifugal chiller. The following data in this subsection is from literature by Zhang et al. (2011a, b). With an increase of the evaporating temperature in the high-temperature water chiller, some significant changes occur in the centrifugal compressor. Table 6.8 lists the compressor design parameters of a 4,000-kW centrifugal chiller when producing 7 and 18 °C chilled water, using R134a as the refrigerant. Compared to the 7 °C condition, the compression ratio of the 18 °C condition decreases to 72 %, while the inlet volume flow rate decreases to 67 %. If a conventional 7 °C water chiller operates at a high evaporating temperature, the working condition will deviate significantly from the design condition because the system flow rate, the compression ratio, and the gas suction state all change considerably. Since the conventional 7 °C water chiller operating at a high evaporating temperature reduces system reliability without significantly improving performance, a high-temperature centrifugal water chiller must be specially designed.

The refrigerating cycle of a high-temperature centrifugal water chiller developed by GREE is shown in Fig. 6.13, the core technology of which is the high-temperature centrifugal compressor. The main features of the water chiller include the following: (1) the compressor effectively combines a variable geometry diffuser and adjustable inlet guide vanes to ensure the efficient and stable operation of the compressor at partial load (Zhang et al. 2011b); (2) a special motor and a lubricating

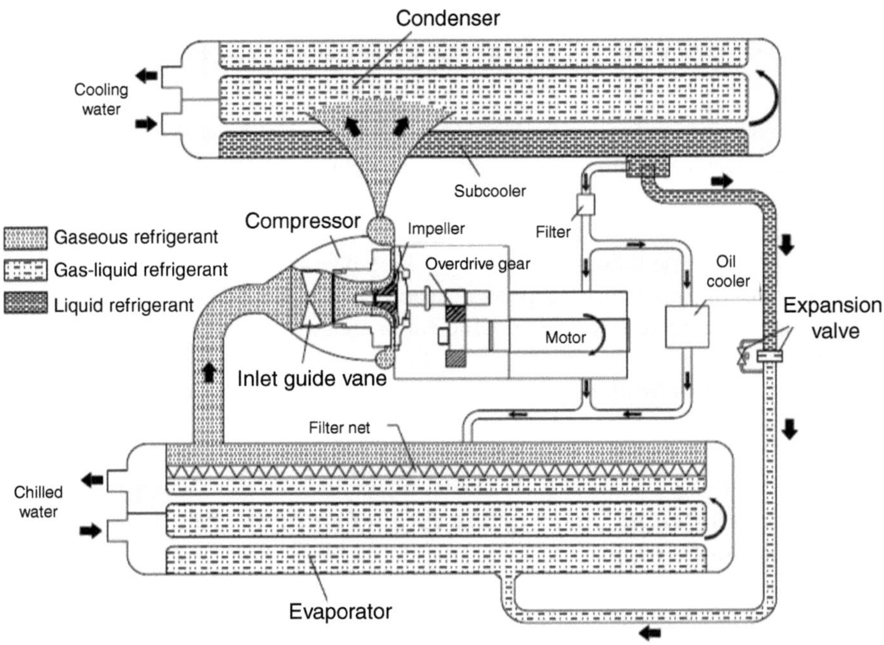

Fig. 6.13 Refrigeration cycle of a centrifugal high-temperature water chiller developed by GREE

Table 6.8 Design parameters of the compressor under different working conditions

Condition	Suction temperature (°C)	Suction pressure (kPa)	Compression ratio	Volume flow rate (m^3/s)
7 °C outlet water	5.0	345	2.7	1.5
18 °C outlet water	16.5	485	1.95	1.0

oil cooling mode ensure that the motor and the lubricating oil are fully cooled and that the machine operates reliably under the small compression ratio condition; (3) the oil return process is designed to be isolated from the influence of the gas suction flow rate and the pressure difference, which ensures the reliability of the oil return under the condition of high evaporation temperature and low compression ratio (Liu et al. 2010); (4) when the chiller operates in surge mode, the control system can make efficient adjustments to keep the chiller away from the surge zone and avoid the occurrence of the surge; (5) a subcooler is introduced into the condenser to improve the subcooling temperature by 3–5 °C, helping the chiller run more efficiently; and (6) a double gearbox design is adopted to reduce the gear noise; the noise of the chiller is only 80–86 db.

Table 6.9 Test results under the 16 °C chilled water condition

Item	Parameter	Unit	Load ratio 100 %	75 %	50 %	25 %
Test conditions	Chilled water outlet temperature	°C	16	16	16	16
	Chilled water flow rate	m^3/h	688	688	688	688
	Cooling water inlet temperature	°C	30	26	23	19
	Cooling water flow rate	m^3/h	860	860	860	860
Test results	Cooling capacity	kW	3,826	2,953	2,034	1,123
	COP	W/W	8.58	10.10	9.52	6.88
	IPLV	W/W	9.47			

Table 6.10 Test results under the 18 °C chilled water condition

Item	Parameter	Unit	Load ratio 100 %	75 %	50 %	25 %
Test conditions	Chilled water outlet temperature	°C	18	18	18	18
	Chilled water flow rate	m^3/h	688	688	688	688
	Cooling water inlet temperature	°C	30	26	23	19
	Cooling water flow rate	m^3/h	860	860	860	860
Test results	Cooling capacity	kW	3,966	3,046	1,978	1,322
	COP	W/W	9.18	10.10	9.80	8.41
	IPLV	W/W	9.77			

Tested Performance

It can be seen that the performance of the high-temperature centrifugal water chiller is improved by optimizing the structure and the working parameters. The performance of a GREE high-temperature centrifugal chiller with a rated cooling capacity of 4,000 kW was tested when producing 16 and 18 °C high-temperature chilled water. According to the test results listed in Tables 6.9 and 6.10, the *COP* of the chiller at 100 % full capacity for producing 18 °C chilled water is 9.18, while the *COP* at full capacity for producing 16 °C chilled water is 8.58.

Performance Comparison

If the conventional 7 °C centrifugal water chiller is used directly to produce high-temperature chilled water, the performance is significantly different compared to the performance-optimized high-temperature water chiller. In order to clarify the performance discrepancy between the two kinds of chillers, a conventional GREE 7 °C chiller with a rated cooling capacity of 4,000 kW was tested when producing 10, 16, and 18 °C chilled water. The results are shown in Table 6.11.

Table 6.11 Test results for a conventional chiller with different chilled water temperatures

Item	Parameter	Unit	Standard condition	10 °C chilled water	16 °C chilled water	18 °C chilled water
Test conditions	Chilled water outlet temperature	°C	7	10	16	18
	Chilled water flow rate	m^3/h	688	688	688	688
	Cooling water inlet temperature	°C	30	30	30	30
	Cooling water flow rate	m^3/h	860	860	860	860
Test results	Cooling capacity	kW	3,963	4,065	4,088	4,095
	Power consumption	kW	683	601.2	601.2	580.8
	COP	W/W	5.78	6.25	6.80	7.05
	η_{ad}	/	0.81	0.79	0.70	0.65
	Compression ratio	/	2.68	2.37	2.00	1.95

As shown in Table 6.11, if the conventional water chiller is used directly to produce high-temperature chilled water, its energy performance is significantly lower than that of the specially designed high-temperature chiller. The *COP* values for the conventional chiller producing 16 and 18 °C chilled water are 6.80 and 7.05, respectively. Thanks to the improvements and optimizations mentioned previously, the performance of the newly designed centrifugal high-temperature water chiller is significantly improved. Compared to the conventional chiller producing high-temperature chilled water, the energy performance of the modified high-temperature chiller producing 16 °C and 18 °C chilled water improved by 26 % and 30 %, respectively.

6.3.2.2 Development Case II: Screw Water Chiller (DunAn)

Optimal Design of the Chiller

For screw compressors with fixed volume ratios, rotary compressors with relatively small internal volumes are required to reduce overcompression loss when producing high-temperature chilled water. Figure 6.14 shows a schematic of the high-temperature screw chiller developed by Zhejiang DunAn Environment (Zhejiang province, China). The following data in this subsection is from literature by Wang et al. (2011). The system works as follows: (1) the refrigerant gas is compressed to a state of high temperature and high pressure in the compressor; (2) the refrigerant gas coming out of the compressor flows into the secondary oil separator so the refrigerant oil can be separated; (3) the refrigerant gas flows into the condenser, is cooled by the cooling water, and condenses to a mid-temperature and high-pressure liquid; (4) after being filtered and dried, the refrigerant liquid flows through the electronic expansion valve and is throttled into a low-temperature and low-pressure two-phase fluid; (5) the fluid enters the evaporator and evaporates by absorbing heat from the chilled water; and (6) the low-temperature and low-pressure gas coming from the evaporator finally flows back to the compressor to finish the refrigeration cycle.

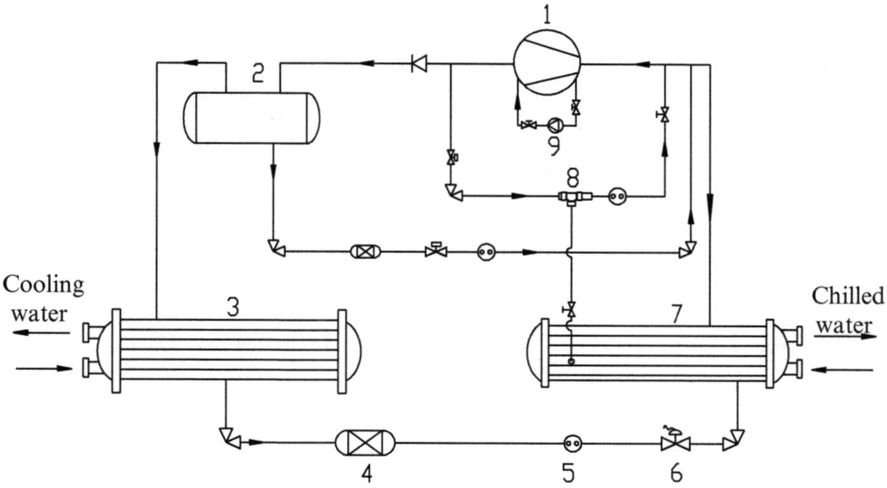

Fig. 6.14 Refrigeration cycle of the DunAn Environment high-temperature screw water chiller (*1* Compressor, *2* Oil separator, *3* Condenser, *4* Drier-filter, *5* Sight glass, *6* Electronic expansion valve, *7* Evaporator, *8* Ejector, *9* Oil pump)

Since the pressure difference of the refrigerant in the high-temperature chiller is lower and the solubility of the lubricating oil in the R134a refrigerant is better when operating at a high evaporating temperature, the oil is difficult to return. To solve this problem, the following modifications are employed: (1) the chiller is designed with a three-stage separator combining a horizontal mechanical separator and an adsorption separator to ensure efficient separation of the lubricating oil from the refrigerant mixture; (2) the ejector and concentrated oil return system are designed to ensure the extended operating duration period of the compressor regardless of the shortage of lubricating oil; and (3) the oil lines are specially balanced to ensure that the oil levels in each compressor are the same, which is beneficial to the safety and reliability of the chiller.

Tested Performance

Table 6.12 lists the tested performance of a typical screw chiller with different inlet and outlet temperatures of the chilled water and cooling water. The cooling capacity of the chiller when producing 7 °C conventional chilled water (Case 1) is 429 kW, and the *COP* is 5.08. The cooling capacity is 606 kW when producing 16 °C high-temperature chilled water (Case 4), and the *COP* is 6.81. Compared to Case 1, the cooling capacity of Case 4 increases by 41 %, and the *COP* increases by 34 %.

When producing 16 °C chilled water, the cooling water parameters of Case 4, Case 6, and Case 7 in Table 6.12 represent (or approach) the 100 % load, 75 % load, and 50 % load conditions, respectively, with tested *COP* values of 6.81, 8.19, and 8.64, respectively.

Table 6.12 Performance of the water chiller with different chilled water outlet temperatures

Parameter		Case 1	Case 2	Case 3	Case 4	Case 5	Case 6	Case 7
Chilled water	Flow rate (m^3/h)	101.1	103.2	105.63	103.1	102.6	104.6	105
	Outlet temperature (°C)	7.02	10.01	13.96	15.97	17.86	15.77	15.5
	Inlet temperature (°C)	10.66	14.09	18.57	21.03	23.28	19.35	18.6
Cooling water	Flow rate (m^3/h)	120.9	120.5	120.9	120	121.78	119.9	122
	Inlet temperature (°C)	29.81	30.0	30	29.84	29.95	25.87	23.4
	Outlet temperature (°C)	33.46	34.11	34.6	34.73	35.07	29.31	26.3
Electricity consumption (kW)		84.45	85.96	87.6	89.01	90.68	53.12	43.6
Cooling capacity (kW)		429.44	490	565.6	605.79	645.13	435.2	377
COP		5.08	5.70	6.46	6.81	7.11	8.19	8.64

6.3.2.3 Development Case III: Centrifugal Refrigerant Chiller Developed by Mitsubishi Heavy Industries, Ltd.

Optimal Design of the Chiller

Figure 6.15 shows a schematic of a micro high-temperature centrifugal water chiller developed by Mitsubishi Heavy Industries (MHI). The following data in this subsection is from material by MHI (Mitsubishi Heavy Industries LTD. 2004, 2005). The cooling capacity is 160 kW, and a "two-stage compression and economizer" vapor compression refrigeration cycle is employed. The centrifugal compressor is a high-speed compressor, and its high rotation speed and strict machining accuracy make it difficult to miniaturize the centrifugal compressor. In fact, the impeller diameters of existing compressors are no smaller than 200–250 mm. Figure 6.16 displays the internal structure of the MHI micro centrifugal compressor, the core technologies of which are the optimized impeller and bearings, as well as its outstanding machining accuracy.

The micro centrifugal high-temperature chiller adopts a flooded evaporator with high-efficiency heat transfer tubes with porous structures and a shell-and-tube condenser with acicular fins to enhance the condensation heat transfer, as seen in Fig. 6.17.

The high-pressure and low-pressure throttling devices are all electronic expansion valves that can accurately control the refrigerant flow rate. In addition, to prevent a reduction in heat transfer performance caused by accumulated refrigeration oil in the evaporator, the liquid refrigerant with high oil content in the evaporator is removed directly by the ejector, which is driven by the high-pressure gaseous refrigerant in the condenser. The refrigerant oil is smoothly returned to the compressor, and the oil in the fuel tank is cooled to ensure effective lubrication of the compressor and the gearbox. The control of the chiller at partial load is achieved by detecting the inlet and outlet temperatures and the flow rate of the chilled water, the temperature of the cooling water, and other system parameters. With these parameters, the motor operating frequency, inlet guide vane angle, and hot gas bypass valve opening are

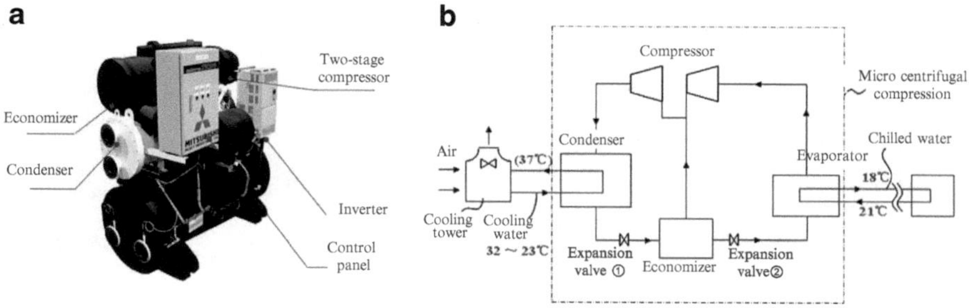

Fig. 6.15 MTWC175 micro centrifugal high-temperature water chiller: (**a**) external view and (**b**) schematic of the refrigeration cycle

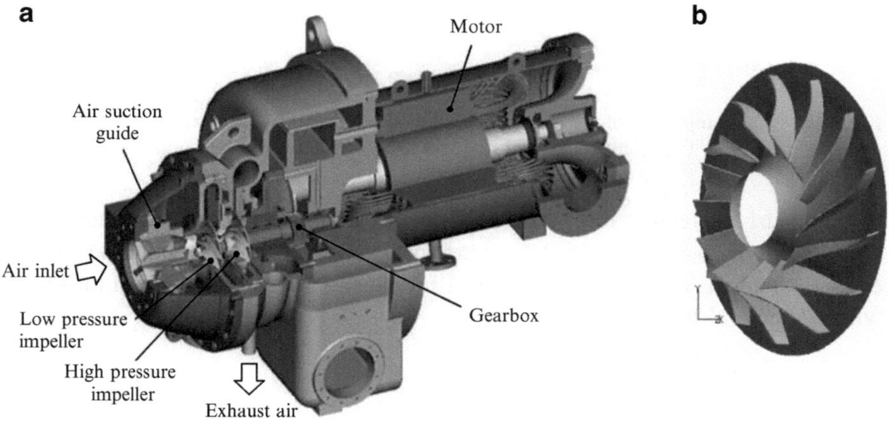

Fig. 6.16 Internal structure of the micro centrifugal compressor: (**a**) micro centrifugal compressor and (**b**) impeller of the compressor

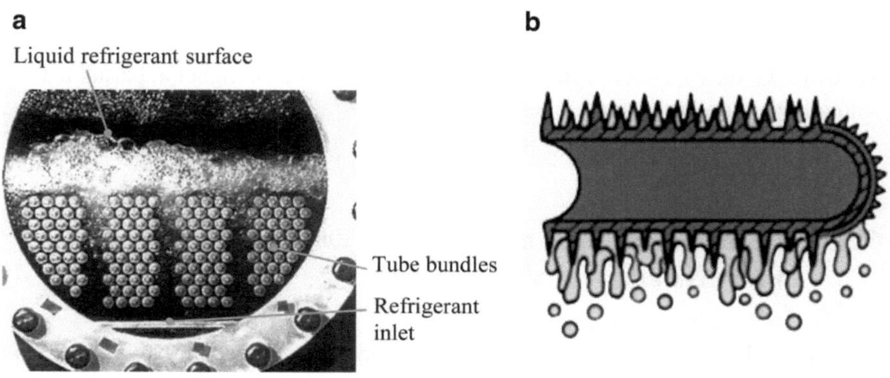

Fig. 6.17 Internal structures of the evaporator and the condenser: (**a**) internal structure of the evaporator and (**b**) heat transfer tube structure of the condenser

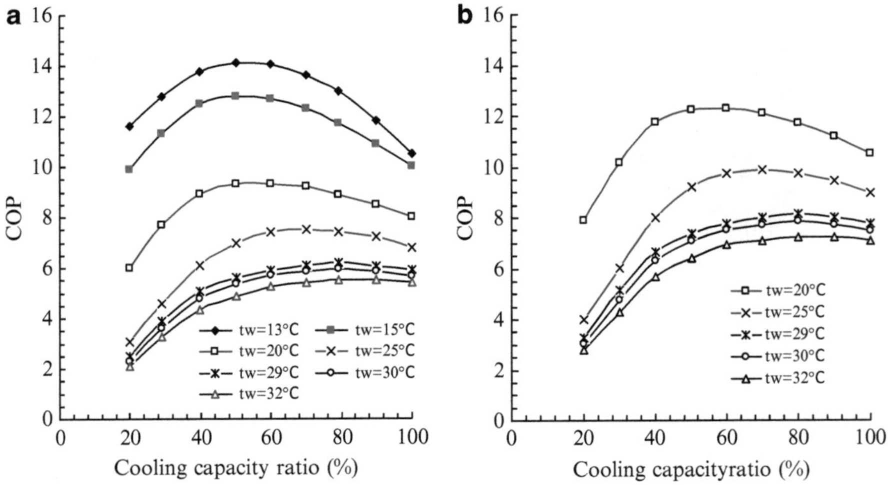

Fig. 6.18 *COP* comparison of the high-temperature water chiller (t_w represents the cooling water temperature): (**a**) 7 °C chilled water condition and (**b**) 18 °C chilled water condition

comprehensively regulated to achieve continuous adjustment of the water chiller with a load range of 20–100 %. The electronic bypass valve between the condenser and the evaporator adopts anti-surge control to ensure safe and stable operation of the chiller with high partial-load operation efficiency.

Tested Performance

As shown in Fig. 6.18a, the rated *COP* of the micro high-temperature centrifugal water chiller producing 7 °C chilled water is 5.4 (cooling water temperature t_w = 32 °C). Figure 6.18b shows the chiller's performance when producing high-temperature chilled water. As shown in Fig. 6.18b, when the chilled water inlet/outlet temperatures are 21 °C/18 °C and the cooling water inlet/outlet temperatures are 37 °C/32 °C, the *COP* of the chiller is 7.1. The *COP* could be higher at partial load or with a lower cooling water temperature.

6.3.3 Development Case of the High-Temperature VRF System

VRF (variable refrigerant flow) air-conditioning systems are becoming more popular in certain small-scale buildings and commercial buildings due to their flexible layout, as well as their ability to utilize refrigerant as the cooling/heating medium directly, which avoids the need to rely on chilled water circulation. In conventional systems, the indoor units of VRF systems have to handle the sensible load and the moisture load at the same time in summer. Thus, the evaporating temperature is constrained by the indoor dew point temperature, which limits the energy performance of the VRF systems. In THIC air-conditioning systems, the indoor moisture load is handled by the humidity control subsystem to regulate indoor humidity, while a temperature control subsystem is responsible for regulating indoor temperature. If a VRF system is used only for indoor temperature control, the system performance could be greatly increased with an increase in the evaporating temperature. In this subsection, the VRF system used to handle the sensible heat load is called the high evaporating temperature VRF (HTVRF) system.

In the HTVRF system, the heat transfer process occurs in the cooling coils of indoor units, and refrigerant is vaporized to cool down the indoor air. The HTVRF system operates with a smaller compression ratio compared to conventional VRF systems. Taking R410A as an example, if the evaporating temperature of a conventional VRF system is 6–8 °C and the condensing temperature is 44–46 °C, then the compression ratio is around 2.6–2.9; if the condensing temperature remains the same and the evaporating temperature increases to 15–17 °C, then the compression ratio of the HTVRF system is reduced to around 2.0–2.2. Thus, the main issues for developing HTVRF systems can be summarized as follows:

- Under a smaller compression ratio, guaranteeing effective and stable operation of the compressor is very important. The structures of compressors should adapt to the operating conditions with a small compression ratio, and overcompression should be avoided. Expansion valves should maintain adequate adjusting ability at the same time. Oil return should be effective, and optimization of the control strategy should be performed.
- As the evaporating temperature increases, the temperature difference between the refrigerant of indoor units and indoor air decreases. Therefore, in order to handle the same sensible load, the air flow rate should be increased; corresponding problems are analyzed in Sect. 3.2 in Chap. 3 (for developing dry FCUs).

A newly developed HTVRF system using R410A as the refrigerant was tested. One outdoor unit is connected with nine indoor units with refrigerant pipes. The system schematic of this HTVRF system is shown in Fig. 6.19, and pictures of the outdoor unit and indoor unit are shown in Fig. 6.20a, b, respectively.

Tables 6.13 and 6.14 list the tested evaporating temperature and cooling capacity of the indoor unit under two typical working conditions:

- *Typical working condition 1*: Outdoor temperature is 27.3 °C, condensing pressure is 2.04 MPa, and condensing temperature is 33.2 °C. Evaporating pressure is 1.29 MPa, and evaporating temperature is around 16 °C. Total cooling capacity of the nine indoor units is 15.0 kW, compressor power input is 2.04 kW, and the *COP* (cooling capacity divided by compressor power input) is as high as 7.4.
- *Typical working condition 2*: Outdoor temperature is 33.2 °C, condensing pressure is 2.29 MPa, and condensing temperature is 37.8 °C. Evaporating pressure is 1.25 MPa, and evaporating temperature is around 15 °C. Total

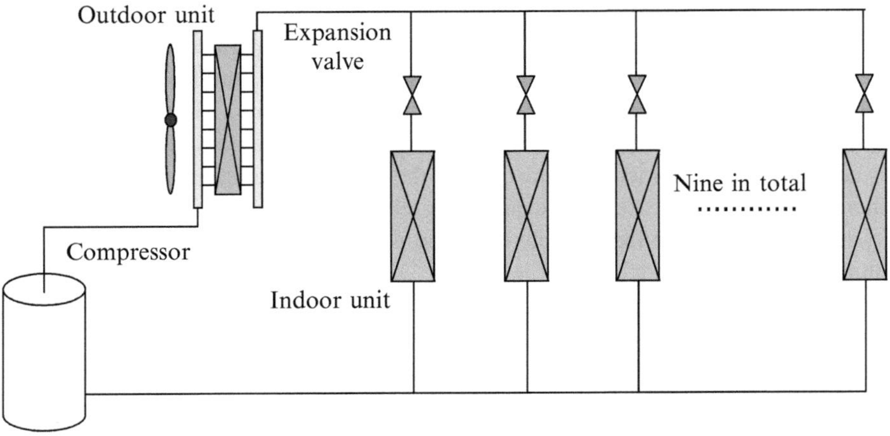

Fig. 6.19 Schematic of the HTVRF system

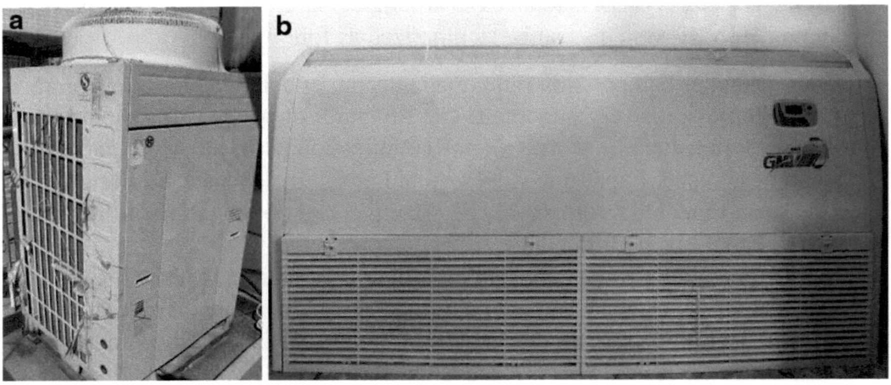

Fig. 6.20 Pictures of the HTVRF system: (**a**) outdoor unit and (**b**) indoor unit

Table 6.13 Experimental results of the HTVRF system (typical working condition 1)

Indoor unit no.	1	2	3	4	5	6	7	8	9
Return air temperature (°C)	25.1	27.9	24.1	26.8	26.1	24.9	25.8	25.4	25.4
Evaporating pressure (MPa)	1.29	1.28	1.29	1.30	1.29	1.28	1.29	1.28	1.29
Cooling capacity (kW)	1.19	3.13	0.81	1.72	1.30	1.41	2.10	2.10	1.26

Table 6.14 Experimental results of the HTVRF system (typical working condition 2)

Indoor unit No.	1	2	3	4	5	6	7	8	9
Return air temperature (°C)	24.7	27.3	23.5	25.2	24.8	24.6	25.8	25.9	26.5
Evaporating pressure (MPa)	1.22	1.24	1.25	1.25	1.27	1.27	1.26	1.24	1.25
Cooling capacity (kW)	1.16	2.67	0.91	1.88	3.17	1.63	2.07	2.15	2.20

cooling capacity of the nine indoor units is 17.8 kW, compressor power input is 2.96 kW, and the *COP* is about 6.0.

Outdoor temperature was relatively low during testing, and therefore, the condensing pressure and temperature were relatively low. The compression ratios of the refrigeration cycle are 1.6 and 1.8 for typical working conditions 1 and 2, respectively. Although the HTVRF system's *COP* was much higher compared to a conventional VRF system, the thermodynamic perfectness of the refrigeration cycle calculated by the evaporating and condensing temperatures is less than 50 %, which implies that there is still a lot of room for improvement for this newly developed HTVRF system. Advanced optimization should be performed under high-temperature working conditions to adjust to the small compression ratio, and this will increase the thermodynamic perfectness. Meanwhile, the important aspects of compressor structure and system control should be examined in detail in order to satisfy the cooling conditions (a relatively small compression ratio) and heating conditions (a relatively large compression ratio) at the same time.

References

Gree Electric Appliances, Inc. of Zhuhai. http://www.gree.com.cn

Jiang Y, Xie XY (2010) Theoretical and testing performance of an innovative indirect evaporative chiller. Sol Energy 84:2041–2055

Jiang Y, Li Z, Xue ZF. An indirect evaporative cooling device. ZL02100431.5, 2002 (Chinese Patent)

Liu H, Zhang ZP, Xie YQ (2010) Development and research on high leaving temperature refrigerating centrifugal compressor. Fluid Mach 38:74–79 (in Chinese)

Mitsubishi Heavy Industries, LTD. MHI Turbo Chiller Microturbo Series "W" MTWC175/350 (for R134a refrigerant) Operating & Maintenance Manual, Issued in November, 2004

Mitsubishi Heavy Industries, LTD. High efficient chiller "MicroTurbo" is the best suited for building energy efficiency, The First Building energy efficiency Forum in Tsinghua University, 22–25 March 2005, Tsinghua University, Beijing, China

Song CL, Zhang GQ, Zhang Q, Chen ZK (1998) Ground-source heat pump – an energy efficient cooling/heating source for air-conditioning system. Energy Conserv 12:7–10

Wang HY, Fang XD, Du LW, Zhang XP, Jiang DL, Jia L (2011) High temperature screw chiller and its experiment. Chin HV&AC 41(1):14–16 (in Chinese)

Xie XY, Jiang Y (2010) Some views on design and thermal performance calculation methods of evaporative cooling air conditioning systems. Chin HV&AC 40(11):1–12 (in Chinese)

Yan QS, Shi WX, Tian CQ (2010) Refrigeration technology for Air conditioning, 4th edn. China Architecture & Building Press, Beijing (in Chinese)

Zhang HQ, Liu XH, Jiang Y (2011a) Performance analysis of periodic heat transfer process of energy-storage heat exchangers. Chin HV&AC 41(3):44–50 (in Chinese)

Zhang ZP, Li HB, Xie YQ, Zhong RX (2011b) Improvement and optimization of a high leaving temperature centrifugal compressor diffuser. Chin HV&AC 41(1):17–20 (in Chinese)

Chapter 7
Design and Operation of THIC Systems

Abstract Temperature and humidity independent control (THIC) system adopts the concept of regulating indoor temperature and humidity separately, and, therefore, the system design process including load calculation method, system principle, as well as equipment selection are all based on this concept. There are many types of THIC systems and a variety of system combinations. In this chapter, THIC systems specifically designed for applications in dry area and humid area are introduced. The performance and applicability of such systems used in different areas are compared, and the annual operating strategy and system regulating methodology are discussed.

Temperature and humidity independent control (THIC) system adopts the concept of regulating indoor temperature and humidity separately, and then system design process including load calculation method, system principle, as well as equipment selecting are all based on this concept. There are many types of THIC systems and a variety of system combinations. In this chapter, THIC systems used in dry area and humid area are introduced. The performance and applicability of such systems used in different areas are compared, and the annual operating strategy and system regulating methodology are introduced.

7.1 Design of THIC System

7.1.1 Overview of System Design

There are two subsystems in THIC system: humidity control subsystem (outdoor air handling system) is used to satisfy indoor requirements for outdoor air and removing indoor moisture load, and temperature control subsystem (indoor sensible terminals) for removing indoor sensible load to control indoor temperature. Based on the fact

X. Liu et al., *Temperature and Humidity Independent Control (THIC) of Air-conditioning System*, DOI 10.1007/978-3-642-42222-5_7,

that indoor moisture load and outdoor air flow rate required could be regarded to be proportional to occupants in most buildings, requirements for indoor humidity control and outdoor air can be both satisfied by adjusting the humidity ratio of supplied outdoor air to a certain value and introducing the handled outdoor air to indoor space with a flow rate proportional to the number of occupants. At the same time, there is another system used to regulate the cooling capacity of sensible terminals to effectively control indoor temperature. With the help of the THIC idea for thermal built environment, indoor requirements for outdoor air, humidity control, and temperature control can be satisfied at the same time.

If the required indoor temperature is 25 °C, theoretically speaking, any cooling source with a temperature lower than 25 °C can be used for the air-conditioning system in summer. In this way, natural cooling sources as well as high-temperature cooling sources with high efficiency could be used in THIC systems. However, in conventional system, under most circumstances, cooling source with a much lower temperature is required, like 7 °C chilled water, because of the need to control humidity and temperature at the same time. If condensation dehumidification method is adopted to dehumidify outdoor air and indoor air, the temperature of the cooling source must be low enough. However, in summer, it is very difficult to find cooling sources with a temperature of around 7 °C naturally existing or cheaply produced. Efficiency of chillers used to produce chilled water with a temperature of around 7 °C is much lower than that of chillers used to produce chilled water with a temperature of around 15–20 °C. Besides, in actual summer conditions, 60–80 % of the cooling capacity of air-conditioning system is used for sensible load to satisfy indoor temperature demand, and the application of THIC system can greatly reduce the energy consumption for indoor temperature control.

In conventional system, indoor temperature and humidity are regulated at the same time. As air exchange is the only solution to realize humidity regulation, introducing cold air to occupants' space is the way commonly used for indoor thermal built environment. However, if the task is only to remove indoor sensible load, it is not necessary to be limited by air circulation any more. Instead, water circulation can be used to directly remove sensible heat to achieve temperature regulation, such as radiant panel for cooling in summer and heating in winter. Because heat capacity of water is much higher than that of air, the profits are not only using the same unit both in summer and winter but also a great reduction of fan power consumption for ventilation and air circulation. Using water circulation, cooling/heating capacity can be transported directly into indoor environment, and the link of forced heat convection between water and air can be omitted, which reduces fan power consumption and space occupied by air ducts.

Based on the analysis above, adopting THIC system can help to greatly reduce energy consumption for air-conditioning from the aspects of cooling sources and distribution system and also makes it possible to effectively use natural cooling sources. It is an energy saving approach, which can also improve indoor thermal environment, and is considered as one of the most promising directions of air-conditioning system in future. There are different types of THIC air-conditioning systems, regarding to different climates, geographical conditions, building functions,

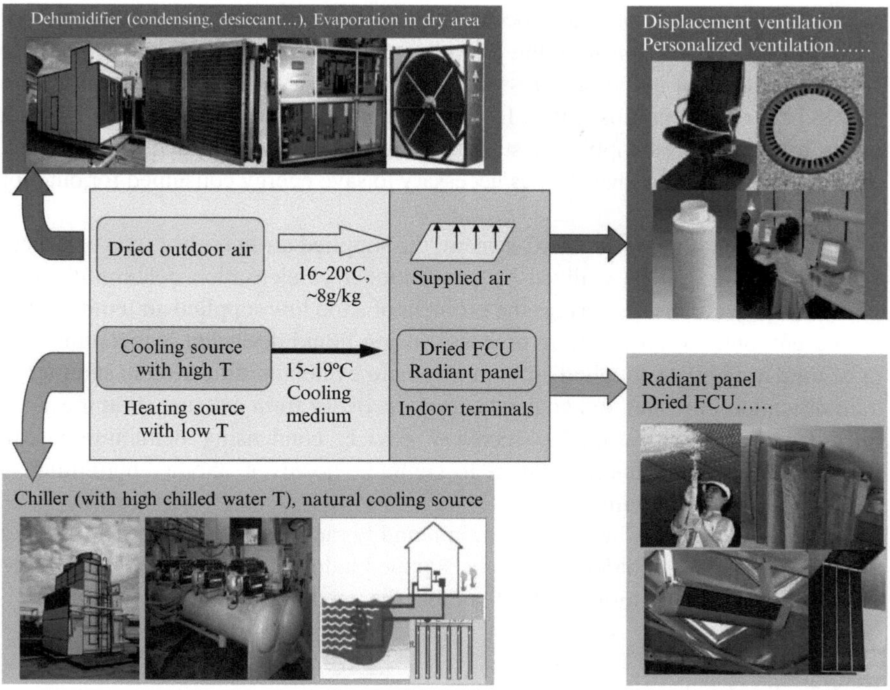

Fig. 7.1 Key components of THIC air-conditioning systems

load characteristics, etc. Referring to Fig. 7.1, it usually consists of outdoor air handling processors and supplied terminals for outdoor air used to control indoor humidity and provide outdoor air, cooling sources used to produce high-temperature chilled water around 15–19 °C, and the corresponding indoor sensible terminals used to control indoor temperature. The main devices are briefly introduced as follows.

7.1.1.1 Outdoor Air Handling Processor

Outdoor air handling processor is used to provide outdoor air with the task of maintaining indoor air quality and extracting indoor moisture load. The air flow rate is designed to satisfy occupants' health requirement, and the indoor humidity requirement can be satisfied with a supplied air humidity ratio lower than the indoor condition (usually around 9 g/kg). During the most humid period, the outdoor air system is running with the lowest air flow rate. If outdoor air humidity ratio is around 10–11 g/kg, the air can be sent directly indoor with a larger flow rate, and the energy consumption of handling outdoor air can be saved. If the actual number of indoor occupants varies a lot, the indoor terminal with variable air volume should be adopted, regulating air flow rate according to the number of occupants, indoor humidity ratio, or CO_2 concentration. In this way, outdoor air flow rate can be reduced if the number of indoor occupants becomes less, resulting in less outdoor air processing power consumption as well as fan power consumption. In order to keep the balance of indoor air pressure, indoor exhaust air with a certain flow rate should be removed in most cases. In summer and winter, exhaust air state is much closer to the required supply air state than that of outdoor air, and efficient heat recovery from indoor exhaust air is necessary to save energy consumed for outdoor air handling process.

As for the outdoor air in THIC system, the supplied air humidity ratio is around 8–10 g/kg. This can be realized by conventional condensation dehumidification system; however, this also brings the problem of a too low supplied air temperature. At present, outdoor air handling processor using liquid desiccant is recommended to be used to regulate supplied air humidity ratio by dehumidification in summer or humidification in winter and achieve energy recovery from exhaust air at the same time. Combined with enthalpy recovery device, condensing dehumidifier can realize outdoor air dehumidification in summer; however, another humidifier is required if applied in winter.

In the northwestern China, outdoor air humidity ratio is rarely higher than 12 g/kg in summer. Thus, the outdoor air can be supplied indoor after being cooled down to realize removing extra moisture load without the need of dehumidification.

7.1.1.2 Production of High-Temperature Cooling Sources

Cooling source's temperature in THIC system is around 15–20 °C, instead of 7 °C in conventional air-conditioning system. The energy saving potential of THIC system can be fully realized if high-temperature cooling sources are taken advantage of.

In the eastern region of China (north of Yangtze River), annual mean temperature is usually lower than 18 °C. And underground water temperature is around 15–20 °C. If allowed by geographical condition, cooling capacity from underground water can be used by drifting wells and recharging. The operating costs are only power consumption of water pumps. As to the small-scale buildings, pipes, with water circulating in, can be buried underground as heat exchanger to get cold water with a temperature of around 20 °C, and there is no need of getting water from and recharging water into underground. If there are sufficient spaces to bury pipes and annual energy balance could be solved, this is an efficient method to obtain high-temperature cooling sources.

By the seashore, lakeside, and riverside, if temperatures of the seawater, lake water, and river water are lower than 18 °C, water can be used as the high-temperature cooling sources through heat exchangers. However, pump power consumption used for transporting water may exceed the power consumption of chillers and then the advantage for energy saving disappears.

Chillers can be adopted when there are no natural cooling sources or it is difficult to be used. To produce high-temperature chilled water, energy efficiency of such chillers could be higher than conventional chillers. At present, centrifugal chiller and screw chiller specially developed for THIC system have been marketed in China, which can be used to produce high-temperature chilled water. To produce 16 °C chilled water, COP of the centrifugal chillers can exceed 8.5, and screw chillers can exceed 7.0, which are much higher than chillers used to produce 7 °C chilled water. Besides, refrigerant could be used as the high-temperature cooling source directly, with no need of secondary water loop. For example, VRF using refrigerant as the intermediate medium is a feasible solution for indoor temperature control of THIC system in small-scale buildings. Compared with conventional VRF for temperature and humidity control, this kind of VRF with an increased evaporating temperature could be more energy efficient, with a COP as high as 6.0 as analyzed in Sect. 5.3.

Actually, in the western part of China, climate in summer is hot but dry, with the dew point temperature lower than 15 °C. Under this condition, indirect evaporating cooler can be used to produce high-temperature chilled water, which could be about 2–3 °C higher than dew point temperature of outdoor air. Without the need of vapor compression refrigeration cycle, this shows up obvious energy saving effects. At present there are already indirect evaporative cooling products widely used in large-scale commercial buildings of Xinjiang, Ningxia, Inner Mongolia, and so on, achieving comfortable indoor environment and energy saving effects.

7.1.1.3 Terminals Used to Regulate Indoor Temperature

High-temperature chilled water is needed to take away sensible heat from indoor to regulate indoor temperature, and corresponding terminals are needed. The terminals are used to deal with sensible heat only, and chilled water temperature is higher than indoor dew point temperature, so there is no condensation phenomenon. Radiant cooling panel and dry FCU are commonly used.

In the last two decades, floor radiant heating has been promoted in the northern of China with fine heating effects. In the same way, chilled water with a temperature of around 15–20 °C can be sent into the floor coils to deal with 30–50 W/m^2 cooling load (if there is solar radiation, the cooling capacity will be increased). With the help of outdoor air handling processor, which undertakes indoor moisture load, under most circumstances, air-conditioning requirements of office buildings can be satisfied. After importing from European countries and domestically developed, capillary radiant terminal has been put forward to realize indoor temperature control in China, which is installed on the ceilings or vertical walls to realize cooling or heating with cold or hot water circulating inside. With radiant cooling method, as panel surface temperature is lower than the surrounding air dew point temperature, there will be condensation risk on the panel surface. So, outdoor air systems are important to realize indoor air humidity control and ensure the normal operation of sensible terminals. Because the density of dry air is higher than that of humid air, humid air stays in the top of occupants' space. If displacement ventilation is used to supply dry air from below, humidity ratio in the lower space will be lower than that in the upper. Thus, the possibility of condensation is higher for ceiling radiant cooling than that of floor radiant cooling. "On-off" control method is regarded as the best way to regulate indoor temperature for radiant cooling/heating, which has been used in many projects with floor heating and shows fine indoor temperature regulating performance.

According to conventional method using forced air circulation for air-conditioning system, air is used as the medium, which exchanges heat with high-temperature chilled water and removes indoor sensible load. The corresponding terminals include dry FCU and passive chilled beam. The former uses fans to realize forced air circulation and heat transfer, and the latter depends on the natural convection caused by the sinking of cold air around the beam. The water temperature in the air-water heat exchanger is higher than indoor dew point temperature, and there is no condensation phenomenon. The condensation pipe or plate is no longer needed, and the problem of condensation water leakage can be avoided, and environment pollution problem caused by mold breeding on the humid surface can be solved. However, dry FCU works at dry conditions; therefore, temperature difference between water and air is significantly reduced, requiring a larger heat transfer area for a same cooling capacity.

In recent years, there have been a number of large-scale office buildings and hospitals applying the THIC systems, located in the climate regions from southern China to northern China, from eastern coastal cities to Xinjiang and Inner Mongolia. After operating for a few years, the new THIC air-conditioning method shows great superiorities in indoor environment control and energy conservation. Chapter 8 of this book introduces part of these applications of THIC systems in China.

7.1.2 *Examples of THIC Systems*

There are different types and plans of THIC systems corresponding to different climates, geographical conditions, building functions, and load characteristics. Typical THIC systems for dry region and humid region are discussed here, respectively.

7.1.2.1 THIC System Used in Dry Region

In dry climate area, humidity ratio of outdoor air is lower than indoor design humidity ratio for occupants' space, which means it can be directly sent into the room, after direct evaporative cooling or indirect evaporative cooling, to realize humidity control. It should be noticed that, after direct evaporative cooling, the humidity ratio is slightly increased, and so it should be checked in the design period whether the supplied outdoor air is sufficient to take away indoor moisture load. High-temperature chilled water, produced by evaporative cooler or indirect evaporative cooler using dry outdoor air, can satisfy the needs of indoor temperature control. Therefore, in dry climate area, the characteristic of dry outdoor air should be fully taken advantage of for constructing thermal built environment according to THIC idea.

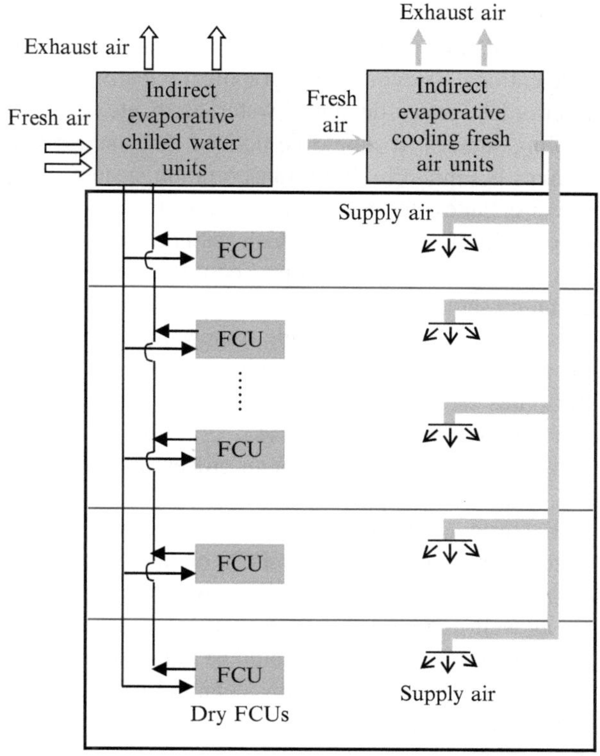

Fig. 7.2 Example of a THIC system for dry climate region

Figure 7.2 shows a typical THIC system used in dry climate area. The outdoor air is cooling down through indirect evaporative cooling device, and indirect evaporative cooler is used to produce 15–20 °C chilled water for indoor temperature control. Compared with conventional vapor compression refrigeration method, THIC system in this case takes full advantage of outdoor dry air, and power is consumed only by fans and water pumps. Thus, energy consumption of compressor is saved, which can greatly increase the system efficiency. Indoor terminal units can be dry FCU or radiant panel. It should be noticed that indirect evaporative cooler and dry FCU is an open loop, and if there is more than one indirect evaporative cooler, water surface balance between coolers should be ensured and water quality should be guaranteed.

7.1.2.2 THIC System in Humid Region

In humid climate area, humidity ratio of outdoor air is high, and the outdoor air needs to be dehumidified before being supplied indoor. The main task of outdoor air handling processor is to dehumidify the outdoor air to the state dry enough to take away indoor moisture load. As indoor temperature and humidity are separately controlled, more natural cooling sources can be used in THIC system than in conventional system. If geographically permitted and with proper temperature, river or lake water, deep well water, earth, etc., can be adopted as high-temperature cooling sources. It should be noticed that the application of such natural cooling sources is limited by fluid transportation system. If the building is far away from rivers or lakes, the pump power consumption caused by long-distance transportation will be increased, which cannot guarantee energy saving. Therefore, in considering to use these natural cooling sources, power consumed for fluid transportation should be assessed. As these natural cooling sources mentioned above are not available, artificial cooling sources including water chiller with high evaporating temperature and VRF with high evaporating temperature can be used. Figure 7.3 illustrates the typical THIC systems using water chiller as cooling sources and different outdoor air processors.

Using independent outdoor air processors and precooling outdoor air processors, Fig. 7.3a, b shows typical THIC systems for humid region. High-temperature water chiller produces chilled water with a temperature of about 15–20 °C, and the chilled water is transported into dry FCU or radiant panel to realize indoor temperature control; outdoor air processors are used to meet the demands of supplying outdoor air as well as indoor humidity control. Independent outdoor air processors with cooling sources in Fig. 7.3a could be liquid desiccant dehumidifiers, solid desiccant dehumidifiers, and condensing dehumidifiers. Figure 7.3b shows the outdoor air processors with precooling module, in which the outdoor air is precooled by high-temperature chilled water (15–20 °C) and then processed by low evaporating temperature chillers as shown in Fig. 7.3b or other outdoor air dehumidification methods before supplied into occupants' space. In this system, high-temperature water chiller accounts for indoor temperature control as well as precooling of outdoor air. Except for using 5–7 °C chilled water produced by the low evaporating temperature chiller to dehumidify the outdoor air, liquid desiccant dehumidifier as well solid desiccant dehumidifier can be used. In this way, the circulating water in THIC system is only high-temperature water, which makes the plumbing system operate much simpler and more convenient than that shown in Fig. 7.3b.

7.2 Load Calculation of THIC System

Loads in the buildings include indoor load (sensible cooling load through envelop, solar radiant, occupant, equipment, lighting, etc., and moisture load from people) and outdoor air load (caused by the temperature and humidity ratio differences between outdoor air and indoor states). The load calculation methods are the same with those of conventional air-conditioning system. In this chapter, indoor load characteristic is introduced, and load splitting is discussed in different THIC systems.

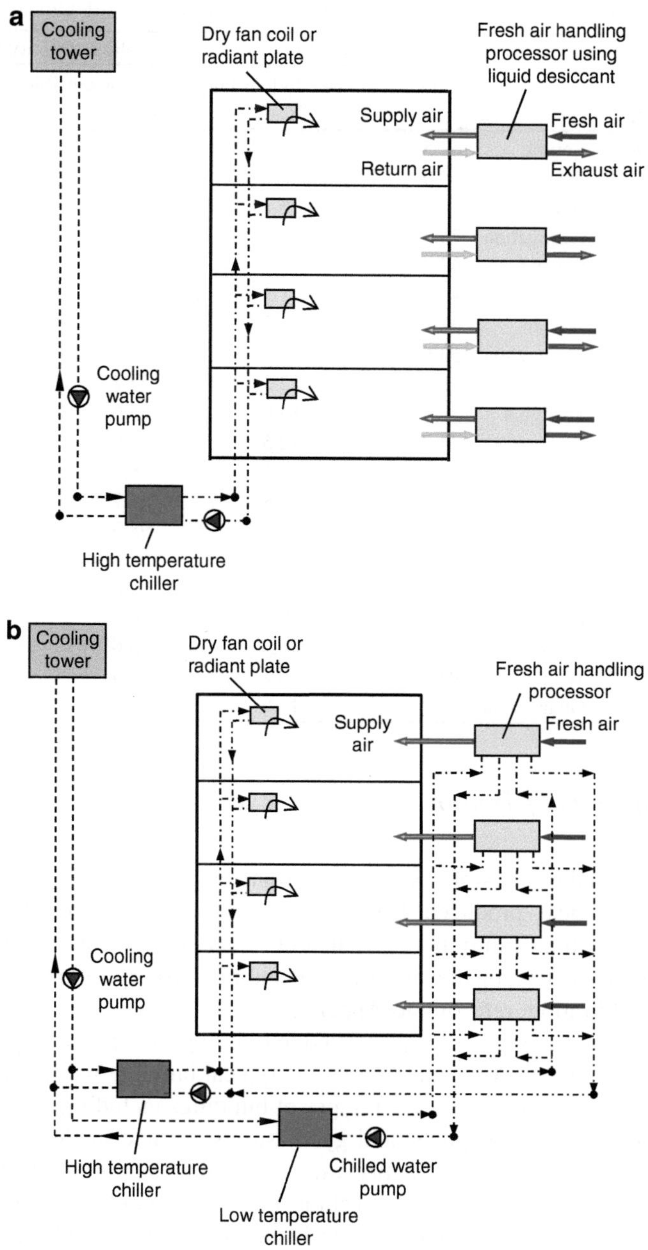

Fig. 7.3 THIC system schematics for humid climate region: (**a**) independent outdoor air processors and (**b**) outdoor air processors with precooling

Table 7.1 Indoor load of buildings with different functions located in different areas

Building function	City	Peak load (W/m^2)		Cooling capacity (kWh/m^2)	
		Indoor sensible load	Indoor latent load	Indoor sensible load	Indoor latent load
Office building	Beijing	62.5	6.7	61.6	3.6
	Shanghai	64.0	6.7	71.3	7.2
	Wuhan	68.1	6.7	74.3	6.9
	Guangzhou	73.6	6.7	93.6	10.2
Hotel	Beijing	48.3	1.5	70.8	1.9
	Shanghai	40.6	1.5	63.1	3.4
	Wuhan	37.7	1.5	67.8	3.4
	Guangzhou	53.6	1.5	81.3	3.6
Shopping mall	Beijing	38.7	12.5	53.2	4.4
	Shanghai	39.6	13.0	65.6	9.2
	Wuhan	44.0	12.5	68.0	8.9
	Guangzhou	37.5	13.1	76.4	10.3
Hospital	Beijing	49.0	3.8	71.7	4.3
	Shanghai	49.8	3.8	98.3	8.7
	Wuhan	55.1	3.8	102.6	8.6
	Guangzhou	51.6	4.9	107.0	10.4
Gymnasium	Beijing	64.2	20.5	30.8	2.3
	Shanghai	65.6	20.7	45.6	5.3
	Wuhan	93.1	20.7	47.9	4.2
	Guangzhou	83.7	18.6	49.5	5.5

7.2.1 Analysis of Indoor Load

The sources of indoor sensible load are heat transferred through building envelope, solar radiation, heat produced from occupants, equipment and lighting, heat brought in through air infiltration, etc. Indoor moisture load comes from moisture generated by occupants, open water surface, and plant transpiration and brought in through air infiltration, referring to Appendix A.

The load characteristic is closely related to building function, local climate, and the operating schedule. Appendix C presents typical building models and parameters set in the models of five different types of buildings including office buildings, hotels, shopping malls, hospitals, and gymnasiums, located in Beijing, Shanghai, Wuhan, and Guangzhou. Table 7.1 lists the calculation results of all the buildings in Appendix C using DeST software. Taking the office building in Beijing as an example, peak values of indoor sensible load and moisture load are 62.5 W/m^2 and 6.7 W/m^2, respectively; during the entire cooling season, the total cooling capacities for indoor sensible heat and indoor moisture load are 61.6 kWh/m^2 and 3.6 kWh/m^2, respectively.

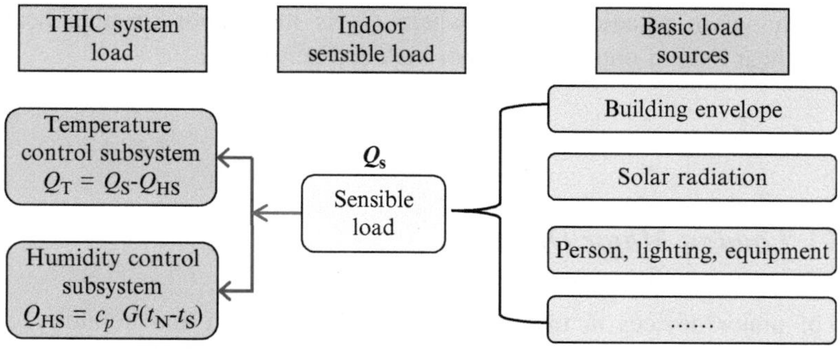

Fig. 7.4 Indoor sensible load appointment of the THIC system ($t_S < t_N$)

7.2.2 *Apportionment of Indoor Sensible Load*

In THIC system, the entire indoor moisture load is removed by the humidity control subsystem. If supplied air temperature of humidity control subsystem is different from indoor air temperature, humidity control subsystem removes or brings in a part of sensible load from (to) the indoor environment. According to the outdoor air handling methods analyzed in Chaps. 4 and 5, supplied air temperature of different methods is different. Thus, the sensible heat taken away (brought in) by the humidity control subsystem will also be different (Zhang 2012).

Indoor sensible load influenced by the supplied outdoor air, Q_{HS}, is calculated by Eq. (7.1), where G is the outdoor air flow rate, m^3/s; c_p is specific heat capacity of air, kJ/(kg · °C); ρ is air density, kg/m^3; t_N and t_S are indoor temperature and supplied outdoor air temperature, respectively, °C.

$$Q_{HS} = c_p \rho G(t_N - t_S) \tag{7.1}$$

- If condensation dehumidification and liquid desiccant dehumidification methods are used to handle outdoor air, supplied air temperature t_S is usually lower than indoor air temperature t_N, and humidity control subsystem undertakes part of indoor sensible load.
- If rotary wheel is used to handle outdoor air, supplied air temperature is usually higher than indoor temperature. Temperature control subsystem has to undertake this part of sensible load caused by the temperature difference between supplied outdoor air and indoor air, as well as entire indoor sensible load.

Figure 7.4 is an example if the supplied outdoor air temperature is lower than the indoor air temperature, indicating the apportionment of indoor sensible heat in THIC system. As the supplied outdoor air temperature t_S is lower than the indoor designed temperature t_N, supplied outdoor air undertakes part of indoor sensible heat, and temperature control subsystem undertakes the rest of indoor sensible heat,

Q_T, as shown in equation below, where Q_S is the indoor sensible heat load (sensible heat load of outdoor air is not included here), kW.

$$Q_T = Q_S - Q_{HS} \tag{7.2}$$

7.2.3 Load of Major Devices

Load of major devices in temperature control subsystem and humidity control subsystem of THIC system is analyzed in the following. In THIC system, outdoor air handling processor is the main device in humidity control subsystem, and high-temperature cooling source (usually high-temperature water chiller), distribution system, and sensible terminal devices are the main devices in temperature control subsystem. According to the plans of THIC air-conditioning systems in design stage, devices to be used for temperature and humidity control subsystems can be determined. And according to the characteristics of different devices and load calculation method mentioned above, the load of key devices can be calculated.

7.2.3.1 Load of Outdoor Air Handling Processor

The main device in humidity control subsystem is outdoor air handling processor, with the major task to produce dry and clean outdoor air and supply it into the room to control indoor humidity. Outdoor air changes from outdoor state to supply air state in outdoor air handling processor, so the load of the processor is calculated by the enthalpy variance of the outdoor air.

If there is no need for precooling or cooling using high-temperature cooling source in humidity control subsystem (seen in Fig. 7.3a), load of outdoor air handling processor Q_{OAP} is calculated as Eq. (7.3), where Q_{OAP} is the product of outdoor air mass flow rate and enthalpy difference between outdoor air and indoor air. The outdoor air handling processor is responsible for all of indoor moisture load $Q_{in,m}$ and outdoor air load Q_{OA} and some of indoor sensible load Q_{HS}.

$$Q_{OAP} = Q_{in,m} + Q_{HS} + Q_{OA} \tag{7.3}$$

If precooling or cooling is required for outdoor air (seen in Fig. 7.3b), the precooling load Q_{pre} needs to be subtracted in calculating Q_{OAP} by Eq. (7.4).

$$Q_{OAP} = Q_{in,m} + Q_{HS} + Q_{OA} - Q_{pre} \tag{7.4}$$

7.2.3.2 Load of High-Temperature Cooling Source

Major devices for temperature control subsystem include high-temperature cooling source, distribution system, and sensible terminal devices. The main task of high-temperature cooling source is for extracting the sensible load, while sometimes sensible cooling is also needed in the outdoor air handling process. Load of high-temperature cooling source is affected by different air handling methods.

If there is no need for precooling or cooling in humidity control subsystem (seen in Fig. 7.3a), load of high-temperature water chiller Q_{HTC} equals to the indoor load extracted by the temperature control subsystem Q_T.

$$Q_{HTC} = Q_T \tag{7.5}$$

If precooling or cooling is required for outdoor air (seen in Fig. 7.3b), precooling load Q_{pre} should be added to the load of high-temperature water chiller Q_{HTC}, as shown in Eq. (7.6).

$$Q_{HTC} = Q_T + Q_{pre} \tag{7.6}$$

Case Study: Taking an office building in Beijing as an example, the building model and relative parameters are given in Appendix C. Indoor sensible peak load Q_S, indoor moisture peak load, outdoor air peak load, and outdoor air flow rate per unit area are 62.5 W/m^2, 6.7 W/m^2, 43.3 W/m^2, and 3.4 m^3/h, respectively. THIC air-conditioning system is adopted in this building. Outdoor air (humidity control subsystem) is responsible for extracting the entire indoor moisture load, and the design temperature of the supplied outdoor air is 20 °C.

The calculation and analysis process is as follows:

1. Calculating the Humidity Ratio of Supplied Outdoor Air

In THIC system, humidity ratio of the supplied outdoor air is lower than indoor design humidity ratio, and the humidity ratio difference is the driving force to remove indoor moisture load. Equation (2.20) of Chap. 2 is used to calculate the humidity ratio of the supplied outdoor air required, and the value is as follows:

$$\omega_s = \omega_{a,in} - \frac{W}{\rho G} = 8.5\,\text{g/kg} \tag{7.7}$$

Therefore, temperature and humidity ratio of the supply air for outdoor air handling processor are 20 °C and 8.5 g/kg, respectively.

2. Appointment of Indoor Sensible Load

In this case, for the supply outdoor air temperature is lower than indoor design temperature, the supply air could also undertake part of indoor sensible load, which is calculated as follows:

$$\begin{aligned} Q_{HS} &= c_p \rho G (t_{a,in} - t_{sa}) \\ &= 1{,}005 \times 1.2 \times 3.4/3{,}600 \times (26 - 20) = 6.8\ \text{W/m}^2 \end{aligned} \tag{7.8}$$

Therefore, indoor sensible load undertaken by indoor sensible terminals is as follows:

$$Q_T = Q_S - Q_{HS} = 62.5 - 6.8 = 55.7\ \mathrm{W/m^2} \quad (7.9)$$

3. Loads of Outdoor Air Handling Processor and High-Temperature Cooling Source

Load appointment of outdoor air handling processor and high-temperature cooling source is closely related to the devices. Here gives the analysis of two typical conditions:

- If the outdoor air handling processor is an independent device (e.g., built-in compressor), high-temperature cooling source just needs to meet the demand of indoor sensible terminals. Loads of the outdoor air handling processor and the high-temperature cooling source are calculated by Eqs. (7.3) and (7.5), respectively. The load of outdoor air handling processor is $Q_{in,m} + Q_{HS} + Q_{OA} = 6.7 + 6.8 + 43.3 = 56.8$ W/m2. The load of high-temperature cooling source is $Q_T = 55.7$ W/m^2. Percentages of load undertaken by outdoor air handling processor and high-temperature cooling source are about 50 and 50 %.
- If precooling by high-temperature chilled water is needed in outdoor air handling process, the high-temperature cooling source has to meet the demands of indoor sensible terminals and outdoor air precooling process at the same time. Loads of outdoor air handling processor and high-temperature cooling source are closely related to the condition of outdoor air after precooling. Then loads of outdoor air handling processor and high-temperature cooling source are calculated by Eqs. (7.4) and (7.6), respectively. Taking the precooling of outdoor air by 16 °C high-temperature chilled water as an example, as the precooling load is 34.1 W/m^2, loads of outdoor air handling processor and high-temperature cooling source are $56.8 - 34.1 = 22.7$ W/m^2 and $55.7 + 34.1 = 89.8$ W/m^2, respectively. Percentages of load undertaken by outdoor air handling processor and high-temperature cooling source are 20 and 80 %.

4. Annual Cooling Consumption Analysis

According to Appendix C, annual cooling loads of indoor sensible load, indoor moisture load, and outdoor air are 61.6, 3.6, and 6.8 kWh/m^2, respectively. As the supplied outdoor air temperature is 20 °C, during the whole cooling season, indoor sensible cooling load undertaken by supplied outdoor air is 6.0 kWh/m^2 and that of sensible terminals is $61.6 - 6.0 = 55.6$ kWh/m^2.

As the outdoor air handling processor is the independent type, annual cooling load is $6.8 + 3.6 + 6.0 = 16.4$ kWh/m^2. And the cooling load of the high-temperature cooling source is 55.6 kWh/m^2. The percentages of cooling loads are then 23 % and 77 %, respectively.

If precooling of outdoor air is needed, under above parameters for precooling, the cooling loads of outdoor air handling processor and high-temperature cooling source are 8.5 kWh/m^2 and 63.5 kWh/m^2 (including precooling outdoor air), respectively.

7.2.4 Efficiency Comparison with Conventional System

Compared to the conventional air-conditioning system, high-temperature cooling source can be used to control indoor air temperature in THIC system, which means that energy efficiency of cooling source can be significantly improved. Terminal devices like radiant panel can be adopted, reducing heat transfer links from chilled water to indoor environment and saving fan power consumption for air circulation. But the heat transfer requirement for devices like FCU is stricter due to a higher chilled water temperature and a smaller temperature difference between chilled water and indoor air. Devices such as cooling source, distribution system, and terminal devices should be taken into consideration when it comes to the comparison between THIC system and conventional system, which means that the comparison is based on the entire air-conditioning system. The THIC system in Fig. 7.3a is taken as an example to introduce the performance comparison.

For the THIC system shown in Fig. 7.3a, high-temperature chilled water generated in high-temperature water chiller is supplied into dry FCU or radiant panel to control indoor temperature. Outdoor air processed by liquid desiccant air handling unit with built-in cooling source is sent into indoor space for humidity control. Here is a typical working condition for analysis, in which the outdoor air parameters are 35 °C, 60 % RH (the corresponding humidity ratio is 21.4 g/kg). Table 7.2 shows the performance indexes of major devices in the THIC system with different terminal devices, including COP_c of the chiller, transport coefficients of chilled water pump TC_{chw}, cooling water pump TC_{cdp}, cooling tower TC_{ct}, outdoor air fan TC_{fan} and FCU TC_{fc}, and COP_a of the liquid desiccant air handling unit (MOHURD, AQSIQ 2004, 2012; Tian et al. 2009). In the THIC system, as the design supply/return temperatures of high-temperature chilled water are 16 °C/21 °C, temperature difference between supply and return chilled water is the same as that in conventional system (7 °C/12 °C), and, thus, transport coefficients of the chilled water pumps in these two systems are the same.

From the performance comparison between different air-conditioning plans in the typical working condition, it can be seen that energy efficiencies of major devices in THIC system are different compared with those in conventional system.

- As to the cooling source, for the conventional system, indoor heat and moisture are removed together. Thus, evaporating temperature of the chiller is restricted, and accordingly energy efficiency of the chiller is restricted. As to the THIC system, indoor temperature and humidity are regulated separately, leading to a higher evaporating temperature, and thus COP_c of the high-temperature water chiller is significantly improved compared with that of the chiller in conventional system.
- As to the sensible terminal, there is a much lower cooling capacity with per unit flow rate for dry FCU used in THIC system compared with the wet FCU in conventional system, because of a higher supply water temperature. That is, transport coefficient of FCU (TC_{fc}) in THIC system is significantly lower than that in conventional system.

Table 7.2 Coefficient of performance (COP) for air-conditioning solutions

Performance indices	Calculation method	Conventional system	THIC system (Fig. 7.3a) Terminal of dry FCU	Terminal of radiant panel
Performance coefficient of the chiller COP_c	Cooling capacity of chiller/energy consumption of the chiller (compressor)	5.5	8.5	8.5
Transport coefficient of the chilled water TC_{chw}	Cooling capacity of chiller/energy consumption of the chilled water pump	41.5	41.5	41.5
Transport coefficient of cooling water pump TC_{cdp}	Heat exhausted by the condenser/energy consumption of the cooling water pumps	41.5	41.5	41.5
Transport coefficient of cooling tower TC_{ct}	Heat exhausted by cooling tower/energy consumption of the cooling tower	150	150	150
Transport coefficient of the outdoor air fan TC_{fan}	Cooling capacity of AHU/energy consumption of the outdoor air fan	20	18	18
Transport coefficient of the FCU TC_{fc}	Cooling capacity of FCU/energy consumption of FCU	50	22	–
COP of liquid desiccant air handling unit COP_a	Cooling capacity of processor/energy consumption of its compressor and pumps	–	5.5	5.5
The whole performance coefficient of cooling plant	Cooling capacity for building/energy consumption of all devices (chiller, water pumps, and cooling tower) in the cooling plant	4.13	5.68	5.68
The whole performance coefficient of the entire system COP_{sys}	Cooling capacity for building/energy consumption of all the devices in air-conditioning system	3.72	4.40	4.94

- As to the distribution devices including pumps and cooling towers, due to the same temperature difference (5 °C) between supply and return chilled (cooling) water in THIC system and conventional system, transport coefficients of the water pumps and cooling tower in these two systems are the same.

According to the performances of major devices in this typical working condition, performance coefficient of the entire system COP_{sys} for different air-conditioning solutions is also shown in Table 7.2. It can be seen that COP_{sys} of the THIC system is significantly higher than that of the conventional system, and COP_{sys} of the THIC system using radiant panel is significantly higher than that of the THIC system using dry FCU. One reason is that performance of the dry FCU is not satisfactory at present, and the other reason is that there is no fan power consumption for radiant panel. However, due to the risk of water condensation, cooling capacity for unit area of radiant panel is limited. So in actual application, it is needed to check whether the area of radiant panel is sufficient to meet the cooling demand in buildings.

As indicated by the analysis above, the idea of THIC provides a feasible solution for energy saving in air-conditioning system. However, it is not that all the THIC systems are more energy efficient than conventional system. Reasonable analysis of the whole system performance is required for achieving a greater energy saving potential.

7.3 Annual Operation for Heating and Cooling

China is a country of vast territory, and there are significant climate differences between different areas. Figure 2.25 in Chap. 2 shows the outdoor climate parameters in typical cities of China in summer. According to whether dehumidification is needed for outdoor air, taking the mean humidity ratio of 12 g/kg in the wettest month as a boundary, dry region and humid region can be distinguished: (a) Dry region in northwest: outdoor air is so dry that the main task of the humidity control subsystem in THIC system is cooling the outdoor air; and (b) Humid region in the east: the main task of the humidity control subsystem is to dehumidify the outdoor air.

According to whether heating is needed in winter and the current winter heating situation in China at present, the following conclusions can be obtained:

- Northern area (heating): Building area of cities and towns in northern China is over 9 billion m^2. Ninety percent of civilian buildings use central heating and the rest use separated heating.
- Yangtze River basin: Building area of cities and towns where the outdoor temperature may be below 5 °C (seen in Fig. 7.5) in winter is about 7 billion m^2. As temperature difference between the outdoor and the required indoor is not large, separated heating is recommended.
- South of the Yangtze River basin: There is almost no heating requirement in winter.

According to the analysis above, central heating is dominated in northern China, and the water from heat grid can be used directly. In Yangtze River basin, separated heating is recommended. As to the south of Yangtze River basin, there is nearly no need for heating in winter. Therefore, only the annual solutions for using THIC system in northern China and Yangtze River basin are stated in the following.

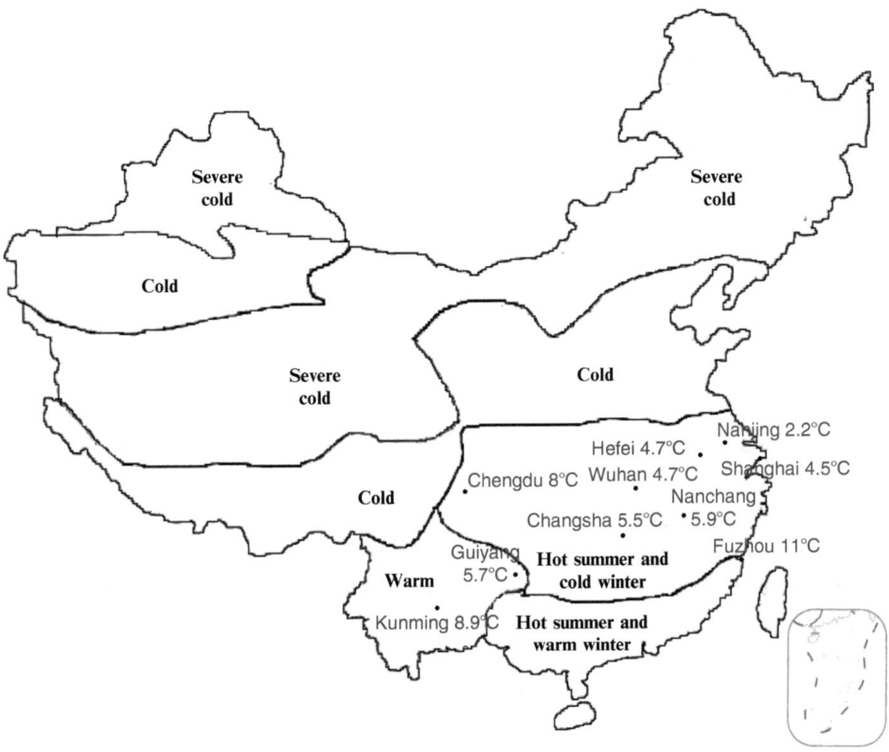

Fig. 7.5 Average outdoor temperature of the coldest month in some cities in southern China

7.3.1 Northern Area

Central heating is dominated in northern China, and the water from the municipal heating systems can meet the building heating requirement in winter. The same terminal device can be used in winter and summer. Sending hot water into the radiant panel or the dry FCU in winter, heating is realized. Figure 7.6a, b illustrates the operating principle of the THIC air-conditioning system in summer and winter, respectively. Here takes the THIC system with condensation dehumidification outdoor air handling processor (with precooling module by high-temperature chilled water and dehumidification by independent heat pump for outdoor air handling process) and dry FCU (or radiant panel) for an example.

As humidity control is required in winter, though there is moisture generated by occupants, the outdoor air still needs to be humidified due to a low humidity ratio of outdoor air in northern China. If cooling coils are used for condensation dehumidification in summer, extra humidifier (circulating water wet membrane humidification, high-pressure spray humidification, and so on) is needed in the air handling processor during winter. If liquid desiccant method driven by heat pump is used for dehumidification in summer, outdoor air can be heated and humidified by switching the four-way valve of the heat pump in winter, and there is no need for extra humidifier.

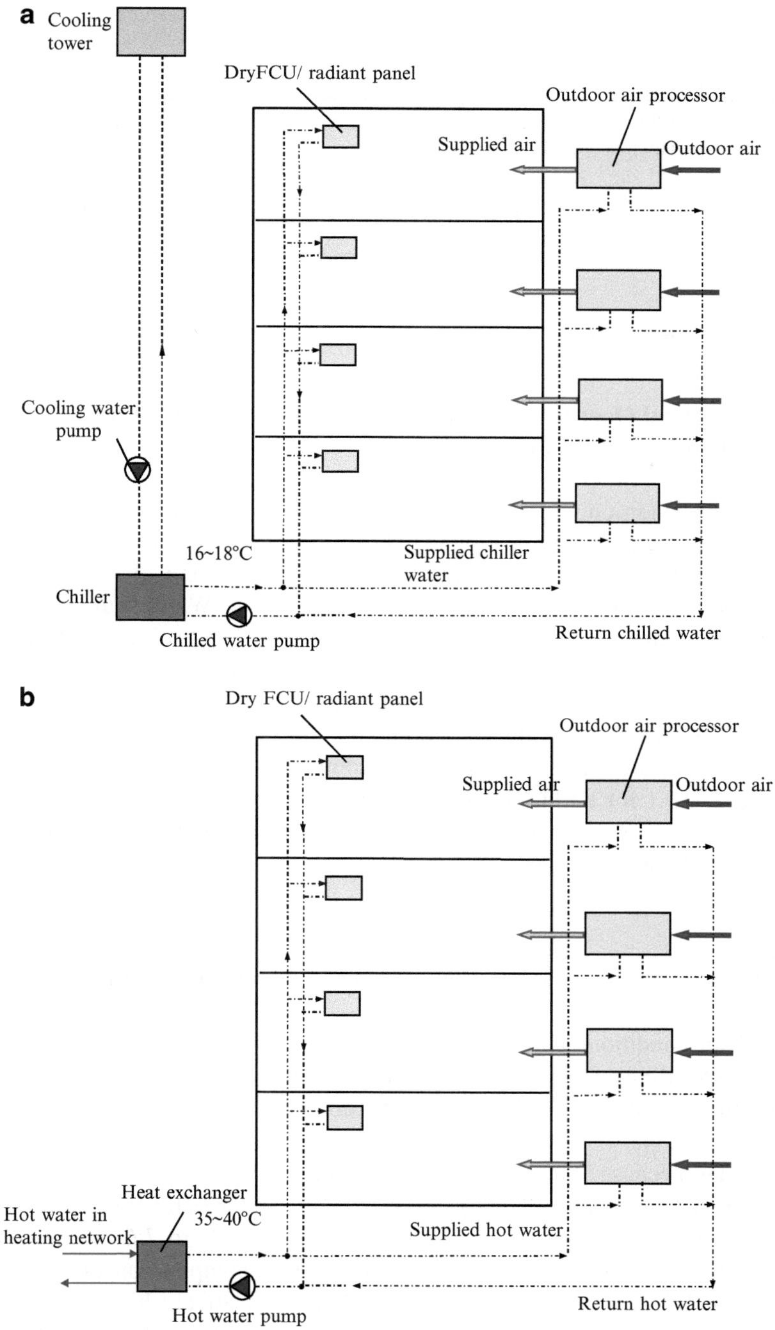

Fig. 7.6 Annual operating principle of the THIC air-conditioning system: (**a**) cooling mode in summer and (**b**) heating mode in winter

Table 7.3 Operating conditions and performance of the single-stage compression water-cooled heat pump (R22)

Operation mode	Cooling			Heating		
Ambient temperature	35 °C			7 °C		
Performance parameters	p_k/p_0	$COP_{cooling}$	φ_0	p_k/p_0	$COP_{heating}$	φ_k
	1.7–2.0	6.3–8.4	0.9–1.0	2.5–2.9	5.1–5.5	0.70–0.75
Notes	18 °C chilled water; SL = 3 °C and SH = 5 °C			35 °C hot water; SL = 3 °C and SH = 5 °C		

7.3.2 *Yangtze River Basin*

7.3.2.1 Annual Operation of Water-Source or Ground-Source Chiller (Heat Pump)

In THIC air-conditioning system, the same terminal device could be used both in winter and in summer. For heating, 35 °C hot water (compared with conventional hot water with a temperature higher than 45 °C for heating, it can be called as low-temperature hot water) meets the heating demand. Water-source heat pump and ground-source heat pump which take groundwater and soil as the heat source have better performance to produce the low-temperature hot water than normal heat pump unit. For heating condition in winter, condensing temperature is around 37–40 °C, and the evaporating temperature is about 3–5 °C due to the good stability of water source and ground source. Then compression ratio is around 2.4–2.8 for R22 and 2.7–3.1 for R134a. If using reciprocating piston compressor, centrifugal compressor or screw compressor, and scroll compressor with proper inner volume ratio, COP of the heat pump is about 4.

7.3.2.2 Annual Operation of Water-Cooled Chiller (Heat Pump)

In THIC air-conditioning system, 16–18 °C chilled water is produced in summer. The corresponding evaporating temperature t_0 is around 14–16 °C, and condensing temperature t_k is around 36–40 °C. Compression ratio is around 1.7–2.0 for R22 and 1.8–2.2 for R134a.

In winter, because of low-temperature hot water (about 35 °C) needed, condensing temperature is around 38–40 °C for the chiller. As the outdoor design condition is that dry-bulb temperature/wet-bulb temperature equals to 7/6 °C, evaporating temperature is usually 2–3 °C. Then corresponding compression ratio is around 2.5–2.9 for R22 and 2.9–3.2 for R134a, which is relatively higher than that in summer condition. Performance of the water-cooled chiller with R22 as the refrigerant is listed in Table 7.3; 18 °C chilled water is provided in summer, and 35 °C hot water is provided in winter. It can be seen that unit ($t_0 = 16$ °C, $t_k = 36$ °C, subcooling temperature SL = 3 °C, superheating temperature SH = 5 °C) with a relative cooling capacity of 1.0 in summer will have a relative heating capacity φ_k of 0.70–0.75 in winter.

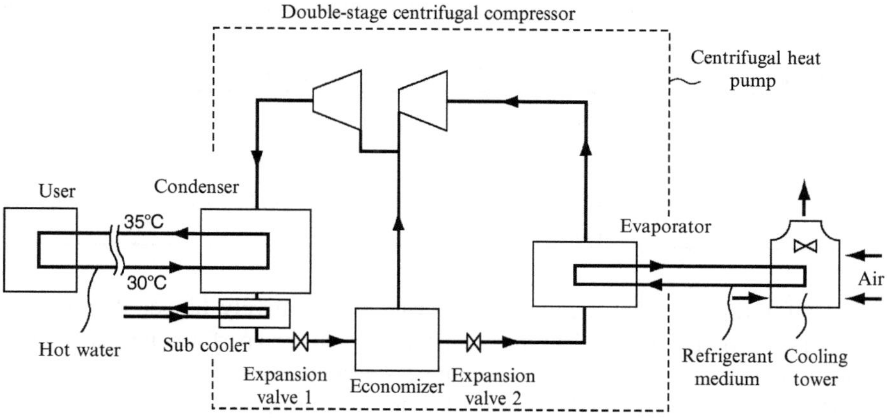

Fig. 7.7 Working principle of a water-cooled two-stage centrifugal compression heat pump in heating mode

Figure 7.7 shows the working principle of a two-stage centrifugal compression heat pump in winter. In summer, cooling water from the cooling tower is taken as heat sink of the chiller, and chilled water produced by the chiller is distributed to indoor terminals for cooling. In winter, the cooling water can be replaced by other fluids (such as glycol solution) which are not easy to freeze. The coolant circulates between the evaporator and the cooling tower, which absorbs heat from air in the cooling tower. With the help of the heat pump system, hot water is generated in condenser and sent to the building. This kind of unit has already been applied in some projects.

7.3.2.3 Annual Operation of Air-Cooled Chiller (Heat Pump)

For air-source heat pump, displacement compressor like reciprocating piston compressor, scroll compressor, and screw compressor are used because of their small capacity. In THIC air-conditioning system, high-temperature chilled water of 16–18 °C is provided in summer. Evaporating temperature is equivalent to that of water-cooled chiller, which is supposed to be $t_0 = 14$–16 °C. While the condenser is an air-cooled heat exchanger with the design condition that dry-bulb temperature and wet-bulb temperature are 35/24 °C, even if the coils with perfect heat transfer performance are adopted, condensing temperature t_k is still higher than that of water-cooled (shell and tube type) condenser. Usually, $t_k = 45$–50 °C. In this case, compression ratio of the air-cooled chiller is a little higher than that of water-cooled chiller. It is 2.1–2.5 for R22 and 2.3–2.8 for R134a.

In winter, hot water with a temperature of about 35 °C is needed, and the corresponding condensing temperature is about 38–40 °C. The design outdoor condition is that dry-bulb temperature/wet-bulb temperature =7/6 °C, and the evaporating temperature is about 2–3 °C. Then compression ratio is around

Table 7.4 Operation conditions and performance of the single-stage compression air-source heat pump (R22)

Operation mode	Cooling			Heating		
Ambient temperature	35 °C			7 °C		
Performance parameters	p_k/p_0	$COP_{cooling}$	φ_0	p_k/p_0	$COP_{heating}$	φ_k
	2.1–2.5	4.3–5.5	0.9–1.0	2.6–2.9	5.0–5.4	0.75–0.80
Notes	18 °C chilled water; SL = 3 °C and SH = 5 °C			35 °C hot water; SL = 3 °C and SH = 5 °C		

2.6–2.9 for R22 and 3.0–3.2 for R134a, which is relatively higher than that in summer condition. Table 7.4 lists the compression ratios and performance parameters of an air-source heat pump with screw compressor (with an inner volume ratio of 2.2) and R22 as the refrigerant. 18 °C chilled water in summer and 35 °C hot water in winter are provided by this unit. It can be seen that the unit (t_0 = 16 °C, t_k = 45 °C, subcooling temperature SL = 3 °C, superheating temperature SH = 5 °C) with a relative cooling capacity of 1.0 in summer has a relative heating capacity φ_k of 0.75–0.80 in winter.

7.4 Operating Parameters of High-Temperature Chilled Water

Transportation system including fans and pumps is an important component of air-conditioning system, and its power consumption takes a great proportion, sometimes over 50 %, of the total energy consumed by the entire air-conditioning system. Reasonable design of transportation system is an important precondition to realize normal operation and energy saving for THIC system. Temperature control subsystem and humidity control subsystem are responsible for different tasks in constructing thermal built environment, and corresponding transportation devices are different. Transportation device for humidity control subsystem is the fan used to supply dry outdoor air into occupants' space. Fans' performance is influenced by the pressure drop of outdoor air handling processors and terminal static pressure needs, which is almost the same as that in conventional air-conditioning system and here is not discussed. This section focuses on the operating parameters of high-temperature chilled water in temperature control subsystem.

7.4.1 Current Operating Parameters of High-Temperature Chilled Water

Indoor temperature is regulated combined with indoor humidity at the same time in conventional air-conditioning system. And to meet dehumidification needs in

Table 7.5 Corresponding dew point temperatures for different indoor design states

t_{db} (°C)	RH (%)	ω (g/kg)	t_{dp} (°C)	t_{db} (°C)	RH (%)	ω (g/kg)	t_{dp} (°C)
26	60	12.6	17.6	25	60	11.9	16.7
26	55	11.6	16.3	25	55	10.9	15.3
26	50	10.5	14.8	25	50	9.9	13.9

Note: t_{db} dry-bulb temperature, *RH* relative humidity, ω humidity ratio, t_{dp} dew point temperature

Table 7.6 Design parameters of high-temperature chilled water in THIC systems

Terminal type	Supply water temperature/°C	Return water temperature/°C	Temperature difference/°C	Resources
Dry FCU	15	20	5	Chen and Liu (2009)
	16	21	5	Lao (2011)
	14	19	5	Xue (2011)
	17.5	20.5	3	Zhao Kang et al. (2011)
Radiant terminal	16	19	3	Lu et al. (2011), Jin (2007), Li et al. (2011)
	17.5	20.5	3	Zhao Kang et al. (2011)

summer, considering energy consumption of transportation system and other economic factors, the designed supplied and return chiller water temperatures are usually 7/12 °C (referring to Appendix B.2). In THIC system, the selection of supply water temperature should, on one hand, consider the limitation of indoor dew point temperature and, on the other hand, consider the influence on performances of cooling sources and terminal devices (Tian et al. 2008).

Table 7.5 lists the dew point temperature under different indoor designed conditions. And the supplied water in temperature control subsystem should ensure that the lowest surface temperature of indoor terminal devices is higher than the dew point temperature of surrounding air to prevent condensation phenomenon. Temperature difference between supplied and return chilled water is usually designed as 5 °C for conventional system and is about 3–5 °C for THIC system at present, seen in Table 7.6.

7.4.2 Key Issues for High-Temperature Cooling

Figure 7.8 shows the schematic of temperature control subsystem commonly used in THIC system. Due to high-temperature cooling source adopted for indoor temperature control in THIC system, temperature difference between air and water is smaller than that of conventional system, which increases energy consumptions of fans and pumps if similar design parameters are adopted. Thus, temperature control subsystem must be designed with different operating parameters from those in conventional system. Taking radiant terminal as an example, the limitation of surface water condensation determines that the lowest surface temperature should be higher

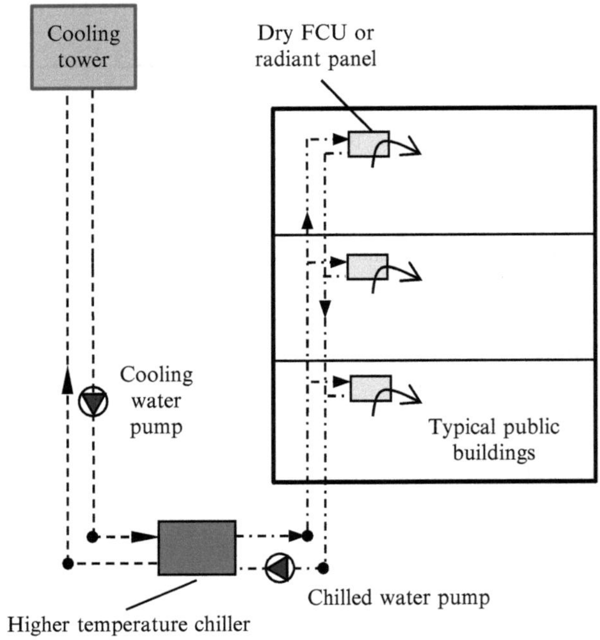

Fig. 7.8 Temperature control subsystem of THIC air-conditioning system

than the dew point temperature of surrounding air. However, the cooling capacity of radiant panel is closely influenced by surface average temperature, the lower the surface mean temperature, the higher the cooling capacity is. The cooling capacity is also influenced by the installation area. In some projects, the installed area is not sufficient to meet the cooling demands. It is the key issue to increase cooling capacity of radiant panel per unit area. As to this kind of water circulation system with "large flow rate and small temperature difference," it can guarantee a uniform surface temperature distribution of radiant panel, i.e., surface lowest temperature is approaching to surface average temperature, and cooling capacity per unit area can be significantly increased. As for chilled water circulation system, if temperature difference between supplied and return water is small (e.g., only about 3 °C), the corresponding circulation flow rate is relatively large. Under this circumstance, adjusting flow rate rarely influences the cooling capacity, and the effect is nonlinear. Thus, instead of regulating water flow rate, "on-off" control method should be adopted for regulating cooling capacity. In one cycle (e.g., 15 min), if the valve is on for 5 min and is off for the other 10 min, the corresponding duty ratio is 0.33, and linear continuous regulation for cooling capacity can be achieved by adjusting the duty ratio. There have been water valves available in market which can reliably work under "on-off" conditions for a long term. With this method, as pipe diameter is appropriately increased to a certain extent, avoiding other control valves except "on-off" valve and adopting water filters with low resistance, water pump pressure head can be limited within 20 m · H_2O, and pump energy consumption for water circulation is also controlled within a reasonable range. This is the new design and regulating

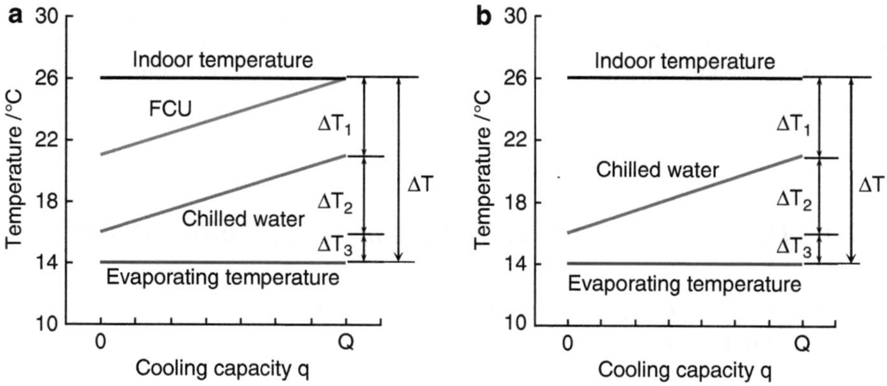

Fig. 7.9 Temperature control subsystem shown in the *T*-*Q* chart (from evaporating temperature to indoor temperature): (**a**) dry FCU and (**b**) radiant panel

method for water transportation system based on the idea of "large flow rate and small temperature difference," which is totally different from the conventional idea of "small flow rate and large temperature difference."

As high-temperature chilled water operates under the idea of "large flow rate small temperature difference," the most appropriate corresponding terminal is the radiant cooling panel. Figure 7.9a, b illustrates the temperature (*T*)-cooling capacity (*Q*) charts showing temperature levels from evaporating temperature to indoor air temperature of the temperature control subsystem when adopting dry FCU and radiant panel, respectively. Taking the *T*-*Q* chart of radiant panel in Fig. 7.9b as an example, temperature difference, ΔT, between indoor air temperature and evaporating temperature consists of three parts: ΔT_1 is the temperature difference between return chilled water and indoor air, reflecting the heat transfer ability of indoor sensible terminals; ΔT_2 is the temperature difference between return and supply chilled water, characterizing the circulating water flow rate; ΔT_3 is the temperature difference between the supply chilled water and the evaporating temperature, which is influenced by the heat transfer ability of the evaporator. The lower the required ΔT, the higher the evaporating temperature and better performance of chiller will be. However, the decrease of ΔT is affected by the increased transportation energy consumption as well as heat transfer ability of the devices in air-conditioning systems. If dry FCU is adopted, as shown in Fig. 7.9a, fans are needed to realize forced convective heat transfer process, resulting in higher energy consumption of the terminal devices.

As to radiant terminals, the forced air circulation process is avoided, helping to reduce the fan power consumption and improve the energy performance of THIC system. If using traditional FCUs, the air flow rate as well as heat transfer area should be increased, which results in difficulty to realize energy saving and reduce initial investment for terminal devices. If forced air circulation is still used to realize cooling, dry FCUs should be used instead of conventional FCUs. In THIC system, it is not just operating the FCU under dry working condition but canceling the condensation water draining system to be a totally different FCU configuration with a significantly lower pressure drop. If air directly flows through the coils circulated by axial fan, the fan head can be reduced significantly. Therefore, the reduced fan head keeps the fan power consumption from rising, although the air flow rate is doubled.

7.4.3 Discussion on Operating Parameters of Chilled Water

7.4.3.1 A Short Distance Between Cooling Source and Terminal Devices

As analyzed above, the key issue for cooling using high-temperature chilled water is the following: to achieve "large water flow rate and small temperature difference" efficiently, power consumed by water pumps could not be significantly increased with larger water circulating flow rate. Furthermore, the relation between the investment of water pipes (e.g., increasing pipe diameter) and operating cost of water pumps should be reexamined.

In the latter half of twentieth century when manufacturing productivity level is still a limiting factor in China, price ratio between steel and coal (with the same mass) is about 40:1; however, this ratio has decreased to as low as 10:1 at present with the rapid development of productive forces. This change forces us to rethink the constitution forms, analyzing method and operating parameters of air-conditioning system, then adapting to current situation. Initial cost of air-conditioning system could be measured by the steel price while operating cost could be reflected by the consumed electricity power. As to China, dominated by thermal power plants, power consumption could be also expressed by coal consumption. As a result, the relation between initial cost and operating cost of air-conditioning system could be reflected by the price ratio between steel and coal. For different times, price ratio between steel and coal could be adopted as an index for comparison on initial cost and operating cost, helping to optimize system operating parameters to achieve an equilibrium relationship between total initial investment and operating cost of the system. The significant decrease of price ratio between steel and coal indicates that the game between initial cost and operating cost is tilting toward the operating cost. Improving the system operating efficiency by increasing initial cost to a certain extent has become an important approach to reduce operating cost. For example, compared with before, manufacturers have been increasing input heat transfer area for evaporators and condensers of chillers to reduce the approach temperature difference (temperature difference between evaporating temperature and outlet chilled water temperature), e.g., the approach temperature difference has been reduced from about 3–5 °C before to within 1 °C at present, even as low as 0.5 °C.

In designing water distribution system for air-conditioning system, water pump head is determined by calculating the pressure drop of index circuit, and then water pump is selected taking water flow rate into account. Resistance of the water transportation system consists of resistance of water chiller, resistance of water pipes, local resistance of cooling plant, resistances of filters and control valves, resistance of terminal device, and so on. A common chosen water pump head could be as high as 30 m · H_2O. In fact, there are usually so many control valves under the traditional design method, and a considerable part of water pump head is consumed by these unnecessary valves. If "on-off" mode is adopted to regulate terminal devices according to duty ratio, these unnecessary valves could be canceled, helping to reduce the pump head consumed by valves. On the other hand, economical specific frictional resistance is usually used for determining water pipe diameter in conventional design procedures, but this original method could not accommodate the interrelation between system initial cost and operating cost any more in this new situation of energy prices. By enlarging water pipe diameter to a certain extent, specific frictional resistance could be reduced, and distribution resistance will be cut down, helping to lower power consumption of water pumps.

Therefore, with the help of appropriately increasing water pipe's diameter, changing the terminal regulating method, and so on, the pressure drop of the water circulation system could be reduced significantly. Taking a pressure drop of 5 m · H_2O of water chiller, a pressure drop of 3–5 m · H_2O for terminal devices, and a pressure drop of 5–10 m · H_2O for pipes and other devices as an example, the total pressure drop for the water circulation system could be controlled within 15–20 m · H_2O, significantly lower than the original pressure drop as high as about 30 m · H_2O in conventional system. Power consumption of water pump P_w (unit: W) could be calculated according to Eq. (7.10) where Δp_w is water pump head, Pa; G_w is water volumetric flow rate, m^3/s; and η_w is efficiency of the water pump.

$$P_w = \frac{\Delta p_w \cdot G_w}{\eta_w} \tag{7.10}$$

As indicated in Eq. (7.10), although circulating water flow rate G_w is increased with the running mode of "large water flow rate and small temperature difference," water pump lift Δp_w could be significantly reduced by enlarging pipe diameter, improving terminal regulating method, and so on. Thus, water pump consumption P_w could be controlled within a reasonable range and without significant increase.

As indicated by the above analysis on water circulation system pressure drop, it could be designed in the light of a constant pressure drop at a certain range (e.g., 15–20 m · H_2O) according to current situation of energy prices. As to the running mode of "large water flow rate and small temperature difference" for high-temperature chilled water, water pump power consumption could be controlled without significant increase with the help of an improved water transportation system. This innovative design thinking could help to reduce water pump head to a great extent and is beneficial to improve system efficiency and reduce operating cost.

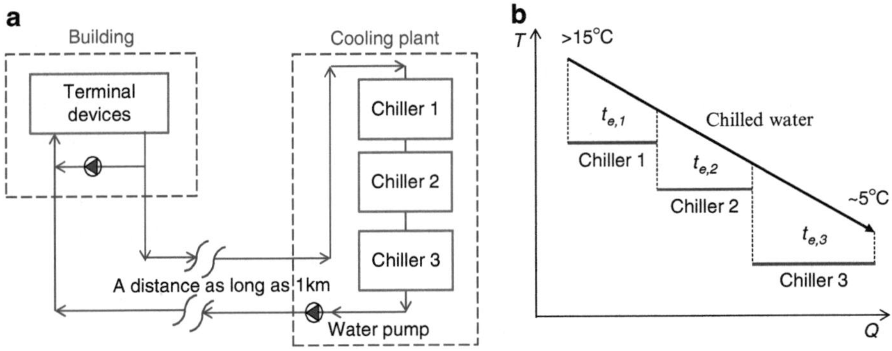

Fig. 7.10 Distributing chilled water for a long distance: (**a**) operating schematic between building and cooling plant and (**b**) heat transfer processes between chilled water and chillers in series

7.4.3.2 Long Distance Between Cooling Source and Terminal Devices

Above analysis focuses in common scenarios where cooling plant is set inside or near the building, indicating that there is a short distance between cooling source and terminal devices. However, in some cases restricted by practical limitations, there is a long distance between cooling plant and terminal devices (e.g., the distance between cooling plant and terminal devices could be as long as 1 km in some buildings, leading to a water pump head as high as 60 m · H_2O). The power consumption for water circulation takes an increasing proportion of the total energy consumption in operation and thus in designing the whole air-conditioning system, reducing this part of water transportation consumption should be taken into account.

Figure 7.10a illustrates an operating schematic for water circulation with a long distance. Due to a long distance between the cooling plant and the building, running mode of the water circulation system with "small water flow rate and large temperature difference" is adopted with a water temperature difference as high as 10–15 °C; in the building, running mode of water circulation system with "large water flow rate and small temperature difference" is chosen with a water temperature difference of about 3–5 °C. Water mixing pumps are used to solve the unmatched flow rates outside and inside the building. Then the heat transfer process between chilled water with a large temperature difference and evaporators of chillers serially connected is plotted in the *T-Q* chart shown as Fig. 7.10b. As temperature difference of supply/return chilled water is as high as 10–15 °C, the running mode to operate chillers in series helps to increase chiller's evaporating temperature and helps to improve chillers' overall COP. Taking the supply chilled water temperature of 5 °C and return water of 20 °C as an example, for these three chillers shown in Fig. 7.10, chiller 1 is to cool the chilled water from 20 to 15 °C, and chiller 2 is from 15 to 10 °C, with chiller 3 from 10 to 5 °C at the end. Considering a 1 °C approach temperature difference, evaporating temperatures are then 14 °C, 9 °C, and 4 °C, respectively. If the condensing temperature is 37 °C and thermodynamic perfectness of the refrigeration cycle is 0.6, corresponding COP for these three chillers are 7.5, 6.0, and 5.0, respectively. Besides, as to the cooling process for chilled water from 20 to 15 °C, cooling water from cooling tower could be adopted to realize this process in some cases, helping to reduce chillers' power consumption further.

7.5 Operating and Regulating Strategy of THIC System

The common points for operating strategies between THIC system and conventional system are regulating methods of the chillers, chilled water pumps, cooling water pumps, cooling towers, and supplying the handled outdoor air from the processor to indoor space. Present section only illustrates the differences between the operation strategies of two systems.

7.5.1 Operating Strategy of the THIC System

Based on the concept of THIC system, a new control mode is proposed for indoor thermal environment. In this control mode, priority is given to the passive approaches that natural cooling sources and waste heat are recommended to be adopted to maintain a comfortable indoor environment. In transition season, "free cooling" by outdoor air could be used to take away indoor heat and moisture, decreasing the operation time of the active air-conditioning system. One thing to note is to check whether the air flow rate of outdoor air ventilation system is sufficient to meet the dehumidification requirements. If natural ventilation is insufficient, active humidity control system is required to meet the dehumidification requirements. The operation modes of natural ventilation are as follows.

- If outdoor temperature and humidity ratio are both lower than these of indoor condition, directly use natural ventilation to extract indoor moisture and sensible load.
- If outdoor temperature is higher than indoor temperature, but outdoor humidity ratio is lower than indoor, natural ventilation could be used to extract indoor moisture load, but sensible terminals are required to extract indoor sensible load and control indoor temperature.
- If outdoor humidity ratio is higher than indoor, stop natural ventilation. In this condition, the passive way couldn't be adopted for extracting indoor moisture and sensible load any more.

There are temperature control and humidity control subsystems in THIC system, regulating indoor temperature and humidity separately. Thus, the logic of the operation strategy for THIC system is easier than that of conventional system, since there is no longer coupled temperature and humidity control. As the outdoor temperature is lower than indoor temperature but outdoor humidity ratio is higher, outdoor air dehumidification processor is required to meet the demands for outdoor air and dehumidification. In summer, condensation phenomenon should be strictly avoided. For the buildings where air-conditioning systems don't need to be in operation 24 h continuously, the devices of THIC system should be turned on in a different order compared with those of conventional air-conditioning system.

Taking a THIC system with the high-temperature water chiller, separate outdoor air handling processor (such as outdoor air handling processor using liquid desiccant or driven by separate heat pumps), and dry FCU as an example, the suggested operating order for the THIC system is as follows:

- Switch on the outdoor air handling processor in advance, and the exact time should be determined according to the realities of situation (shown as Fig. 7.11).
- Get the dew point information with the help of indoor temperature and humidity sensors. If indoor dew point temperature is lower than the supply chilled water temperature which is set about 16–18 °C, switch on the FCU and terminal water valve. Then the high-temperature water chiller could be in operation.
- The order to switch on the high-temperature water chiller is as follows successively: cooling pumps → cooling towers → chilled water pumps → chillers.
- After normal operation of the THIC system, motor damper of outdoor air branch should be automatically regulated based on the monitoring data of temperature and humidity sensor, and the water valve of FCU should be automatically regulated and switched based on the monitoring data of temperature sensor for regulating water temperature.
- The order to switch off the THIC system is as follows: high-temperature water chiller → chilled water pumps → cooling water pumps → cooling towers → dry FCU → outdoor air handling processor.

The basic concept for controlling and regulating of the entire THIC system is as follows:

- Outdoor air processor: regulating the outdoor air handling processor based on the difference between the measured value of humidity ratio (which could be calculated by the dry-bulb temperature and relative humidity) and the set point. One solution is to adopt a variable air flow rate and a fixed humidity ratio of the supply air. Another solution is to adopt a variable humidity ratio and a fixed outdoor air flow rate.
- Indoor sensible terminals (dry FCUs and radiant terminals): regulating the devices based on the difference between measured indoor temperature and the set value. The air flow rate and the water valve of the dry FCUs could both be regulated. For radiant terminals, variable duty ratio with fixed water flow rate, variable supply water temperature with mixed water pumps, and variable water flow rate could be used to regulate indoor temperature.

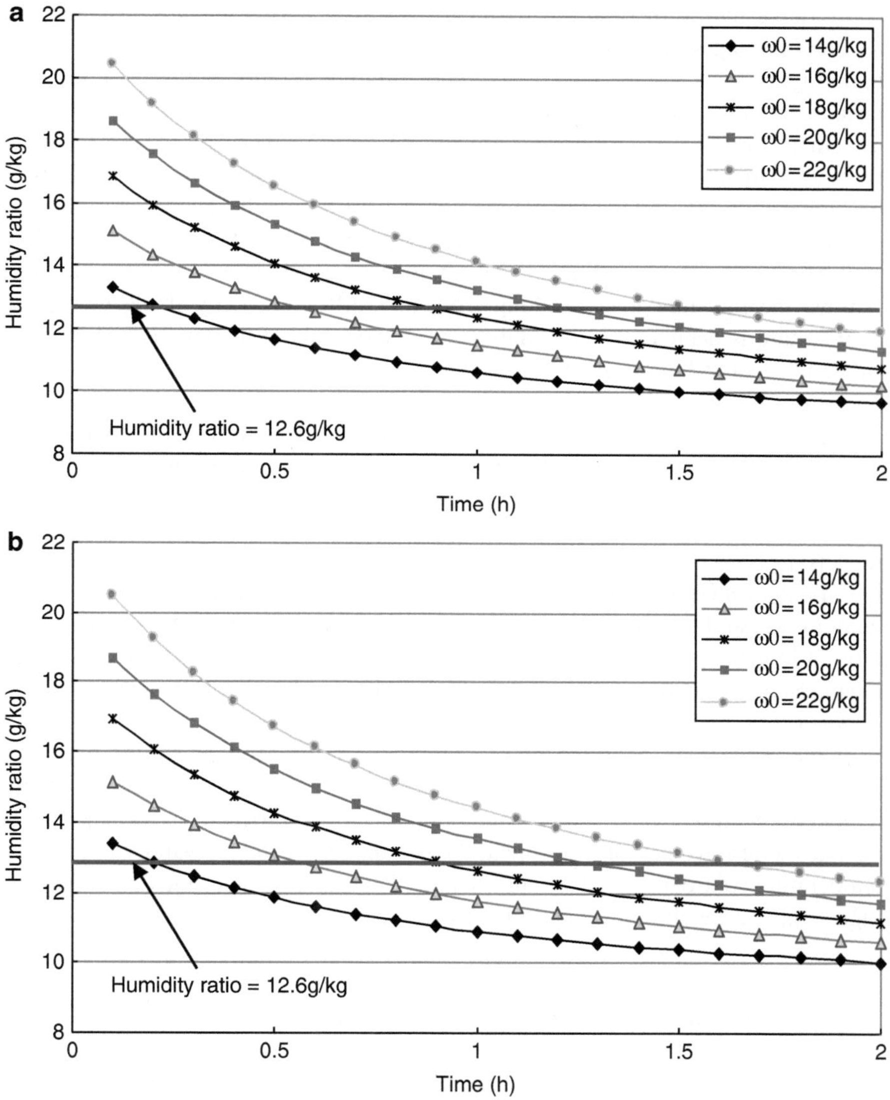

Fig. 7.11 Influence of indoor initial humidity ratio on the advanced operating time of the outdoor air handling processor: (**a**) humidity ratio of the supply outdoor air is 8 g/kg and (**b**) humidity ratio of the supply outdoor air is 8.5 g/kg

7.5.2 Regulating Strategy for Supplied Outdoor Air

7.5.2.1 Operating Time in Advance of Outdoor Air Handling Processor

In most actual buildings, air-conditioning systems always operate discontinuously. During working time, air-conditioning system is in operation to maintain a comfortable indoor thermal environment. However, during nonworking time, the air-conditioning system isn't in operation anymore. In summer, if the outdoor air humidity ratio is higher than that indoor, humidity ratio indoor will increase because of air infiltration through opening windows while the air-conditioning system is turned off during the nonworking period. Then for the next day, if radiant terminals are to start again, the dew point temperature of indoor air could be higher than the surface temperature of radiant terminals. Then water condensation occurs. Thus, the outdoor air handling processor in THIC system should be switched on in advance. Supply dry air to decrease indoor humidity ratio and remove the moisture load resulting from air infiltration. Only if the dew point temperature is decreased to a certain degree, terminals of the temperature control subsystem could be switched on, and then the THIC system could be in normal operation.

Figure 7.11 shows the variances of indoor humidity ratio with different indoor initial humidity ratios (ω_0) and supplied outdoor air humidity ratios in a typical office room, varying with operating time in advance of the outdoor air handling processor. As indoor design state is 26 °C and 60 %RH, the corresponding humidity ratio is 12.6 g/kg. If humidity ratio of the supplied outdoor air is 8 g/kg and the initial indoor humidity ratio is 20 and 16 g/kg, the time in advance to switch on the outdoor air handling processor is 1.2 h and 0.6 h, respectively. After that humidity ratio of indoor air could be approaching to the design value. For the discontinuous operation of air-conditioning system, a higher initial indoor humidity ratio results in a longer time to switch on the outdoor air handling processor in advance.

If the outdoor air handling processor with high-temperature chilled water for precooling is adopted in the THIC system, outdoor air should be precooled by high-temperature chilled water, and the outdoor air handling processor is switched on in advance to remove indoor moisture at the beginning. However, water valves of the temperature control terminals should be closed during this stage, and for this period, high-temperature chilled water is just used to precool the outdoor air.

7.5.2.2 Regulation of the Outdoor Air Handling Processor

Outdoor air handling processor is regulated according to the indoor requirement for outdoor air and dehumidification demand. There are humidity sensors and CO_2 sensors measuring indoor humidity ratio and indoor air quality, respectively. There are also infrared sensors in some buildings monitoring the room and judging whether there are occupants in the room or not. According to the measured data, outdoor air handling processor could be regulated.

In this section, a case is given and analyzed, where a quantity of sensors are installed in the building. This building located in Beijing is an office building with a total area of 20,000 m^2. The supplied outdoor air terminals are pressure independent VAV boxes. Raised floor ventilation is used, and the height of the raised floor is 350 mm. Outdoor air is supplied to the raised floor cavity and then

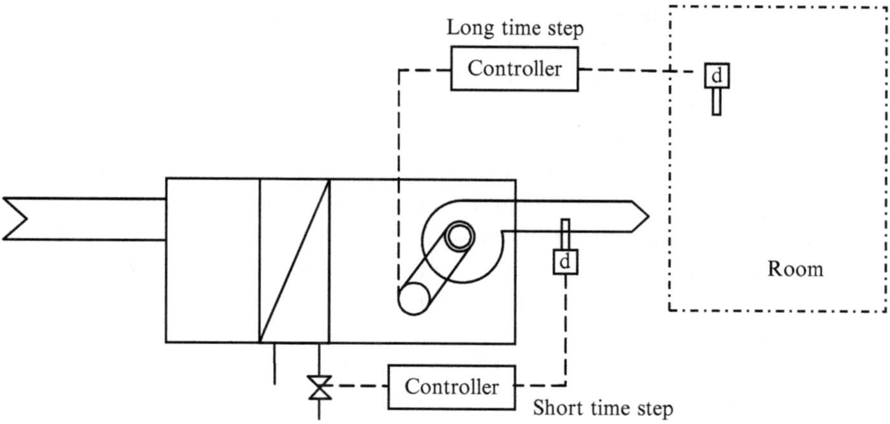

Fig. 7.12 Operating strategy of the condensation dehumidification air handling processor

supplied indoor through air vents on floor. There are three types of VAV boxes: (1) the maximum outdoor air flow rate is 382 m^3/h and the minimum is 44 m^3/h; (2) the maximum outdoor air flow rate is 594 m^3/h and the minimum is 72 m^3/h; and (3) the maximum outdoor air flow rate is 764 m^3/h and the minimum is 105 m^3/h. The outdoor air VAV boxes are regulated according to the measured data of the CO_2 sensors in exhaust air duct and the infrared sensors in the rooms. The infrared sensors monitor the room and determine whether there are occupants in the room or not. If there are no occupants, air flow rate of the VAV box is set as the minimum flow rate to remove indoor pollution and maintain a good indoor air quality. The minimum outdoor air flow rate is 15 % of the maximum one. If there are occupants indoor, air flow rate of the VAV boxes is regulated according to the CO_2 sensors. There are also sensors monitoring the windows and determining whether the windows are closed or not. If not, the VAV boxes will be switched off and supplied outdoor air is also stopped.

Operating strategy of the outdoor air handling processor in partial load could use the method with a fixed supplied air humidity ratio and a variable outdoor air flow rate or the method with a fixed outdoor air flow rate and a variable supplied air humidity ratio. Taking outdoor air handling processors using condensation dehumidification and liquid desiccant as examples, operating strategies are introduced as follows.

Condensation Dehumidification Air Handling Processor

Operating strategy of the condensation dehumidification outdoor air handling processor is shown in Fig. 7.12, using the temperature and humidity sensors to measure supplied air humidity ratio. According to the difference between the measured value and the set point, chilled water flow could be regulated, and the time step is around 10 s. According to the difference between indoor measured humidity ratio and the set value, outdoor air flow rate or the set point of supplied air humidity ratio could be regulated. The time step is around 15 min which is much longer than the time step of regulating the flow rate of chilled water. If the outdoor air handling processor includes the exhaust air heat recovery equipment, fans in outdoor air duct and exhaust duct should be linkage controlled.

If outdoor air needs to be humidified in winter, separate humidification device should be set up in the outdoor air handling processor, and hot water should be supplied to the coils to heat and humidify the outdoor air. Then operation strategy is also to control the humidity ratio of the supplied air. Water flow rate of the coils and the humidification device could be regulated according to the difference between the measured humidity ratio and the set value. Supplied air flow rate or the set value of supplied air humidity ratio could be regulated according to difference between indoor measured humidity ratio (or CO_2 concentration) and the set point.

If the humidity control subsystem in winter is turned off and only temperature control subsystem is in operation, operation strategy of the outdoor air handling processor could be as follows: controlling supplied air temperature and regulating opening of the water valve for coils according to the difference between measured supplied air temperature and the set value. If the air flow rate of outdoor air handling processor is variable, supplied air flow rate could be regulated according to the difference between measured indoor CO_2 concentration and the set point.

Liquid Desiccant Outdoor Air Handling Processor

Operating strategy of the liquid desiccant outdoor air handling processor is similar to that of condensation dehumidification outdoor air handling processor. The control logic consists of the longtime step method and the short-time step method. The control logic of the longtime step is the same, but that of the short-time step is different. For the liquid desiccant outdoor air handling processor, the number of heat pumps in operation or the frequency of heat pump could be regulated, and liquid concentration could be regulated by adjusting the adding water rate to meet the requirement of controlling supplied air humidity ratio. And the time step is as short as 10–15 s. Operating strategy of the longtime step method is to regulate the outdoor air flow rate or the set value of supplied air humidity ratio according to the difference between indoor measured humidity ratio and the set value. The longtime step is as long as about 15 min.

As to the liquid desiccant outdoor air handling processor, through switching the four-way valve of heat pumps, outdoor air could be heated and humidified in winter, and the operating strategy is the same as that in summer.

7.5.3 *Regulating Strategy of Sensible Terminals*

Sensible terminals commonly used for THIC system consist of two types, which are the dry FCUs and the radiant panels. These two kinds of terminals could operate both in winter and summer to meet the requirements of heating and cooling,

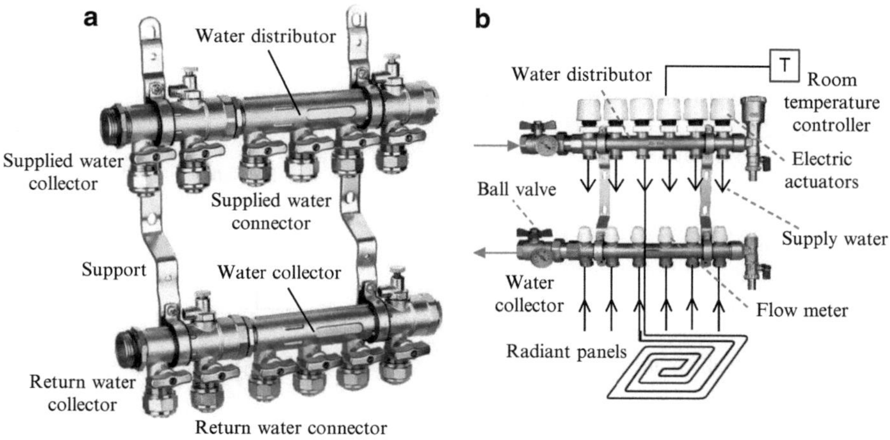

Fig. 7.13 Variable flow rate control mode: (**a**) water distributor and collector and (**b**) schematic of the variable flow rate control mode

with similar operating strategies. The operation strategy of dry FCU is similar to that of wet FCU, including the variable air speed, the indoor temperature controller, and the on-off control of the supplied water by the solenoid valves (Liu et al. 2010). Here focuses on the regulating strategy of radiant panels. There are three main regulating methods for radiant panels. The first is the variable flow control mode. The second is variable duty cycles of the water valve in fixed flow rate system. The third is mixing water control mode. As to the first two methods, temperature of the supplied water for the radiant panels is fixed while the third is to regulate the supplied water temperature. These three methods are introduced as follows.

7.5.3.1 Variable Flow Control Mode

In the design handbook of *Practical Design Manual for Heating and Air-conditioning* (2nd edition, 2008), this control mode is introduced in details. Here quotes a typical control mode from that handbook, which consists of indoor temperature controllers, electro-sensitive (thermosensitive) actuators, and water segregators with internal valve core. Structure maps of the manifold and segregator for radiant panels and the scheme of the control mode are illustrated in Fig. 7.13a, b, respectively. According to the difference between the measured indoor temperature and the set point, the electro-sensitive (or thermosensitive) actuators are controlled to change the valves' opening. And the supplied water flow rate is changed to main indoor temperature.

7.5.3.2 Variable Duty Cycles of the Water Valve in Fixed Flow System

As to the first regulating method, supplied water flow decreases in partial load, which would result in nonuniformity of surface temperature distribution of radiant

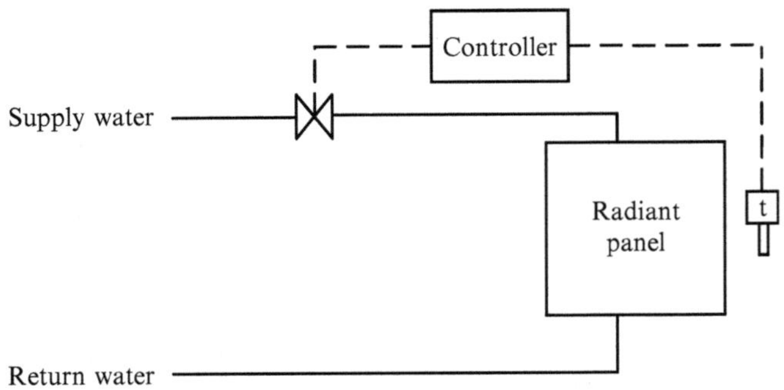

Fig. 7.14 Schematic of the on-off control mode

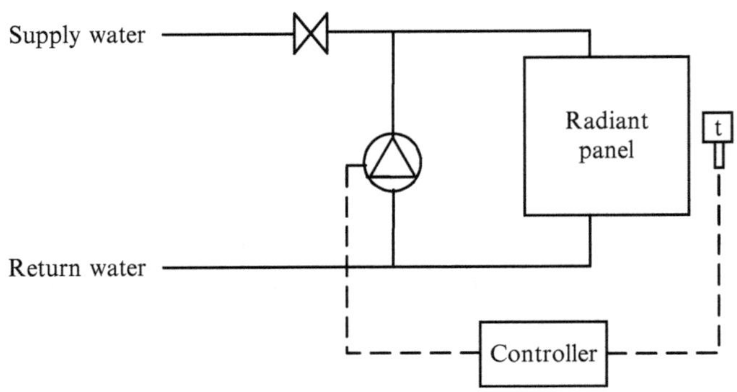

Fig. 7.15 Supply water temperature control mode by mixing pump

panel. The core principle of the control mode introduced here is to fix the water flow if the water valve is on and to change the duty cycle of the water valve to regulate the heating/cooling capacity of radiant panels. The schematic of this regulating method is shown in Fig. 7.14. On-off valves according to measured indoor temperature are installed in each branch. Duty cycle of the on-off valve is changed about every half an hour to regulate cooling/heating capacity of the radiant panel.

7.5.3.3 Mixing Water Pump Control Mode

The mixing water mode driven by terminal pumps could be used to regulate the supplied water temperature, as shown in Fig. 7.15. As water pump's speed reaches the maximum, no chilled water is supplied to the radiant terminal. With the speed of mixing water pump decreasing, the ratio of mixed water decreases, indicating that the mean water temperature in radiant terminals becomes lower and cooling capacity is increasing.

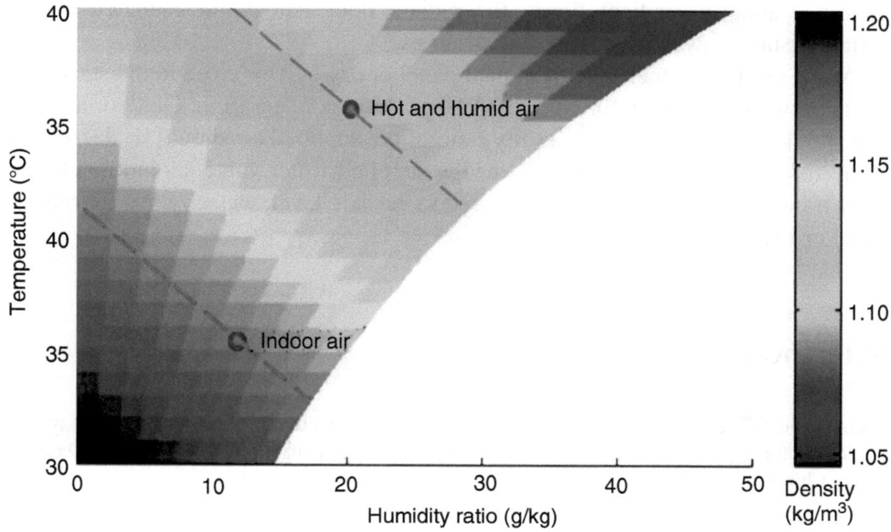

Fig. 7.16 Influence of air temperature and humidity ratio on air density

7.5.4 Anti-sweat Measures and Regulating Strategy

Avoiding water condensation on terminal's surface is the precondition for normal operation of THIC system in summer. To avoid water condensation, temperature sensor should be installed on the coolest surface in the room and guarantee the lowest temperature of surfaces is higher than indoor dew point temperature. Based on the experience, the coolest point indoor is always far away from the windows and adjoins the supplied water pipe near the inner corner. Theoretically, if the lowest temperature of the cold surfaces (not the supplied chilled water temperature) is higher than the dew point temperature, water condensation is impossible. As suggested in ASHRAE handbooks, supply water temperature should be 0.5 °C higher than indoor dew point temperature to ensure normal operation. And according to other references, temperature of the radiant panels should be 1–2 °C higher than indoor dew point temperature.

Besides, the places where there are external windows and doors with risk of outdoor hot and humid air infiltration should be properly taken into account, especially in humid region. Figure 7.16 shows the air density varying with temperature and humidity ratio. A higher outdoor temperature or humidity ratio leads to a lower air density. Outdoor hot and humid air infiltrated gathers in the top of indoor space; thus, the risk of water condensation for radiant ceiling is higher compared with radiant floor. Condensation water pans are suggested to be designed for dry FCUs near external windows and doors in case of moisture condensation. In some buildings, there are detectors monitoring the opening of windows. If the windows are open, radiant panels or the water valve of the dry FCUs will be closed. For the buildings adopting radiant floors, the radiant floors should be installed with a certain distance away from the external doors.

As the coldest temperature of radiant panel or dry FCU is close to the surrounding air dew point temperature, there is risk of water condensation. Outdoor air flow rate should be increased, or humidity ratio of the supplied air should be decreased. If the risk still exists, radiant panel or the supplied water valve should be closed. Only if indoor humidity ratio is decreased to the safe level, water valves for radiant panel or dry FCU can be opened again.

References

Chen P, Liu SQ (2009) Temperature and humidity independent control air-conditioning system design for inpatient building of central hospital of Sanmenxia. Refrig Air Cond 9:66–71 (in Chinese)

Jin Y (2007) Design for the Sino-Italian ecological and energy efficient building of Tsinghua University. Chin HV&AC 37:73–75 (in Chinese)

Lao YM (2011) Feasibility analysis on cold source energy efficiency based on temperature and humidity independent control with refrigeration dehumidification. Build Sci 27(8):48–50 (in Chinese)

Li Y, Hu R, Feng TT (2011) Application of temperature and humidity independent control air conditioning system to the second phase of Qingdao xiangxi villa. Chin HV&AC 41(1):42–47 (in Chinese)

Liu LB, Fu L, Jiang Y (2010) Maintaining uniform hydraulic conditions with intelligent on-off regulation. Build Environ 45(12):2817–2822

Lu B, Chen M, Zhang YN (2011) Energy-saving design for air-conditioning system in ECO building. Refrig Air Cond 11(1):77–81 (in Chinese)

MOHURD, AQSIQ (2004) GB 50019-2003 Code for design of heating ventilation and air conditioning. China Architecture & Building Press, Beijing (in Chinese)

MOHURD, AQSIQ (2012) GB50736-2012 Code for design of heating ventilation and air conditioning. China Architecture & Building Press, Beijing (in Chinese)

Tian XD, Shi M, Zhou JC, Li JG, Lu HY (2008) Research about the value of chilling water temperature difference in independent humidity control AC system. Fluid Mach 36(12):75–78 (in Chinese)

Tian XD, Liu H, Zhang ZP, Li HB (2009) Research on centrifugal chiller which has high-leaving temperature and its performance. Fluid Mach 37(10):53–56 (in Chinese)

Xue ZY (2011) Energy saving design on temperature and humidity independent control system in a subway control center. Energy Conserv 11–12:113–116 (in Chinese)

Zhang T (2012) Design methodology and performance analysis of temperature and humidity independent control (THIC) air-conditioning system. Master thesis, Tsinghua University, Beijing (in Chinese)

Zhao K, Liu XH, Zhang T, Jiang Y (2011) Performance of temperature and humidity independent control air-conditioning system applied in an office building. Energy Build 43:1895–1903

Chapter 8
Application Cases of THIC Systems

Abstract Recently, THIC system has been paid more attention to and successfully commercialized. Based on the idea of THIC for air-conditioning system, there are various schemes and devices for constructing a THIC system. Application cases with various types of high-temperature cooling sources, outdoor air handling units, and indoor terminals have been discussed in literatures. This chapter focuses on some typical examples of THIC air-conditioning systems applied in commercial buildings with different functions (such as office building, hospital building, large space building, and industrial factory) that are located in different regions and corresponding to different outdoor climates.

In recent years, THIC system has been paid more attention to and applied in many buildings (Liu et al. 2006; Xie 2009; Zhu et al. 2010; Fan et al. 2011). Based on the idea of THIC for air-conditioning system, there are various schemes and devices for constructing a THIC system. Table 8.1 lists some application cases analyzed in literature, with various kinds of high-temperature cooling sources, outdoor air handling units, and indoor terminals. This chapter mainly introduces the applications of THIC air-conditioning systems in buildings with different functions (such as office building, hospital building, large space building and industrial factory) and located in different regions corresponding to different outdoor climates.

X. Liu et al., *Temperature and Humidity Independent Control (THIC) of Air-conditioning System*, DOI 10.1007/978-3-642-42222-5_8,

Table 8.1 Application cases of THIC systems

Building function	Location	Building area/m^2	Temperature control subsystem	Humidity control subsystem	Indoor terminals	Resources
Restaurant	Xinjiang	1,000	15–20 °C chilled water with indirect evaporative cooling	Outdoor air cooled by the chilled water returned from the fan coil units	Dry fan coil unit	Xie et al. (2007)
Office building	Beijing	20,000	16 °C chilled water from LiBr absorption chillers	Condensation dehumidification with 5 °C chilled water from LiBr absorption chillers	Radiant ceiling and floor air distribution	Jin (2007)
Office building	Beijing	94,000	High-temperature chilled water from centrifugal chillers	Condensation dehumidification with low-temperature chilled water from centrifugal chillers	Dry fan coil unit	Lin et al. (2009)
Hospital	Henan	48,000	15 °C chilled water from underground-source heat pumps	Liquid desiccant outdoor air handling unit driven by heat pumps	Dry fan coil unit	Chen and Liu (2009)
Pavilion	Shanghai	3,470	Return air is mixed with handled outdoor air and then cooled by chilled water	Rotary solid desiccant outdoor air handling unit with heat recovery and precooling	All-air system	Hu (2010)
Archive	Sichuan	3,600	Underground water (18 °C) as the cooling source	Condensation dehumidification with chilled water from underground-source heat pumps	All-air system	Fang (2010)
Office building	Beijing	30,000	16 °C chilled water from screw chillers	Condensation dehumidification with low-temperature chilled water from centrifugal chillers	Chilled beam	Lu et al. (2011)
Office building	Shenzhen	21,000	17.5 °C chilled water from centrifugal chillers	Liquid desiccant outdoor air handling unit driven by heat pumps	Dry fan coil unit and radiant panel	Zhao et al. (2011)
Airport terminal	Xi'an	47,000	High-temperature chilled water from the cooling plant	Liquid desiccant outdoor air handling unit driven by heat pumps with precooling	Radiant floor	Zhou (2011)
Villa	Qingdao	508	16–19 °C chilled water from underground-source heat pumps	Condensation dehumidification with chilled water from underground-source heat pumps	Capillary radiant panel	Li et al. (2011)

8.1 Application in an Office Building (Humid Region)

8.1.1 Description of the THIC System in an Office Building

8.1.1.1 Basic Information

A 5-story office building, as shown in Fig. 8.1, is located in Shenzhen, China, with total building area of 21,960 m^2 and the areas of 5,940, 5,045, 3,876, 3,908, and 3,191 m^2 for the 1st to 5th floor, respectively. The main functions of the 1st floor are mess hall, archive, and carport, while the 2nd to 4th floors are the office rooms, with the 5th floor as the meeting room. And there is a vestibule vertically through up the 2nd to 4th floors in the north of this building. As to the vestibule, curtain wall with ventilation shutters on the top is applied on its north facade, as shown in Fig. 8.2.

The outdoor condition in Shenzhen is rather hot and humid all through the year as shown in Fig. 8.3. The annual outdoor air relative humidity is about 80 %, and humidity ratio in summer is as high as 20 g/(kg dry air). The building requires both cooling and dehumidification in a long period of time and no heating and humidification requirement in winter. Therefore, how to handle the moisture efficiently is the key issue in such a subtropical area.

8.1.1.2 THIC Air-Conditioning System

The THIC system serves from 1st to 4th floor with the net air-conditioning area of 13,180 m^2 (the total area of 18,769 m^2), and the 5th floor is served by several stand-alone air conditioners and is not within the scope of our discussion. The schematic of the THIC system is represented in Fig. 8.4 with the main devices' parameters listed in Table 8.2.

The right side of Fig. 8.4 is the humidity control subsystem, including 9 liquid desiccant outdoor air handling units that supply adequate dry outdoor air into the occupied spaces. As the flow rate of the supplied outdoor air is proportional to the number of people, the pollutants, CO_2, and latent heat produced by human bodies can be removed by outdoor air. The schematic of the heat pump-driven outdoor air processors using liquid desiccant is illustrated in Fig. 5.12, which is composed of a two-stage enthalpy recovery device and a two-stage air handling device coupled with refrigeration cycles. LiBr aqueous solution is employed as liquid desiccant in these air processors. The enthalpy recovery device is used to recover the energy from indoor exhaust air to decrease the energy consumption of the outdoor air handling process. And in the heat pump-driven air handling device, the diluted solution from the dehumidification modules is heated by the exhaust heat from the condenser and concentrated in the regeneration modules. Then the hot concentrated solution is cooled by passing through the heat exchanger and evaporator before it enters the dehumidification modules and lastly used to remove moisture from the outdoor air.

Fig. 8.1 The tested office building in Shenzhen

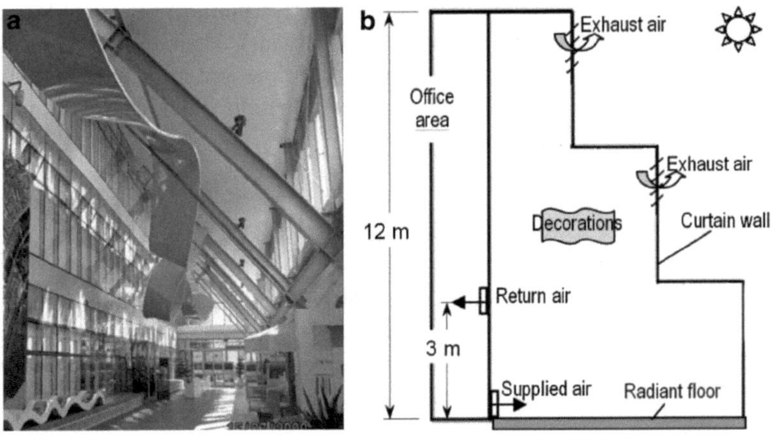

Fig. 8.2 Vestibule in the office building: (**a**) photo and (**b**) schematic figure

The left side of Fig. 8.4 is the temperature control subsystem that undertakes the rest sensible load to control indoor temperature, including a high-temperature chiller, a cooling tower, a cooling water pump, a chilled water pump, and indoor terminal devices (radiant panels and dry fan coil units). The high-temperature chiller is a centrifugal chiller with the rated COP of 8.3 (designed condition: the inlet and outlet temperature of the chilled water and cooling water are 20.5 °C/17.5 °C and 30.0 °C/35.0 °C, respectively), which is much higher than the conventional chiller operating at the chilled water temperature of 12 °C/7 °C. As for indoor terminal devices, as shown in Fig. 8.5, FCUs operating in "dry condition" are set up in the restaurant, archive, and office regions which serve about 81 % of the entire cooling load of the temperature control subsystem, while radiant floor and radiant ceiling panels are applied in vestibule and some office rooms which serve the rest 19 %.

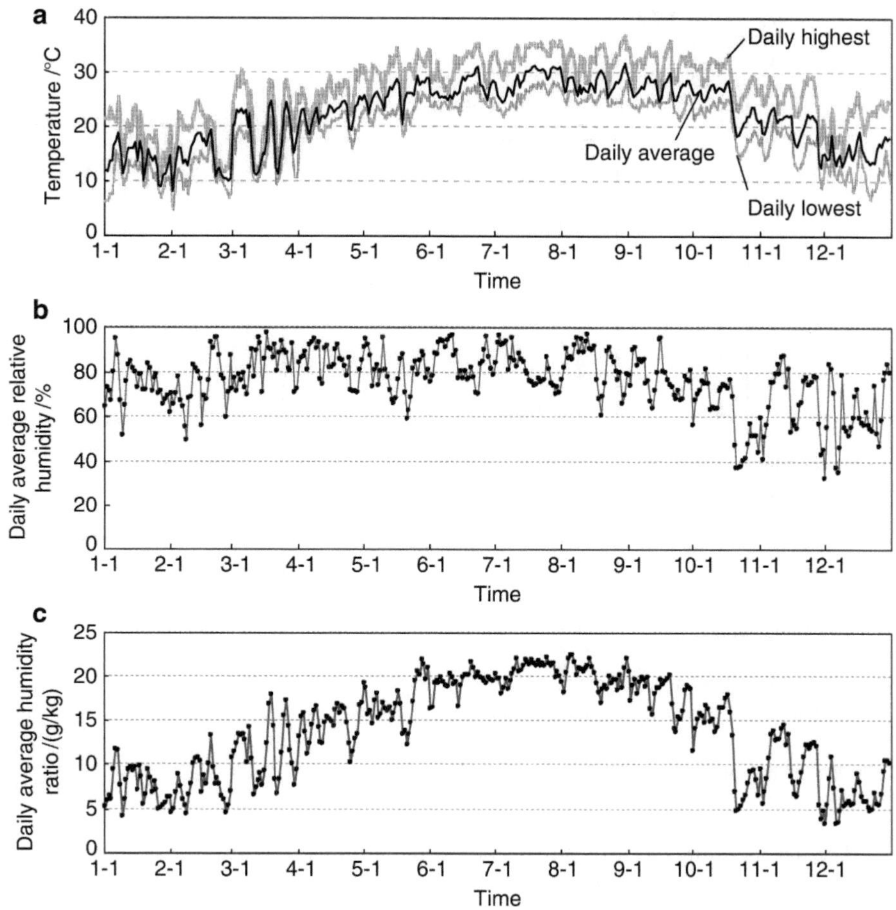

Fig. 8.3 Annual weather data of Shenzhen: (**a**) daily highest, average, and lowest temperatures; (**b**) daily average relative humidity; and (**c**) daily average humidity ratio

The whole THIC system layout has been introduced briefly as analyzed above. Particularly, stratified air-conditioning, a key design principle of large space, is selected in the air-conditioning design of the vestibule as shown in Fig. 8.2b. Specifically, in the occupied zone (indicating the area with a height within 2 m), chilled water with a temperature of 17.5 °C is pumped and distributed into radiant floor for cooling, and the handled dry outdoor air is supplied in the bottom with indoor exhaust air expelled in the middle of the space, as shown in Fig. 8.6a. This configuration forms a "dry air layer" to protect the cold floor surface from condensation. Solar radiation that enters through glass curtain wall is absorbed by the ornamental decorations in the higher space above occupied zone, and the heat is then carried away by natural ventilation through the shutters (shown in Fig. 8.6b) directly.

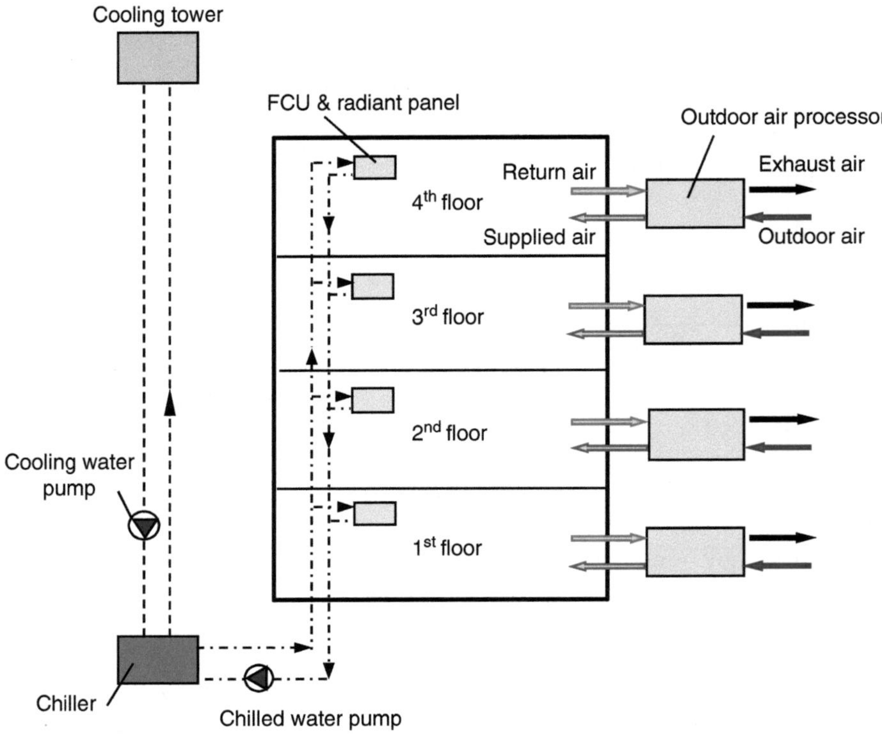

Fig. 8.4 Schematic of the THIC air-conditioning system

The temperature control subsystem and humidity control subsystem can be operated separately according to ambient condition and indoor requirement:

- The two subsystems operate together at hot and humid outdoor climate.
- Only the humidity control subsystem operates at cold but humid ambient condition.
- Outdoor air is directly introduced into occupied spaces after filter when outdoor air is dry enough, such as with a humidity ratio of 11 g/kg.

8.1.2 Performance Test of the THIC System

The field test of the energy efficiency of the THIC system was conducted under both the partial load condition and design condition. The former one was on May 27, 2009, with ambient temperature of 29.3 °C and relative humidity of 79 %, and the latter one was on July 16, 2009, with ambient temperature of 34.9 °C and relative humidity of 61 %. The measurement was divided into two parts: humidity control subsystem and temperature control subsystem.

Table 8.2 Main devices of the THIC system in the office building

Subsystem	Device	Quantity	Rated parameters
Temperature control subsystem	Chiller	1	Cooling capacity: 893 kW; power consumption: 107 kW; rated COP = 8.3
	Cooling tower	1	Flow rate: 200 m^3/h; power consumption: 7.5 kW
	Cooling water pump	1	Flow rate: 180 m^3/h; head: 29 m; power consumption: 22 kW
	Chilled water pump	1	Flow rate: 262 m^3/h; head: 32 m; power consumption: 37 kW
	Indoor terminal devices	–	Radiant panels and dry FCUs, serving 19 % and 81 % of the entire sensible cooling, respectively
Humidity control subsystem	Outdoor air handling units	1	Fresh air flow rate: 2,000 m^3/h; cooling capacity: 39 kW; power consumption[a]: 10 kW; for offices in the 4th floor
		2	Fresh air flow rate: 4,000 m^3/h; cooling capacity: 83 kW; power consumption[a]: 25 kW; for offices in the 4th floor
		4	Fresh air flow rate: 5,000 m^3/h; cooling capacity: 103 kW; power consumption[a]: 28 kW; 2 units for the 2nd floor and 2 units for the 3rd floor
		1	Fresh air flow rate: 8,000 m^3/h; cooling capacity: 166 kW; power consumption[a]: 45 kW, for vestibule in the 2nd floor
		1	Fresh air flow rate: 10,000 m^3/h; cooling capacity: 196 kW; power consumption[a]: 45 kW, for restaurant in the 1st floor

[a]Including power consumption of compressors, solution pumps, and fans inside the outdoor air units

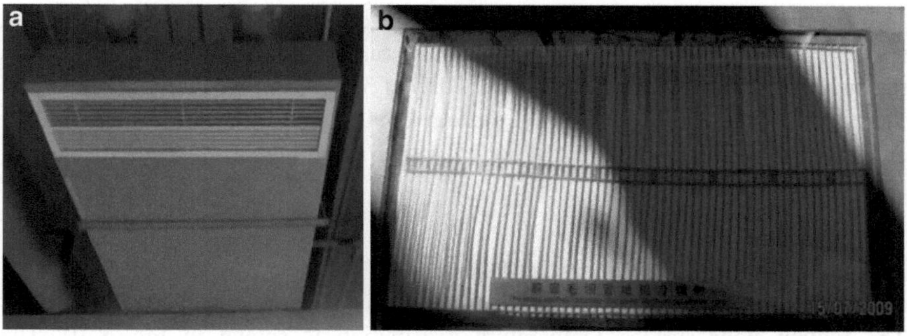

Fig. 8.5 Indoor terminal devices: (**a**) dry fan coil unit and (**b**) capillary radiant panel

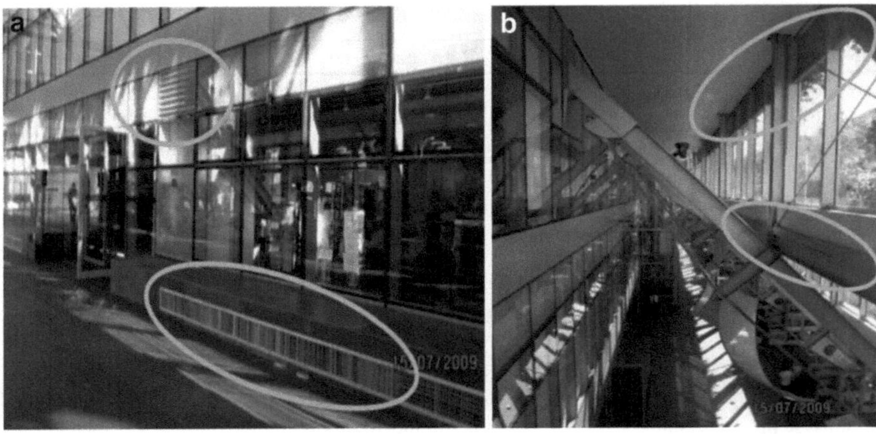

Fig. 8.6 Supply, return air vents, and top shutter: (a) supply and return air vents and (b) top shutter

8.1.2.1 Indoor Thermal Environment

Outdoor Condition

The outdoor condition in Shenzhen is rather hot and humid, as shown in Fig. 8.7. The peak air temperature was 35 °C, and the humidity ratio was as high as 20 g/kg. The solar radiation was measured by a portable solar pyranometer. The peak solar radiation was as high as about 1,000 W/m^2 at noon, as indicated in Fig. 8.8.

Indoor Thermal Environment

This system has been brought into operation since July 2008. Figure 8.9 shows the tested results of indoor temperatures, humidity ratios, and CO_2 concentrations with the outdoor temperature and relative humidity of 34.9 °C and 61 %, respectively. As indicated by the figure, the THIC system could provide a comfortable indoor environment with suitable thermal condition and good indoor air quality.

Radiant Floor Surface Temperature Distribution

The radiant floor surface temperature distribution is shown in Fig. 8.10. The air-conditioning system was switched on at 6:30 in the morning, and the radiant floor surface temperature reached steady at 9:30. During the afternoon (16:00–17:00), the radiant floor surface temperature rose when direct solar radiation reached the floor surface.

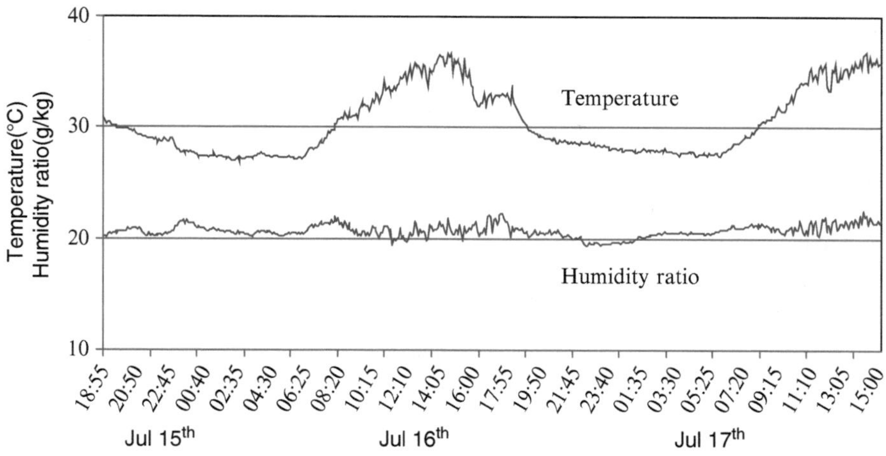

Fig. 8.7 Ambient temperature and humidity ratio during the test

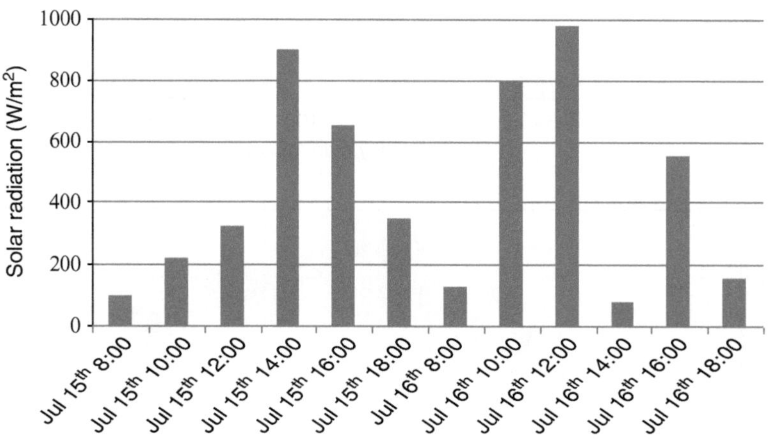

Fig. 8.8 Solar intensity during the test

The effect of the indoor furniture shelter on the surface temperature of the radiant floor is shown in Fig. 8.11. As the radiation heat transfer with the surrounding environment of the sheltered radiant floor is weaken by the furniture shelter, the floor surface temperature is about 1 °C lower than the unsheltered radiant floor. Hence, the surface of the sheltered radiant floor could have a risk of condensation. The temperature and humidity parameters of the air near the floor were measured, as shown in Fig. 8.12. It can be seen that the dew point temperature of the surrounding air was about 15 °C, lower than the lowest temperature of the radiant floor surface. Therefore, there was no risk of condensation during the test.

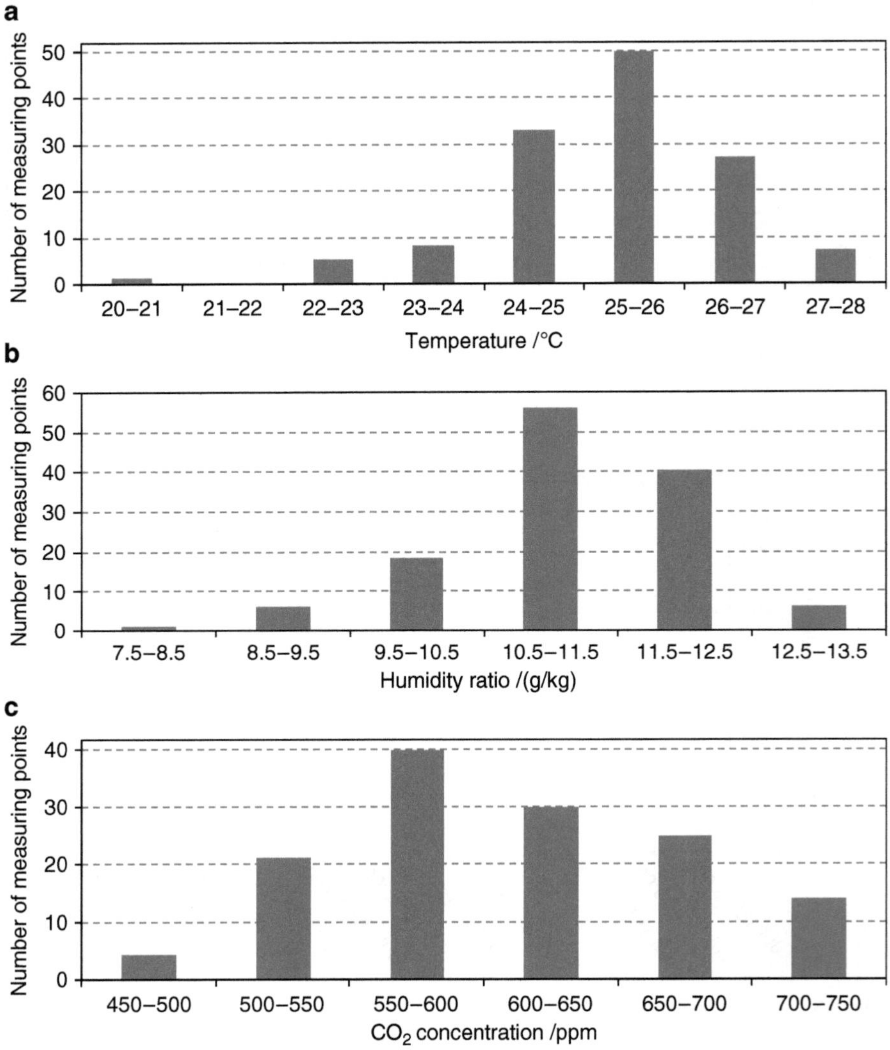

Fig. 8.9 Indoor air parameters of the office building: (**a**) temperature, (**b**) humidity ratio, and (**c**) CO_2 concentration

Air Temperature and Humidity Ratio Distribution in the Vertical Direction

Figure 8.13 depicts the tested temperatures and humidity ratios along the vertical direction in the vestibule. In the occupied zone (the height within 2 m), the temperature and humidity ratio were about 26 °C and 12 g/kg, respectively. The results meet human thermal comfort well.

Both the temperature and humidity ratio increased fast along the vertical direction, which reached about 30 °C and 20 g/kg at the height of 10 m. And the peak

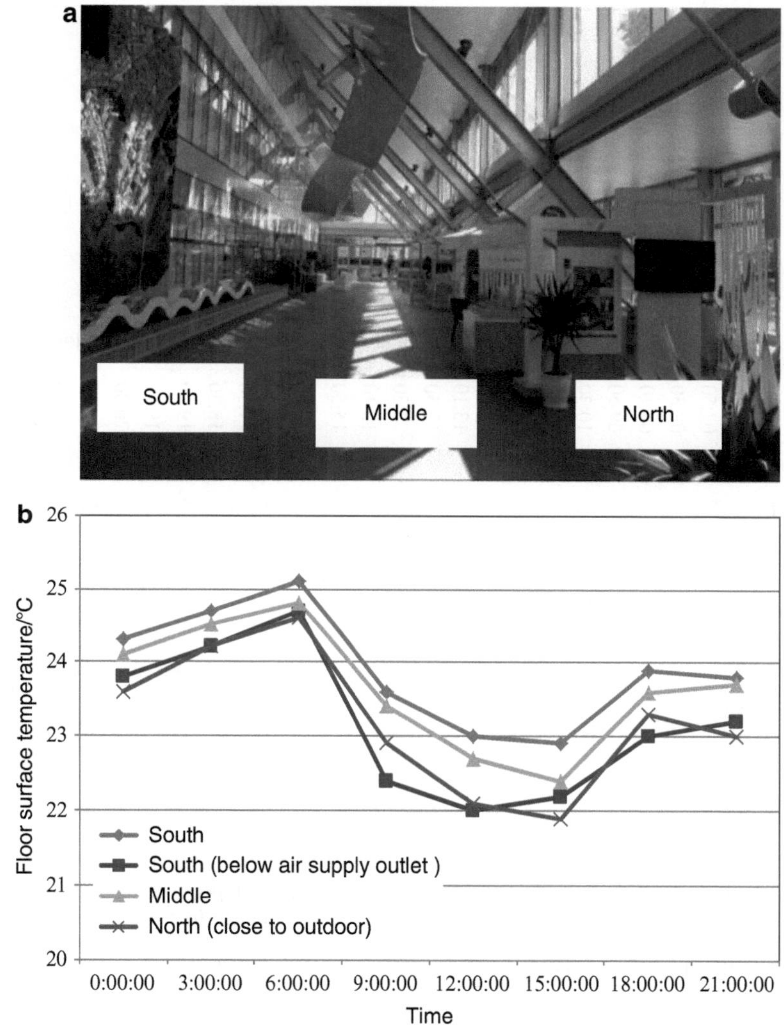

Fig. 8.10 Radiant floor surface temperature profile: (**a**) field test points and (**b**) radiant floor surface temperature profile

temperature at top space (above 7 m) occurred at noon, due to the strong solar radiation and high ambient temperature. According to the test results, the THIC system applied in large space is effective on energy saving, which only keeps the occupied zone in comfort condition and forms apparent stratification of indoor temperature and humidity ratio along the vertical direction. Moreover, the natural ventilation from the shutters contributed to removal of the absorbed heat from decorations to outdoor environment.

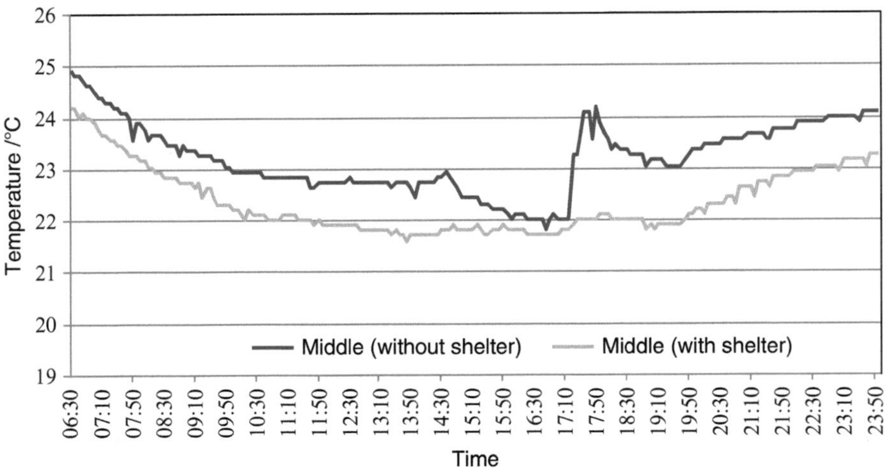

Fig. 8.11 Effect of the indoor furniture shelter on the radiant floor surface temperature

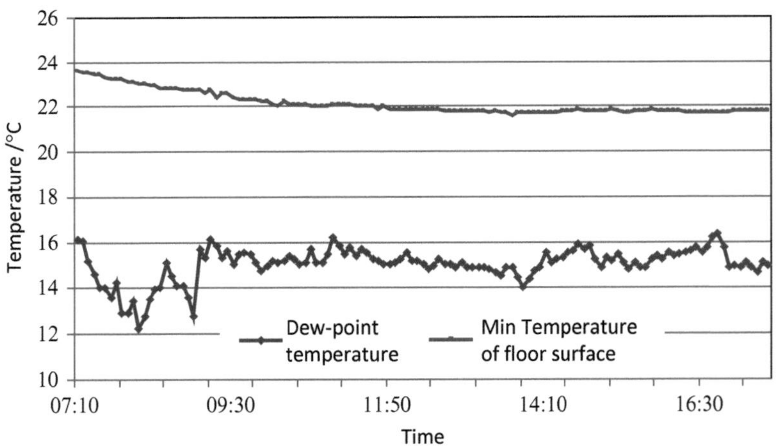

Fig. 8.12 Test results of radiant floor minimum temperature and surrounding air dew point temperature

8.1.2.2 Energy Efficiency of Humidity Control Subsystem

The performances of the liquid desiccant outdoor air units were tested one by one on May 27, 2009, according to the measured flow rates and air inlet and outlet parameters through the processor and input power of compressors, solution pumps, and fans. The tested results of seven outdoor air units are summarized in Table 8.3; the other two processors are neglected due to the difficulty of installing the sensors.

The east side processor of the 2nd floor is a typical example for the outdoor air unit, and its specific operation information is shown in Table 8.4. The outdoor air

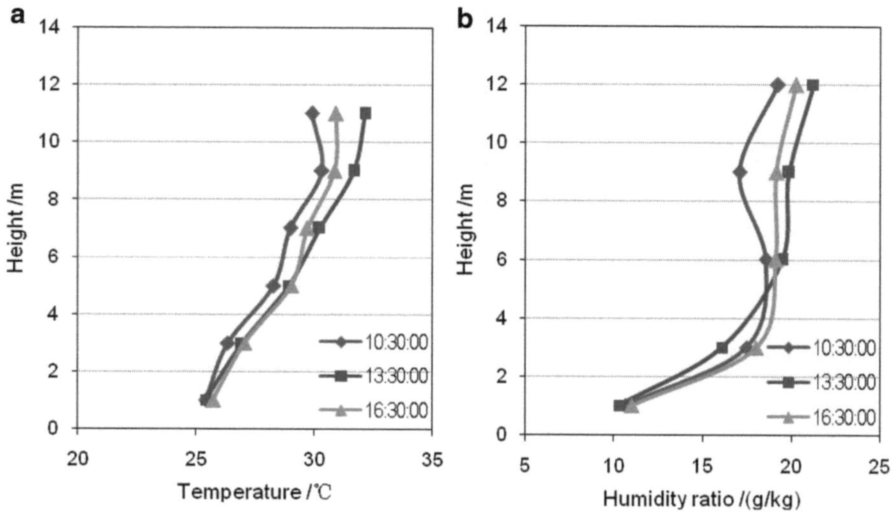

Fig. 8.13 Temperature and humidity ratio distribution in the vertical direction of the vestibule: (**a**) temperature and (**b**) humidity ratio

flow rate was 5,059 m^3/h, the outdoor air parameters were 29.3 °C and 20.3 g/kg, and the supply air parameters were 17.1 °C and 6.2 g/kg. So the cooling capacity (Q_{air}), calculated by energy balance equation, was 82.6 kW. The power consumption of compressors together with solution pumps (P_{air}) inside the processor was 17.8 kW, and the power consumption of the supply air and exhaust air fans (P_{fan}) was 2.2 kW. Therefore, the performance of the outdoor air unit (COP_{air}), the transport coefficient of fans (TC_{fan}), and the performance of the entire humidity handling process (COP_{hum}), as shown in Eqs. (8.1), (8.2), and (8.3), are 4.7, 37.5, and 4.2, respectively.

$$COP_{air} = \frac{Q_{air}}{P_{air}} \tag{8.1}$$

$$TC_{fan} = \frac{Q_{air}}{P_{fan}} \tag{8.2}$$

$$COP_{hum} = \frac{Q_{air}}{P_{air} + P_{fan}} \tag{8.3}$$

As shown in Table 8.3, the COP_{air} of the tested seven outdoor air units were in the range of 4.4–4.9, the TC_{fan} of the fans are 35–40, and the COP_{hum} of the entire humidity handling processes were 4.0–4.4.

According to the tested data on May 27, 2009, and rated parameters of the outdoor air units and fans, the calculated cooling capacity of the entire humidity control subsystem was 773.0 kW with total inside compressors and solution pumps

Table 8.3 Performance of the outdoor air handling units (outdoor condition: 29.3 °C and 20.3 g/kg)

Location	Supply outdoor air			Cooling capacity/kW	Power consumption/kW				
	Flow rate/(m^3/h)	Temperature/°C	Humidity ratio/(g/kg)		Compressors and solution pumps	Fans	COP_{air}	TC_{fan}	COP_{hum}
East side of 2nd floor	5,059	17.1	6.2	82.6	17.8	2.2	4.7	37.5	4.2
West side of 2nd floor	5,195	16.7	6.1	86.0	17.6	2.3	4.9	37.4	4.3
East side of 3rd floor	4,972	16.8	6.5	80.4	18.2	2.2	4.4	36.5	4.0
West side of 3rd floor	5,215	16.6	6.2	86.4	17.6	2.2	4.9	39.3	4.4
East side of 4th floor	4,261	16.7	6.4	69.5	15.0	1.7	4.6	40.9	4.2
Middle side of 4th floor	1,940	16.5	6.2	32.1	7.1	0.9	4.5	35.7	4.0
West side of 4th floor	4,307	16.3	6.1	72.0	15.3	1.8	4.7	40.0	4.2

Table 8.4 Operation condition of a typical outdoor air handling unit (east side of 2nd floor)

Outdoor condition	Supplied outdoor air	Stage	Evaporating temperature/°C	Condensing temperature/°C	Solution parameter (temperature and concentration)	
Temperature: 29.3 °C	Temperature: 17.1 °C	I	11.0	50.8	Dehumidification module	Inlet: 15.8 °C, 34.6 %
Humidity ratio: 20.3 g/kg	Humidity ratio: 6.2 g/kg					Outlet: 20.5 °C, 34.4 %
	Flow rate: 5,059 m^3/h				Regeneration module	Inlet: 44.2 °C, 34.9 %
						Outlet: 38.3 °C, 35.2 %
		II	4.4	51.3	Dehumidification module	Inlet: 9.0 °C, 44.1 %
						Outlet: 14.0 °C, 43.9 %
					Regeneration module	Inlet: 44.3 °C, 44.8 %
						Outlet: 37.9 °C, 45.1 %

input power of 166.9 kW and total fans input power of 20.0 kW, so the coefficient of performance of the humidity control subsystem (COP_{HUM}), shown in Eq. (8.4), is 4.1.

$$COP_{HUM} = \frac{\Sigma Q_{air}}{\Sigma(P_{air} + P_{fan})} \tag{8.4}$$

Similarly, on the basis of the tested data on July 16, 2009, the COP_{HUM} under the design condition was 4.1, with the calculated cooling capacity of 915.0 kW, total inside compressors and solution pumps input power of 194.4 kW, and total fans input power of 25.1 kW.

8.1.2.3 Energy Efficiency of Temperature Control Subsystem

The performances of the chiller, cooling tower, cooling and chilled water pumps, and indoor dry FCUs were measured, based on the measured flow rates and water inlet and outlet parameters through the chiller and input powers of above facilities. The tested results of each device, under both partial load condition and design condition, are listed in Table 8.5.

According to the tested data under the partial load condition, the calculated cooling capacity of the temperature control subsystem (Q_{CH}) was 446.1 kW with total input power of 120.8 kW. So the COP of the temperature control subsystem (COP_{TEMP}), as shown in Eq. (8.5), was 3.7. The COP_{TEMP} under the design condition was 4.1, with the calculated cooling capacity of 543.4 kW and overall power consumption of 133.6 kW.

$$COP_{TEMP} = \frac{Q_{CH}}{P_{CH} + P_{CT} + P_{CTP} + P_{CWP} + P_{FCU}} \tag{8.5}$$

8.1.2.4 Energy Efficiency of the Entire THIC System

The entire load of the building is composed of the latent load and sensible load. The former consists of both the moisture introduced by outdoor air and produced from occupants, and the latter is caused by the ventilation, heat transfer, solar radiation, heat dissipation of devices, activity of occupants, etc., as illustrated in Fig. 8.14. In the THIC system, processed outdoor air is used to remove the entire latent load and part of the sensible load, while cooling plant provides high-temperature chilled water to remove the rest sensible load. By calculation, under the partial load condition, the humidity control subsystem removed 63 % of entire cooling load with 61 % environmental control energy consumption, and the temperature control subsystem removed the rest 37 % load with the rest 39 % energy consumption. The same conclusion also can emerge from the test of design point.

Table 8.5 Performance of the temperature control subsystem

	Cooling capacity	Device	Power consumption	Tested performance	Comments
Partial load condition (May 27)	$Q_{CH} = 446.1$ kW	Chiller	$P_{CH} = 52.5$ kW	$COP_{CH} = \frac{Q_{CH}}{P_{CH}} = 8.5$	Outdoor: 29.3 °C, 20.3 g/kg
		Cooling tower	$P_{CT} = 3.7$ kW	$TC_{CT} = \frac{Q_{CH}+P_{CH}}{P_{CT}} = 134.8$	Chilled water flow rate: 239 m^3/h
		Cooling water pump	$P_{CTP} = 14.6$ kW	$TC_{CTP} = \frac{Q_{CH}+P_{CH}}{P_{CTP}} = 34.2$	Cooling water flow rate: 180 m^3/h
		Chilled water pump	$P_{CWP} = 30.6$ kW	$TC_{CWP} = \frac{Q_{CH}}{P_{CWP}} = 14.6$	Chilled water temperatures: 17.5 °C/19.1 °C
		Fan coil units (serve 81 % of the Q_{CH})	$P_{FCU} = 19.4$ kW	$TC_{FCU} = \frac{0.81 \times Q_{CH}}{P_{FCU}} = 18.6$	Cooling water temperatures: 31.0 °C/33.3 °C
Very hot and humid condition (July 16)	$Q_{CH} = 543.4$ kW	Chiller	$P_{CH} = 63.8$ kW	$COP_{CH} = \frac{Q_{CH}}{P_{CH}} = 8.5$	Outdoor: 34.9 °C, 21.6 g/kg
		Cooling tower	$P_{CT} = 3.7$ kW	$TC_{CT} = \frac{Q_{CH}+P_{CH}}{P_{CT}} = 147.0$	Chilled water flow rate: 260 m^3/h
		Cooling water pump	$P_{CTP} = 14.4$ kW	$TC_{CTP} = \frac{Q_{CH}+P_{CH}}{P_{CTP}} = 37.7$	Cooling water flow rate: 180 m^3/h
		Chilled water pump	$P_{CWP} = 29.4$ kW	$TC_{CWP} = \frac{Q_{CH}}{P_{CWP}} = 18.5$	Chilled water temperatures: 17.5 °C/19.3 °C
		Fan coil units (serve 81 % of the Q_{CH})	$P_{FCU} = 22.3$ kW	$TC_{FCU} = \frac{0.81 \times Q_{CH}}{P_{FCU}} = 24.4$	Cooling water temperatures: 32.9 °C/35.7 °C

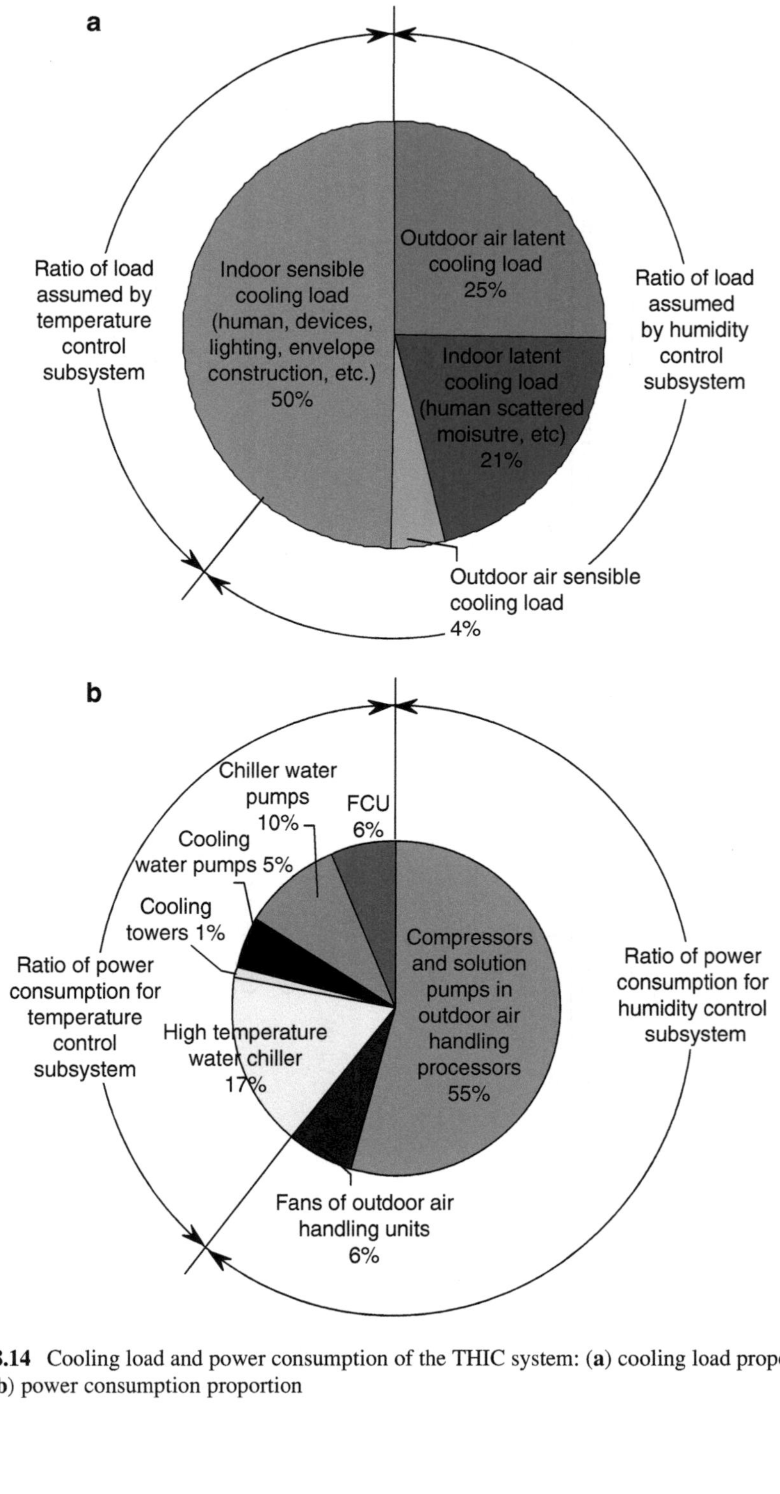

Fig. 8.14 Cooling load and power consumption of the THIC system: (**a**) cooling load proportion and (**b**) power consumption proportion

Calculated from the above tested data, the overall COP of the THIC system (COP_{SYS}) under partial load condition and design condition were 4.0 and 4.1, respectively, as shown in Eq. (8.6).

$$COP_{SYS} = \frac{Q_{CH} + \Sigma Q_{air}}{(P_{CH} + P_{CT} + P_{CTP} + P_{CWP} + P_{FCU}) + \Sigma(P_{air} + P_{fan})} \tag{8.6}$$

Based on the tested results of these two typical operating conditions, it is convinced that the THIC system in this office building has achieved a high efficiency with its total COP over 4.0. By comparison, the measured average coefficient of performance of whole system in traditional fan coil systems is usually around 3.0. Therefore, there is a remarkable energy efficiency improvement of the THIC system comparing with the conventional system.

8.1.3 Energy Consumption of the THIC System

Energy consumption of the THIC system was measured by the power metering monitoring system. Figure 8.15 shows the monthly power consumption of the system from April 15, 2009, to October 15, 2009 (apart from weekends and statutory holidays). The total energy consumption was 425,000 kWh during the cooling season, and the humidity control subsystem occupied 61 % of the total power consumption which was proportional to the ratio of cooling load that the humidity control subsystem undertook.

The annual energy consumption in unit building area and unit net air-conditioning area of the tested THIC system were 22.6 kWh/($m^2 \cdot$ year) and 32.2 kWh/($m^2 \cdot$ year), respectively. However, the average energy consumption levels of office building of the similar building envelope and occupant density during the same time in Shenzhen were around 42–49 kWh/(m^2·year) according to the investigated results by Zi (2007). Therefore, the THIC system in this office building achieved noticeable energy saving in operation compared with the conventional air-conditioning system, and the added cost can be recalled within 2 years.

8.1.4 Discussions

According to our knowledge, cooling air can be realized more easily than dehumidification by condensation, since the latter one requires lower temperature of cooling source than the former. However, the COP of the tested temperature control subsystem is lower than that of the humidity control subsystem in present THIC system. Thus, this section focuses on how to improve the performance of the temperature control subsystem.

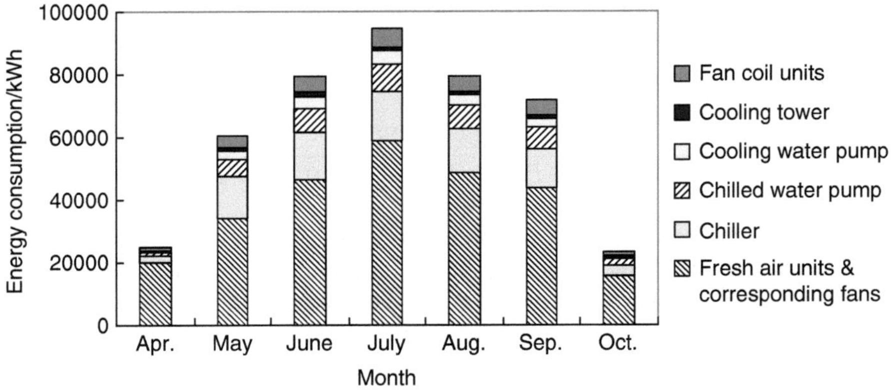

Fig. 8.15 Monthly power consumption of the THIC system

According to the performance of each component in the temperature control subsystem shown in Table 8.5, three main improvements of the temperature control subsystem are recommended: (1) reducing the frequency of the chilled water pump, (2) improving the cooling tower performance by tightening the strap of the fan's belt, and (3) improving the performance of FCUs under dry working condition. The first two methods can be easily achieved in the building, while the third one depends on the improvement of new FCU products.

8.1.4.1 Improve the Performance of Chilled Water Pump

As indicated in Fig. 8.14b and Table 8.5, the power consumption of the chilled water pump was near 60 % of the chiller. It was because the variable-speed pump has operated at 50 Hz in the past that the chilled water temperature difference between the inlet and outlet of chiller was only 1.6–1.8 °C, while the designed value was 3.0 °C. Therefore, the power consumption of the chilled water pump can be considerably cut down after reducing the chilled water flow rate by frequency regulation. For instance, the energy savings of the chilled water pump could be up to 42 %, if the transport coefficient of chilled water pump (TC_{CWP}) is increased from the current 14.6 to an empirical value of 25.

8.1.4.2 Improve the Performance of Cooling Tower

According to Table 8.5, the transport efficient of cooling tower (TC_{CT}) was near 140, which was lower than the empirical index of 200. Detailed performance test was then launched. For the cooling tower, the mass flow rate ratio of air to water was 0.55, with air flow rate of 99,620 kg/h and water flow rate of 180,000 kg/h, and the temperatures of inlet water, outlet water, and outdoor air wet bulb were 35.6 °C, 32.9 °C, and 28.3 °C, respectively. Hence, the efficiency of the cooling tower (the temperature variance of the cooling water divided by the possible maximal temperature variance) was (35.6–32.9)/(35.6–28.3) = 37 %. By field survey, the low air flow rate, which led to low efficiency of the cooling tower, was caused by the loose strap of fan.

The energy consumption of the cooling tower in the entire air-conditioning system is relatively small, but it seriously influences the chiller performance by affecting the condensing temperature. The expected efficiency of the cooling tower can be higher than 55 % if the air flow rate increases to about 180,000 kg/h, and thus the condensing temperature can decrease at least 2 °C. As a result, the COP of the chiller (COP_{CH}) can be increased from current 8.5 to 9.3.

8.1.4.3 Improve the Performance of FCU

As the chilled water inlet temperature is higher than the indoor dew point temperature in this THIC system, fan coil units that operate under dry working condition are put into use in this building to remove large part of the sensible load of the building. At present, the configuration of the adopted dry FCUs is similar to the originally wet FCUs (with standard chilled water inlet temperature of 7 °C and transport coefficient of 50). However, the tested transport coefficient of the dry FCUs (TC_{FCU}) was around 20, as shown in Table 8.5. Factually, the configuration of dry FCUs should be rather different from wet FCUs, since condensing water no longer exists in dry FCUs, and some dry FCUs with new configuration have appeared with TC_{FCU} close to 50. So there is a huge potential to improve the performance of dry FCU with the transport coefficient increased to around 40.

The contribution of the above three improvements to the temperature control subsystem is summarized in Table 8.6. And the expected COP_{TEMP} and COP_{SYS} can be increased to 4.8 and 4.4 after these alterations, respectively. Therefore, the energy consumption of the altered temperature control subsystem can save 23 % compared to present situation, and the entire THIC system can save 9 % compared to its performance at present.

8.1.5 Conclusion

The operating performance of the THIC system in an office building in Shenzhen is presented in this section. Liquid desiccant outdoor air units are used to supply dry outdoor air to control indoor humidity, and chilled water with the temperature of 17.5 °C is pumped and distributed into radiant panels and dry fan coil units to control indoor temperature. The following are the conclusions based on the tested results:

1. The THIC system can provide a comfortable indoor environment that indoor temperatures, humidity ratios as well as CO_2 concentrations were all within the comfortable ranges.

Table 8.6 Improvement of the temperature control subsystem

		Chiller	Cooling tower	Cooling water pump	Chilled water pump	Fan coil units
Partial load condition	Present power consumption/kW	52.5	3.7	14.6	30.6	19.4
	Expected power consumption/kW	47.7 (COP_{CH} increases from 8.5 to 9.3)	3.7	14.6	17.8 (TC_{CWP} increases from 14.6 to 25)	9.0 (TC_{FCU} increases from 18.6 to 40)
Very hot and humid condition	Present power consumption/kW	63.8	3.7	14.4	29.4	22.3
	Expected power consumption/kW	58.4 (COP_{CH} increases from 8.5 to 9.3)	3.7	14.4	21.7 (TC_{CWP} increases from 18.5 to 25)	13.6 (TC_{FCU} increases from 24.4 to 40)

2. The COP of the entire THIC system can reach 4.0, with the COP of the temperature control subsystem and humidity control subsystem of 3.7–4.1 and 4.1, respectively. The energy consumption of the THIC system in the tested office building was 32.2 kWh/(m^2 · year) (net air-conditioning area), that is, the energy efficiency was much higher than that of the conventional air-conditioning system available in literature.
3. Possible improvements of the temperature control subsystem are provided, including modification on the chilled water pump, cooling tower, and FCUs. Thus, the expected system COP can be further increased to 4.4, which can save 9 % compared to present air-conditioning system.

8.2 Application in a Hospital Building (Dry Region)

8.2.1 Basic Information

A hospital building, located in Urumchi of Xinjiang province, has been put into use since 2007, shown in Fig. 8.16. The data in this subsection is from the PhD thesis by Xie XY (tutor: Jiang Y). The total area of this hospital building is about 46,000 m^2. Hereinto, the basement is used as garage, laundry room, storeroom, etc. And the floors over ground are consulting rooms, treatment rooms, wards, clinics, and operating rooms. The area of ward and clinic from 5th to 18th floor is about 38,000 m^2, and the THIC air-conditioning system with indirect evaporative cooling technology is adopted for regulating the thermal built environment.

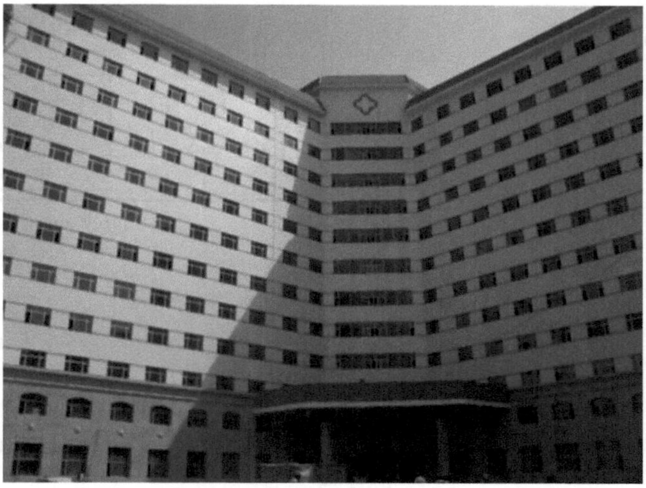

Fig. 8.16 A hospital in Urumchi

8.2.1.1 Indirect Evaporative Cooling Air-Conditioning System Based on THIC Idea

As indicated by the outdoor climate listed in Fig. 4.1, Urumchi is located in the northwestern dry and hot climate zone. Thus, outdoor dry air can be used for removing the indoor moisture load directly without dehumidification in summer, and chilled water produced by the indirect evaporative cooling technology can be used to remove indoor sensible load. Then the THIC system for this building could be constructed, with its operating schematic shown in Fig. 8.17.

8.2.1.2 Key Devices of the THIC System

Schematic and actual unit of the indirect evaporative chiller are shown in Fig. 8.18a, b, respectively. An indirect evaporative chiller and a total number of 374 dry FCUs are adopted for the temperature control subsystem; their rated power consumptions are listed in Table 8.7. As to the humidity control subsystem, the indirect evaporative outdoor air cooling unit is adopted, with its operating schematic shown in Fig. 8.19. The corresponding air flow rate for the indirect evaporative outdoor air cooling unit is 100,000 m^3/h, with rated total power consumption for fans of 56 kW.

Fig. 8.17 Schematic of the THIC air-conditioning system in the hospital

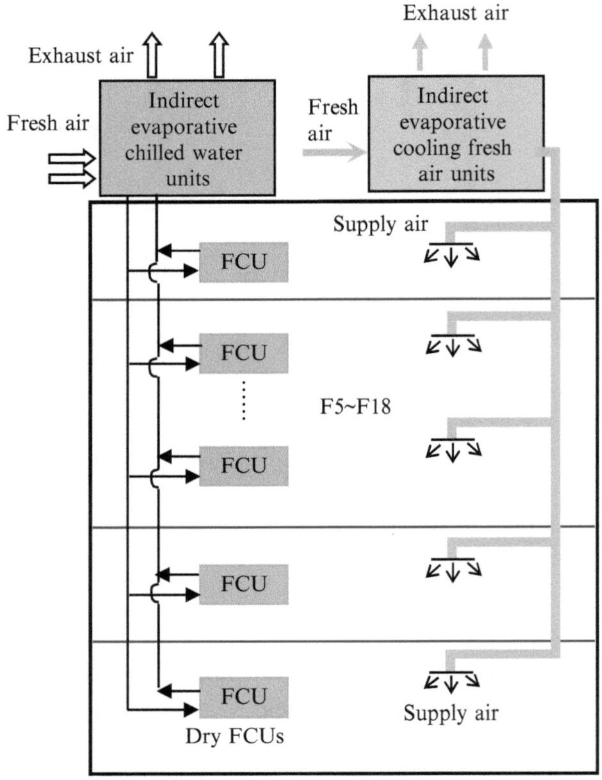

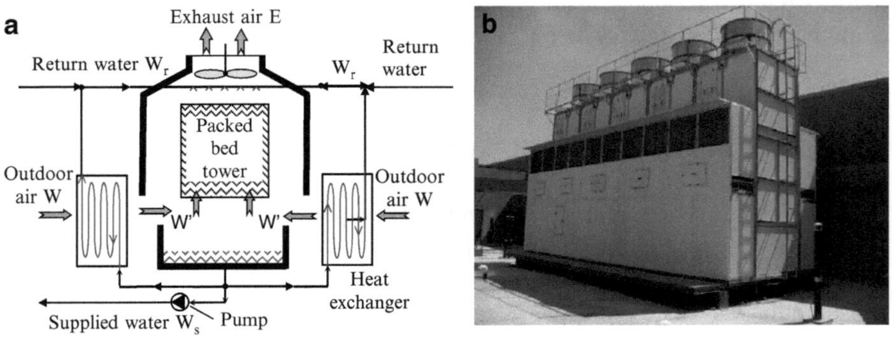

Fig. 8.18 Indirect evaporative chiller: (**a**) schematic and (**b**) photo

Table 8.7 Indirect evaporative chiller and dry fan coil units

Indirect evaporative chiller			Dry fan coil units
Nominal cooling capacity (kW)	Exhaust fan (kW)	Chilled water pump (kW)	(kW)
700	17.5	20	24.9

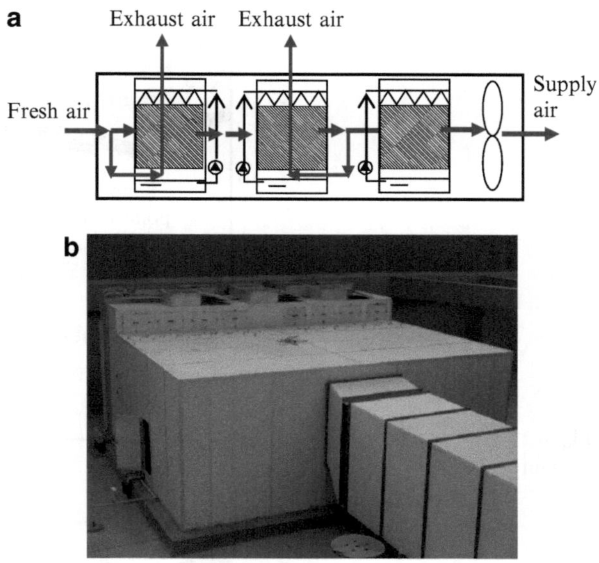

Fig. 8.19 Indirect evaporative cooling unit: (**a**) schematic and (**b**) photo

8.2.2 Performance Test Results of the THIC System

8.2.2.1 Indirect Evaporative Water Chiller and Outdoor Air Handling Unit

Figures 8.20 and 8.21 show the performance of the indirect evaporative chiller and the outdoor air cooling unit in a typical day of summer, respectively. As indicated in Fig. 8.20, chilled water temperature of the indirect evaporative chiller was about 15–17 °C, lower than the corresponding outdoor wet-bulb temperature. As indicated in Fig. 8.21, supply air temperature of the indirect evaporative outdoor air handling unit was 18.21 °C, approaching the outdoor wet-bulb temperature. And humidity ratio of the supply air was 10–11 g/kg.

Based on the indirect evaporative cooling method for this THIC air-conditioning system, the measured indoor temperature and relative humidity are shown in Fig. 8.22. As outdoor temperature varied, indoor temperature could be maintained at 24–26 °C, and indoor corresponding relative humidity could be about 50–60 %, satisfying thermal comfort requirements of human well and creating a comfortable indoor thermal built environment.

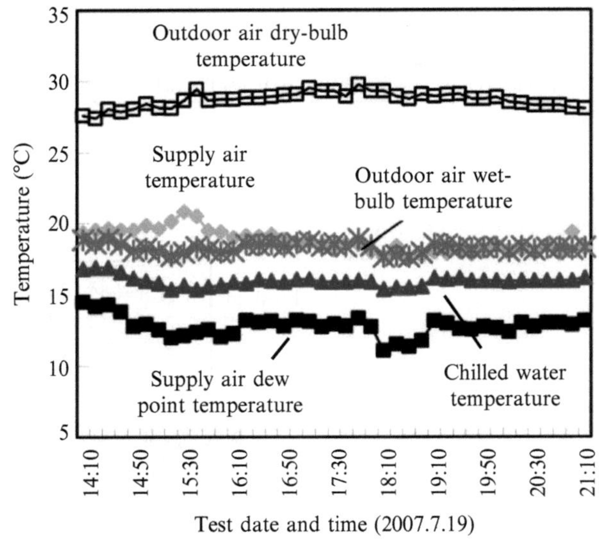

Fig. 8.20 Supply chilled water and supply air temperatures

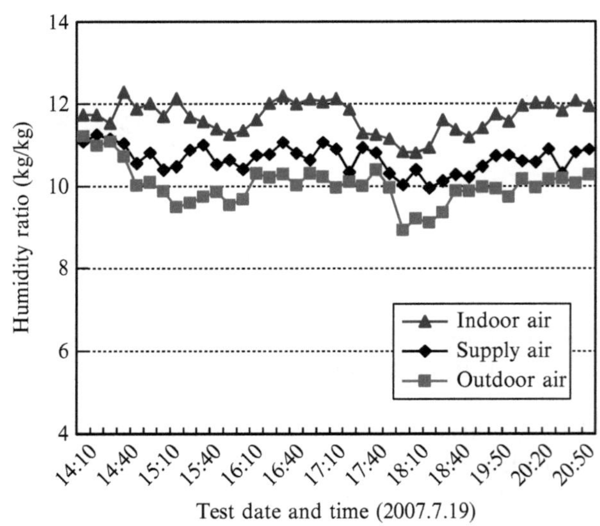

Fig. 8.21 Test results of air humidity ratios

8.2.2.2 Cooling Capacity of the THIC System

In this THIC air-conditioning system using indirect evaporative cooling method, chilled water produced by the indirect evaporative chiller is the main cooling source for cooling, responsible for extracting most of the indoor sensible heat, and the rest part of indoor sensible load could be removed by the handled dry air supplied by the indirect evaporative outdoor air cooling unit. Then the total indoor sensible load

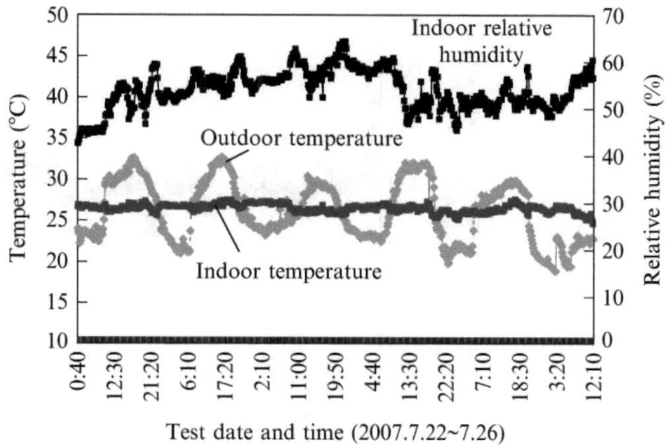

Fig. 8.22 Indoor air temperature and humidity ratio

removed by the indirect evaporative chiller and the outdoor air cooling unit can be calculated as follows:

$$Q_r = G_w \times c_{pw} \times (t_{w,r} - t_{w,out}) + G_f \times c_{pa} \times (t_{r,a} - t_{f,a}) \tag{8.7}$$

The total sensible heat removed from indoor space is shown in Fig. 8.23 during the test period, and the ratio of the heat removed by the indirect evaporative chiller is given as Fig. 8.24. As indicated by these two figures, sensible heat removed by the THIC air-conditioning system is about 400–500 kW, and sensible heat removed by the indirect evaporative chiller accounts for about 40–60 % of the entire sensible load. For this THIC air-conditioning systems, the ratio of sensible load extracted by the evaporative chiller mainly depends on outdoor air parameters and heat and moisture load ratio of the building. For this hospital building, outdoor air required for the air-conditioning system is too large, so that the sensible heat removed by the dry outdoor air can reach 50 %.

8.2.3 Energy Consumption Analysis

Power consumption of this THIC air-conditioning system using indirect evaporative cooling is given in Table 8.8, and the performance of this air-conditioning system is listed in Table 8.9. The coefficient of performance (COP) for the indirect evaporative chiller COP_C was 15.9, higher than that of the conventional vapor compression water chiller, and COP for the indirect evaporative outdoor air handling unit (COP_{air}) was 12.6. For the entire temperature control subsystem using chilled water, including chillers, chilled water pumps, and dry FCUs, the overall COP (COP_{TEMP}) was 4.6. For the entire humidity control subsystem system including outdoor air handling unit and fans, the overall COP (COP_{HUM}) was 4.0.

Fig. 8.23 Total removed sensible load

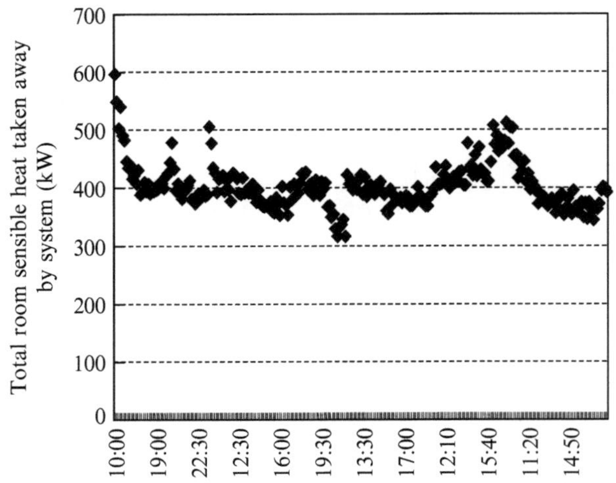

Fig. 8.24 Removed sensible load by indirect evaporative chiller

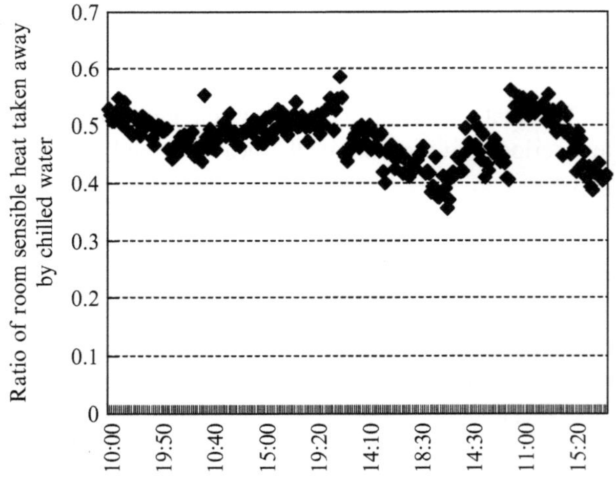

Table 8.8 Energy consumption of each component (kW)

Indirect evaporative chiller P_C	Indirect evaporative cooling units P_{air}	Chilled water pumps P_{chp}	Supply air fan P_{fan}	Dry fan coil units P_{fc}	Total energy consumption ($P_C + P_{air} + P_{chp} + P_{fan} + P_{fc}$)
13.8	13.4	14.9	28.3	19.2	89.6

Table 8.9 Energy coefficient of the air-conditioning system

	Entire performance	Performance of each component
Temperature control subsystem	$COP_{TEMP} = 4.6$	Indirect evaporative chiller: $COP_C = Q_T/P_C = 15.9$
	$Q_T = 219$ kW	Chilled water pumps: $TC_{chp} = Q_T/P_{chp} = 14.7$
		Dry fan coil units: $TC_{fc} = Q_T/P_{fc} = 11.4$
Humidity control subsystem	$COP_{HUM} = 4.0$	Indirect evaporative cooling units: $COP_{air} = Q_H/P_{air} = 12.6$
	$Q_H = 170$ kW	Supply air fan: $TC_{fan} = Q_H/P_{fan} = 6.0$

8.3 Application in a Large Space Building (Airport)

8.3.1 Description of the THIC System in an Airport

8.3.1.1 Basic Information

Terminal 3 of Xi'an Xianyang International Airport (Fig. 8.25), located in Shanxi Province, was designed by China Northwest Architectural Design and Research Institute. The terminal has been in use since May 2012, and it has a total building area of 258,000 m^2. Terminal 3 has three floors: one underground floor and two that are above ground. As a typical airport terminal, the building includes a check-in hall, a departure hall, a baggage claim area, and offices. Figure 8.25b gives an outlook of the check-in hall. The maximum height of the terminal is 37.0 m, and the underground depth is 8.6 m.

The outdoor environmental conditions in Xi'an are shown in Fig. 8.26, including annual outdoor air temperature, relative humidity, and humidity ratio. The average relative humidity of the outdoor air in summer is around 60 %, and the humidity ratio is 15–20 g/kg. The cooling season for Terminal 3 is from May 15 to September 15; according to the daily operating schedule of the airport terminal, the air-conditioning system is from 6:00 a.m. to 12:00 a.m. midnight.

Because of the extremely large spaces in this airport terminal, radiant floors and displacement ventilation were selected for the THIC system to conserve air distribution energy during operation. Moreover, since electricity prices in Xi'an vary tremendously (the off-peak price is only about one-third of the peak price), ice storage technology was utilized in order to maximize operating cost savings.

8.3.1.2 THIC Air-Conditioning System

The building envelope is dominated with large glass walls with the internal shade. There are several typical large open spaces, including check-in hall and departure waiting hall on the second floor. The height of check-in hall is 26.5 m and its area is 32,000 m^2, and the height of departure waiting hall is 11.0 m and its area is

Fig. 8.25 Terminal 3 in Xi'an Xianyang International Airport: (**a**) building outlook and (**b**) check-in hall

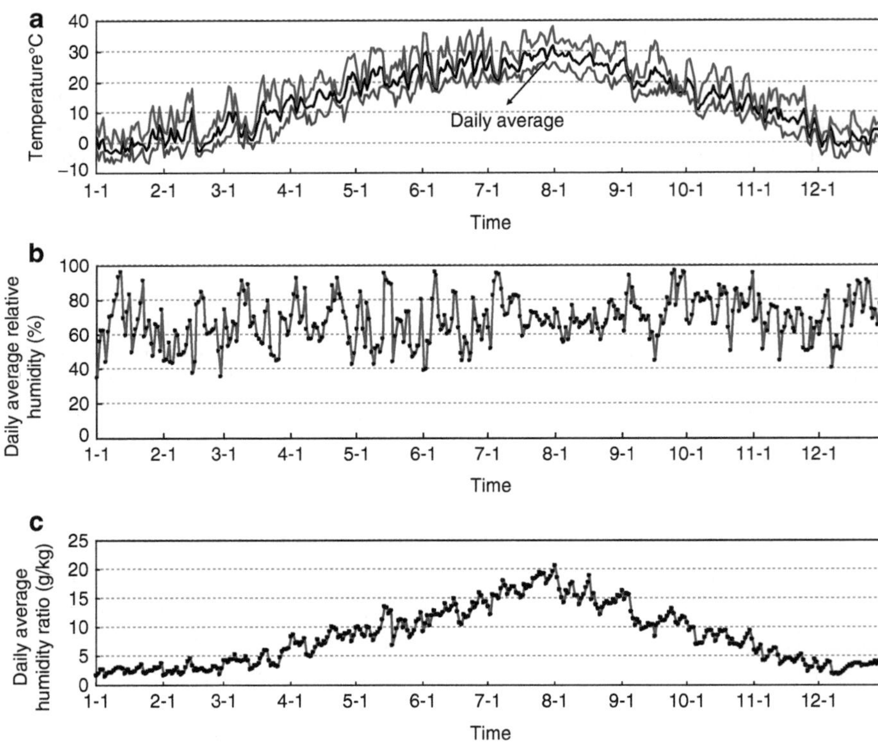

Fig. 8.26 Annual weather data of Xi'an: (**a**) daily maximum, average, and minimum temperatures; (**b**) daily average relative humidity; and (**c**) daily average humidity ratio

15,000 m^2. The THIC air-conditioning system is employed in the large spaces of the airport terminal (i.e., the check-in hall and the departure hall), as shown in Fig. 8.27, with a total area of about 47,000 m^2.

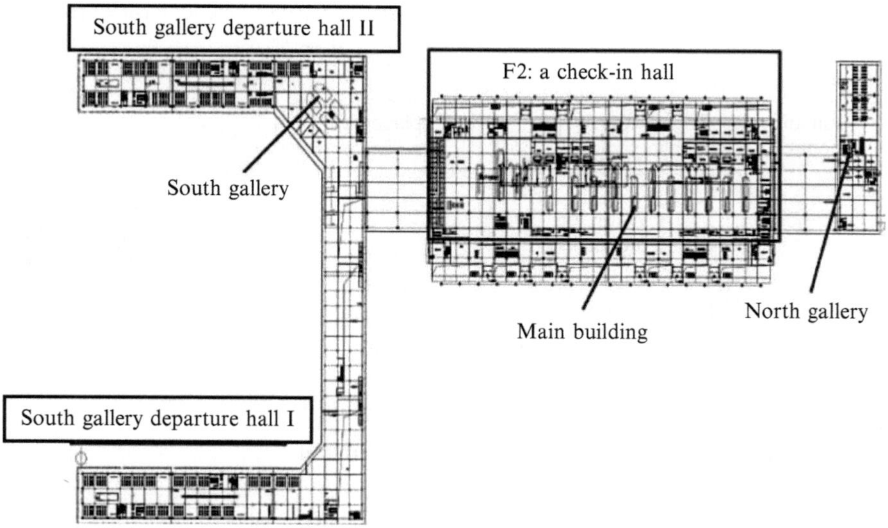

Fig. 8.27 THIC air-conditioning system application region

Figure 8.28 shows the operating schematic of the THIC system in the airport terminal. In this THIC system, outdoor air is dehumidified by a heat pump-driven liquid desiccant processor, and high-temperature chilled water is used to precool the air. Displacement ventilation is utilized as the air supply terminal (Jiang et al. 2010), and dry air is supplied into the indoor environment for humidity control. Dry FCUs (fan coil units) and radiant floors are used for temperature control in the THIC system, and dry FCUs with condensing plates are placed near the doors and windows instead of radiant floors to avoid condensation.

Figures 8.29 and 8.30 show the operating schematic of the terminal devices in the check-in hall and the departure hall, respectively. There are a variety of combinations of the terminal devices in the THIC air-conditioning system, as listed in Table 8.10. The radiant floors are designed for cooling with supply and return water temperature of 14 °C/19 °C in summer (designed cooling capacity of 60 W/m^2) and heating with supply and return water temperature of 40 °C/30 °C in winter. To be mentioned, the air is returned through the grilles located in the middle of the building in summer, and the top skylight is used to expel the high-temperature air dirty; in winter, the air is returned through at the top of the building, weakening the temperature gradient of indoor air in the vertical direction, and in the transition season, top skylight is open for mechanical and natural ventilation.

8.3.1.3 Cooling Plant

The basic task of the cooling plant is to supply chilled water to Terminal 3 and the other nearby buildings, and Fig. 8.31 illustrates the operating principle of the

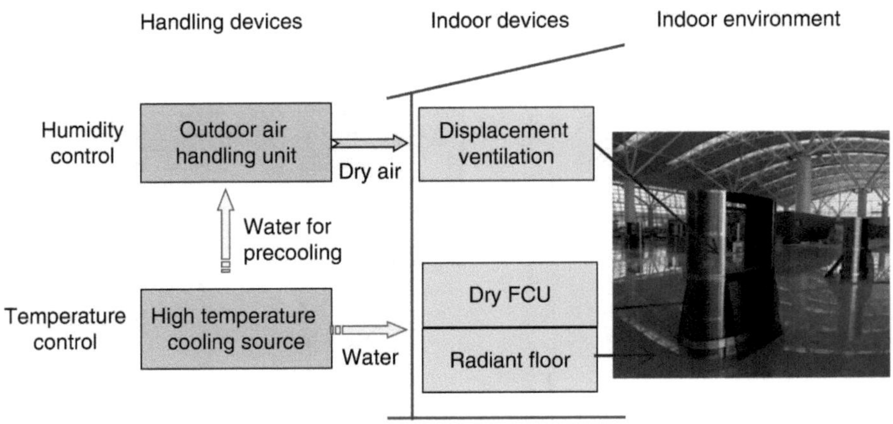

Fig. 8.28 Operating schematic of the THIC system in Terminal 3

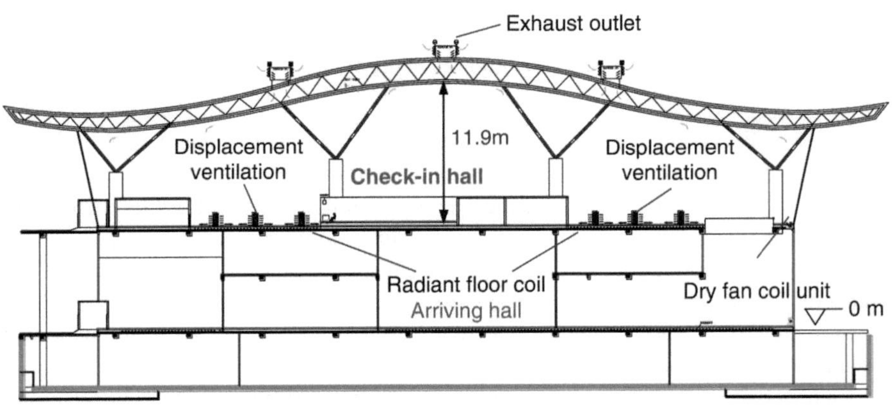

Fig. 8.29 Indoor terminals of the check-in hall

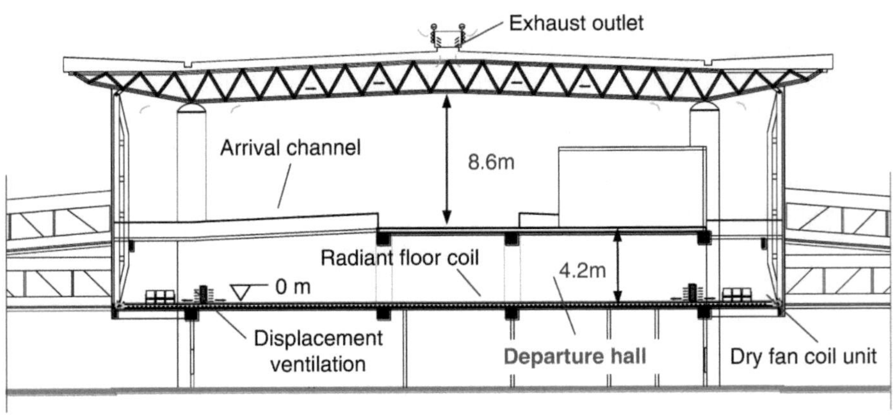

Fig. 8.30 Indoor terminals of the departure hall

cooling plant. There are three series of chillers, and the basic parameters are listed in Table 8.11. Chillers A1–A3 work in both air-conditioning and ice storage conditions. In contrast to conventional water chillers, glycol is the refrigerating medium for chillers A1–A3, and three plate heat exchangers are adopted for heat transfer between the chilled water and glycol. Chillers B1–B3 and chiller C only work in air-conditioning mode, with water as the heat transfer fluid. Because of the differences in operating conditions, the rated energy performances of chillers B1–B3 and chiller C are superior to those of chillers A1–A3.

According to the running time of Terminal 3 and peak/off-peak electricity prices, the cooling plant utilizes ice storage technology from 11:00 p.m. to 5:00 a.m. During the ice storage period, chillers A1–A3 operate in ice storage mode, and the tank is used for storing ice. In the daytime, the cooling plant runs in melting ice mode to produce chilled water for the airport terminal. If the melting ice cannot satisfy the cooling capacity requirement of the chilled water, the chillers then operate as supplements. The handling devices in the airport terminal are designed to operate in series, and a large Δt (greater than 10 °C) of the chilled water is expected. The operation of the cooling plant is designed to accommodate this requirement and improve energy efficiency. Return water from the terminal devices flows out of the collector (shown in Fig. 8.31) and is cooled step-by-step. The water first flows through chiller C and chillers B1–B3 to be cooled and then flows through the plate heat exchangers cooled by the glycol from chillers A1–A3. Then, the water flows out of the heat exchangers into the ice storage tank and is cooled to the required temperature, achieving a more matched heat transfer process.

The chillers are controlled according to the outlet chilled water temperature. If the melting ice is not sufficient, chillers B1–B3 or C should be turned on with priority because their energy efficiencies are higher than those of chillers A1–A3, as indicated in Table 8.11. Various chilled water pumps have different functions in the cooling plant. As shown in Fig. 8.31, pumps B1–B3 and C correspond to the water circulation of chillers B1–B3 and C, respectively; pumps A1–A6 correspond to water circulation between the cooling plant and Terminal 3.

8.3.2 Performance On-Site Test in Summer

8.3.2.1 Indoor Thermal Environment

Terminal 3 has been in use since May 2012, and its indoor environment was examined in terms of the large spaces adopting the THIC system (Zhang et al. 2013). The measured indoor parameters of the check-in hall are shown in Fig. 8.32. The temperatures of these areas were mainly between 22 and 24 °C, while the humidity ratios were concentrated in a range of 10–11 g/kg. These suitable thermal conditions showed that the indoor environment was comfortable for occupants.

Table 8.10 Components and functions of the THIC system

	Terminal devices	Air supply	Functions
Temperature control subsystem	Radiant floor	–	(1) Undertake the basic sensible cooling load; (2) increase the longwave radiation heat transfer, reduce the indoor cooling load, and improve the thermal comfort; (3) reduce air-conditioning investment; (4) increase the return water temperature and reduce the transportation energy consumption and cooling energy consumption
	Radiant floor + dry fan coil units	Bottom ventilation beside the curtain wall	(1) Provide supplemental cooling capacity; (2) reduce the outer zone cooling load and improve the thermal comfort; (3) reduce air-conditioning investment; (4) increase the return water temperature and reduce the transportation energy consumption and cooling energy consumption
	Dry fan coil units	Bottom displacement ventilation	(1) Provide supplemental cooling capacity; (2) make full use of the replacement air for free cooling in transition season and reduce the cooling energy consumption; (3) increase the return water temperature and reduce the transportation energy consumption and cooling energy consumption
Humidity control subsystem	Liquid desiccant outdoor air handling units	Bottom displacement ventilation	(1) Undertake the latent cooling load for indoor humidity control; (2) provide fresh air for air quality; (3) make full use of replacement air for free cooling in transition season and reduce the cooling energy consumption; (4) reduce transportation energy consumption

8.3.2.2 Operating Performance of Outdoor Air Processor

A heat pump-driven liquid desiccant outdoor air processor with a precool module was adopted for humidity control in the THIC system. Figure 8.33 shows the operating schematic of the processor and a psychrometric chart of the air handling process. Outdoor air (OA) first flows through the precool coils and is precooled to OA_1 by the high-temperature chilled water. Then, the air flows into the dehumidifier using liquid desiccant and is handled to the supply air state with the required humidity ratio. In the liquid desiccant dehumidification module, the concentrated solution is cooled by the evaporator before flowing into the dehumidifier. The diluted solution then leaves the dehumidifier, heated by the condenser, and is concentrated in the regenerator. Outdoor air is used for desiccant regeneration, and the humid regenerated air is exhausted to the ambient.

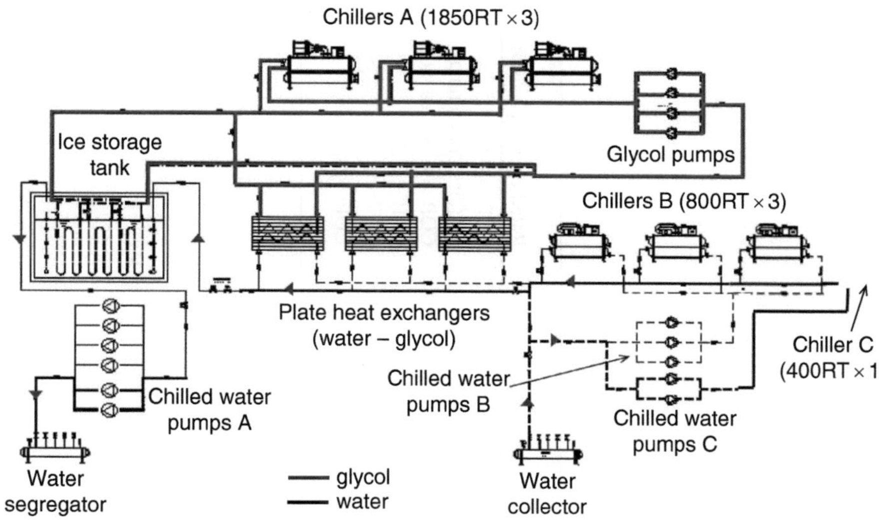

Fig. 8.31 Operating schematic of the cooling plant: melting ice mode combined with chillers for air-conditioning

Figure 8.34 illustrates the measured hourly performance (including air temperature and humidity ratio) of one outdoor air processor in the departure hall. As indicated in Fig. 8.34a, the outdoor air temperature was about 30–32 °C and the supply air temperature stabilized at around 19 °C. The inlet water temperature for precooling was about 14.5 °C, with a Δt of 3.6 °C. The outdoor air humidity ratio was about 18.0 g/kg, while the supply air humidity ratio for extracting indoor moisture was 9.6 g/kg.

8.3.2.3 Operating Performance of Radiant Floors

Two kinds of radiant floors are utilized in Terminal 3: a marble type and a plastic cement type. The basic components of the two types are identical: insulating layer (30 mm), pea gravel concrete (70 mm), and cement mortar (25 mm) over floor. Although the marble type (marble: 25 mm) is five times thicker than the plastic cement type (plastic cement: 5 mm), marble has superior thermal conductivity. Consequently, the conduction thermal resistance of the marble type floor (0.1 W/(m^2·°C)) is about half of that of the plastic cement type floor (0.21 W/(m^2·°C)).

A heat flow meter with a measuring accuracy of 5 % was used to test the cooling capacity of the radiant floors. As the marble type floor is mainly present in the check-in hall and the plastic cement type floor is mainly used in the departure hall, different measuring points were selected to investigate the performance of these two kinds of radiant floors. Figures 8.35 and 8.36 show the measured performances of the marble type and the plastic cement type radiant floors, respectively, in a cloudy day. It can be seen that the surface temperature of the marble type floor is

Table 8.11 Types and rated energy performances of key devices in the cooling plant

Chiller	Type	Rated cooling capacity (kW)	Number	Heat transfer fluids	Rated energy performance
A	Centrifugal; for both ice storage and air-conditioning modes	6,506 (air-conditioning)	3	Glycol (25 % concentration)	COP = 4.84 (air-conditioning)
		4,238 (ice storage)			COP = 4.06 (ice storage)
B	Centrifugal; only for air-conditioning mode	2,813	3	Water	COP = 5.52
C	Centrifugal; only for air-conditioning mode	1,407	1	Water	COP = 5.21

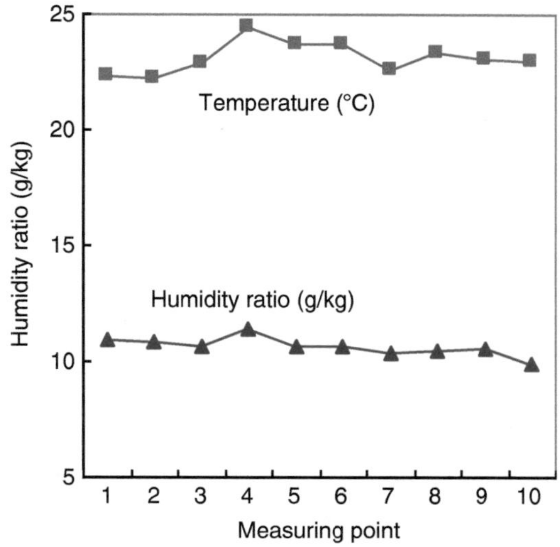

Fig. 8.32 Indoor air parameters of Terminal 3 (the check-in hall)

about 22–23 °C, with a cooling capacity of about 25–40 W/m^2. As shown in Fig. 8.35a, the temperatures of floor surfaces under seats are all lower than those of surfaces without seats (by about 1 °C). Furthermore, the temperatures of the floor surfaces are all higher than the indoor dew point temperatures, indicating that there is no risk of condensation.

As shown in Fig. 8.36, the measured surface temperature of the plastic cement type floor in the departure hall is only 19–20 °C, which is lower than that of the marble type floor. This surface temperature discrepancy exists because the average supply and return water temperature of the plastic cement type floor is only 15 °C in improper operation, while it is approximately 20 °C for the marble type floor. Thus, the hourly cooling capacity of the plastic cement type floor, which is around 30–50 W/m^2 according to Fig. 8.36b, is higher than that of the marble type floor. If the average supply and return water temperature is similar for these two types, the surface temperature of the marble type would be lower than that of the plastic cement type due to the former's lower thermal resistance, and the cooling capacity of the former would be higher than that of the latter under similar application conditions.

8.3.2.4 Energy Performance of the Chillers

The cooling plant operates in either ice storage mode or melting ice mode. In the system design, a large Δt (about 13 °C) of the chilled water is expected. The actual Δt of the chilled water in operation was about 10 °C, as the supplied chilled water was 2–3 °C. The operating performances of certain chillers are listed in Table 8.12 (both air-conditioning mode and ice storage mode). For the air-conditioning mode, the outdoor air temperature is 32.2 °C, with a relative humidity of 46.3 %. Chillers A1 and A2 are in operation, acting as auxiliary chillers to the melting ice.

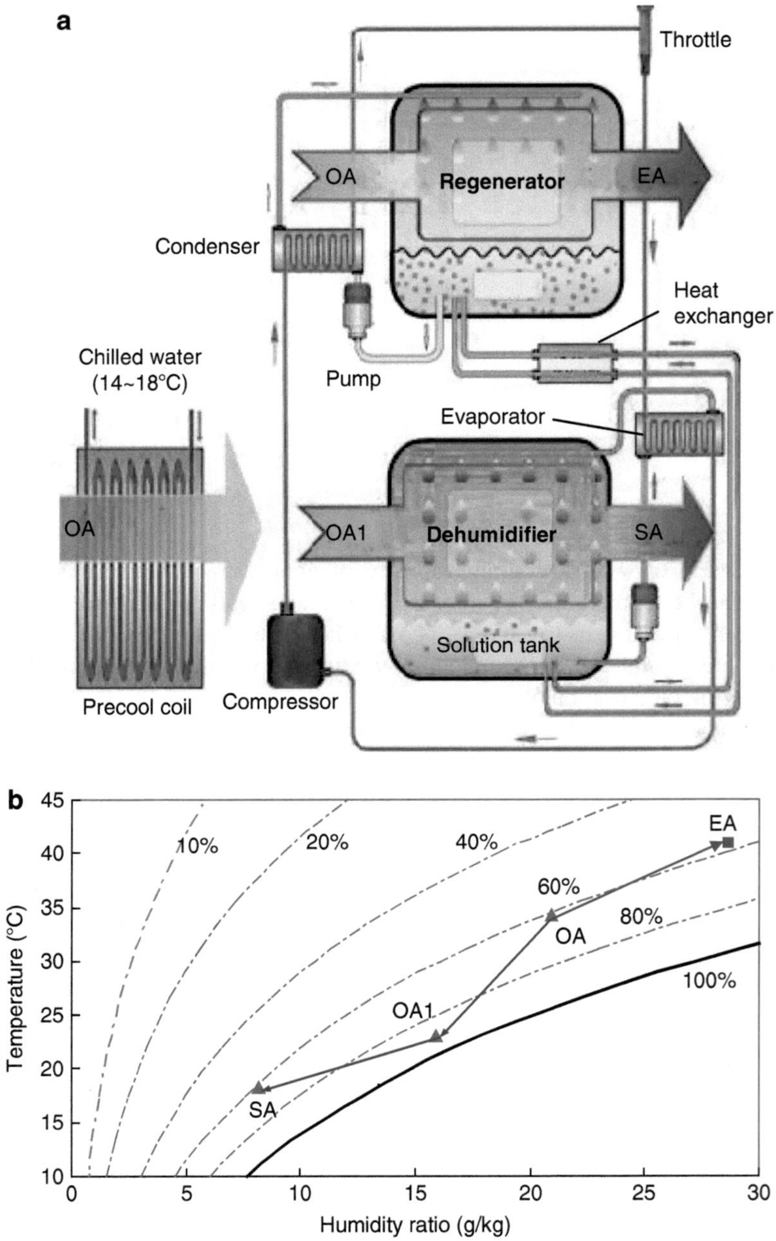

Fig. 8.33 Heat pump-driven liquid desiccant outdoor air handling unit with precooling module: (**a**) operating schematic and (**b**) typical air handling process in the psychrometric chart

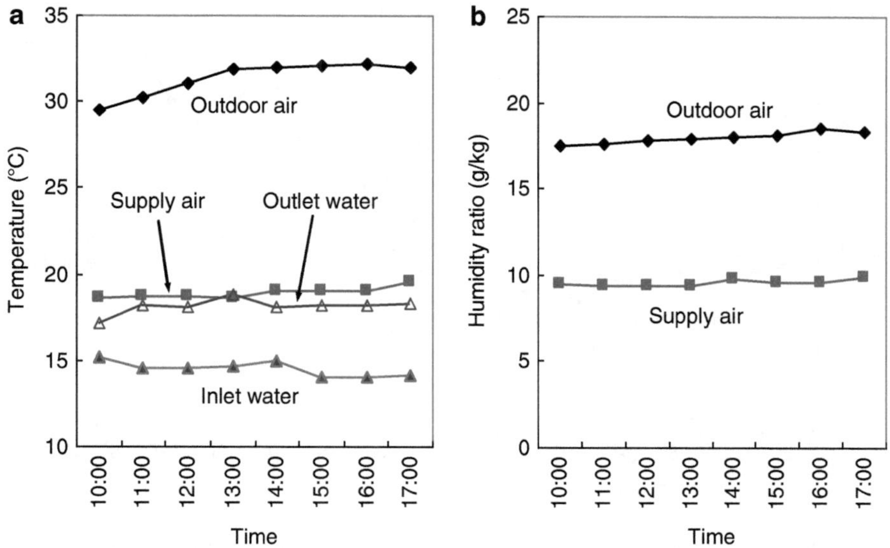

Fig. 8.34 Operating parameters of the liquid desiccant outdoor air processor: (**a**) temperature and (**b**) humidity ratio

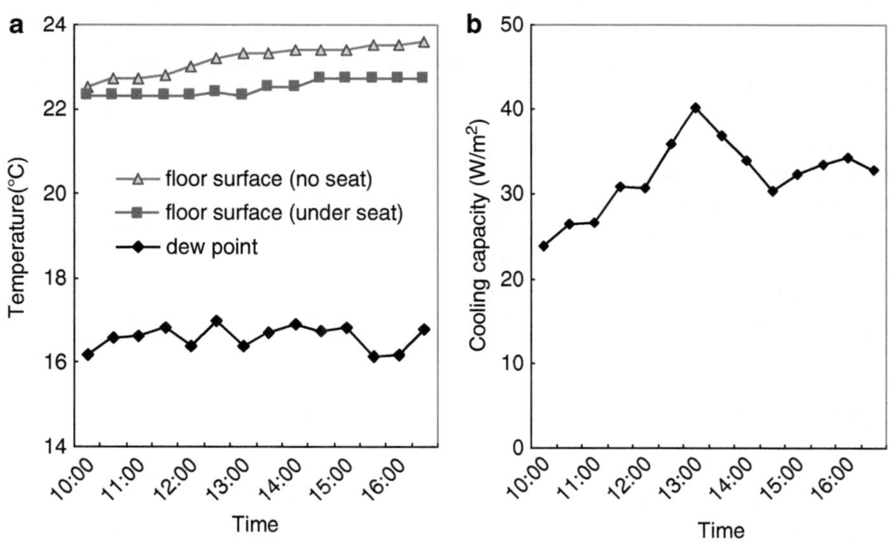

Fig. 8.35 Measured performance of the marble type radiant floor in the check-in hall: (**a**) temperatures and (**b**) cooling capacity

As indicated in Table 8.12, the temperature differences between the outlet glycol temperature and evaporating temperature t_e are only 0.8 and 0.7 °C for chillers A1 and A2, respectively. And their corresponding evaporator heat exchange efficiencies are 0.84 and 0.85, respectively, indicating good heat transfer performance of

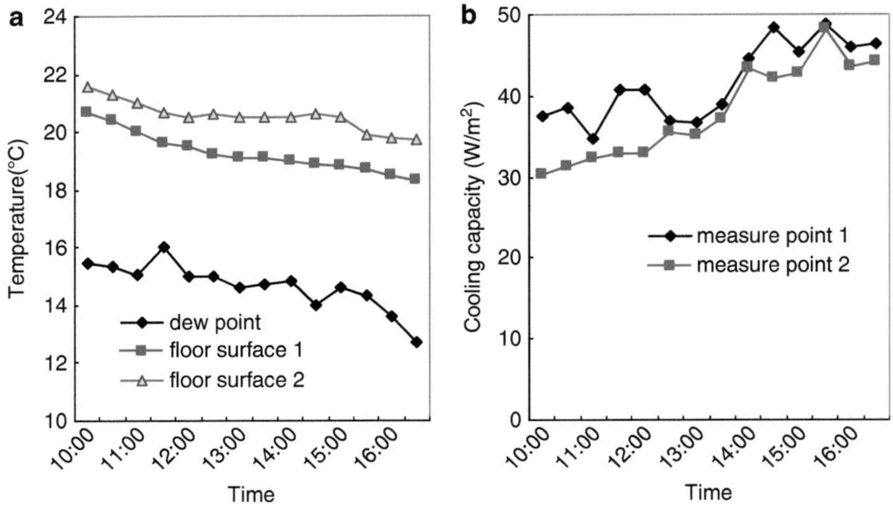

Fig. 8.36 Measured performance of the plastic cement type radiant floor in the departure hall: (**a**) temperature and (**b**) cooling capacity

Table 8.12 Operating performance of chillers in typical working conditions

	Chiller A1 (air-conditioning)	Chiller A2 (air-conditioning)	Chiller A1 (ice storage)
Ambient condition	32.2 °C, 46.3 %	32.2 °C, 46.3 %	31.1 °C, 49.4 %
Glycol inlet temperature (°C)	7.3	7.3	−1.8
Glycol outlet temperature (°C)	3.1	3.2	−5.5
Evaporating temperature (°C)	2.3	2.5	−6.2
Cooling water inlet temperature (°C)	28.3	28.3	28.1
Cooling water outlet temperature (°C)	32.5	32.8	31.9
Condensing temperature (°C)	33.5	34.5	32.8
Cooling capacity of the chiller (kW)	4,675	4,563	4,124
Power consumption of the chiller (kW)	1,009	1,076	1,045
Ideal COP of the chiller	8.82	8.61	6.84
Actual COP of the chiller COP_{ch}	4.54	4.15	3.95
Thermodynamic perfect degree η	0.51	0.48	0.58

the evaporators. Compared with the rated performances in Table 8.11, the cooling capacities of these two chillers in this condition are about 70 % of the rated cooling capacity. COP_{ch} is a little lower than the rated performance, and thermodynamic perfect degree η (actual COP divided by the ideal COP of a Carnot chiller operating at the same evaporating and condensing temperatures) is about 0.5. Chiller A1 is the chiller in operation in ice storage mode, and the outdoor condition is 31.1 °C, 49.4 %. The measured cooling capacity and COP_{ch} are close to the rated performance in this condition, and η is 0.58, which is higher than in air-conditioning mode.

Table 8.13 Energy performances of cooling plant devices in typical working conditions (melting ice mode)

Devices	Number	Power (kW)	Measured performance	Comments
Chiller	2	$P_{ch} = 2{,}085$	$COP_{ch} = \frac{Q_{ch}}{P_{ch}} = 4.43$	Outdoor condition: 32.2 °C, 46.3 %; $Q_{ch} = 9{,}238$ kW
Glycol pump	2	$P_{gl} = 451$	$TC_{gl} = \frac{Q_{ch}}{P_{gl}} = 20.5$	Glycol flow rate = 2,067 m^3/h; glycol temperatures: 7.3 °C/ 3.1 °C
Chilled water pump	2	$P_{chw} = 593$	$TC_{chw} = \frac{Q_{ch}+Q_i}{P_{chw}} = 26.6$	Chilled water flow rate = 1,560 m^3/h; temperatures: 11.6 °C/ 1.9 °C[a]
Cooling water pump	2	$P_{cwp} = 378$	$TC_{cwp} = \frac{Q_{ch}+P_{ch}}{P_{cwp}} = 30.0$	Cooling water flow rate = 2,257 m^3/h; temperatures: 28.3 °C/ 32.7 °C
Cooling tower	3	$P_{ct} = 135$	$TC_{ct} = \frac{Q_{ch}+P_{ch}}{P_{ct}} = 83.9$	

[a]Inlet and outlet temperatures of the chilled water for the ice storage tank are 5.5 °C and 1.9 °C, respectively

8.3.2.5 Energy Performance of the Cooling Plant

In addition to the chillers, there are chilled water pumps, glycol pumps, cooling water pumps, and cooling towers in the cooling plant. The typical performances of these devices in melting ice mode are listed in Table 8.13. The inlet and outlet temperatures of the chilled water for the ice storage tank are 5.5 °C and 1.9 °C, respectively, indicating that cooling capacity of melting ice Q_i is 6,521 kW. As the cooling capacity of the two chillers is 9,238 kW (from Table 8.12), the cooling capacity from melting ice accounts for 41 % of the total cooling load. Transport coefficient TC is adopted to evaluate the energy performances of different distribution devices (GB/T 17981). Limited by the higher viscosity of glycol, TC_{gl} is only 20.5, which is significantly lower than the case if water were chosen as the transfer medium. TC_{chw} is equal to the total cooling capacity divided by the power consumption of chilled water pumps in the cooling plant. As the distance between the cooling plant and Terminal 3 is quite long (about 1 km), TC_{chw} is only 26.6, which is lower than the recommended value in China's design standards (GB 50189).

For ice storage mode, the chiller's performance is listed in Table 8.12, with a cooling capacity of 4,124 kW. The glycol pumps, cooling water pumps, and cooling towers are in operation in this condition (i.e., everything except the chilled water pumps). The energy performances of the cooling plant devices in ice storage mode, including the power consumption and the TC of the distribution devices, are listed in Table 8.13. Energy efficiency ratio EER is equal to the cooling capacity divided by the total power consumption, and for the ice storage mode, EER_i is 2.74.

Not taking the cooling capacity loss in the ice storage process into account, EER_i is also assumed to be 2.74 for the cooling capacity from the ice storage tank in melting ice mode. Then, equivalent power consumption P_i for melting ice is the quotient of Q_i divided by EER_i, written as Eq. (8.8):

$$P_i = \frac{Q_i}{EER_i} = 2,381 \text{ kW} \tag{8.8}$$

Thus, the EER of the cooling plant in typical melting ice mode can be derived as Eq. (8.9):

$$EER = \frac{Q_{ch} + Q_i}{P_i + P_{ch} + P_{gl} + P_{chw} + P_{cwp} + P_{ct}} = 2.62 \tag{8.9}$$

However, as the terminal has been in use less than 2 years, there is still insufficient data for detailed analysis on the energy performance. The detailed analysis of the energy performance will be investigated in further research.

8.3.3 *Performance On-Site Test in Winter*

8.3.3.1 Indoor Thermal Environment

In winter, as to this THIC system for Terminal 3, radiant floor is responsible for indoor heating where hot water with a relative low temperature (about 35–40 °C) is adopted to realize "low temperature heating." The performance test of this THIC system with radiant floor for heating was carried out in December, 2012. Figure 8.37 gives the outdoor temperature during the test, indicating that the mean outdoor temperature of Dec. 19 was about 2 °C while of Dec. 20 was lower than 0 °C.

Taking the check-in hall as an example, Fig. 8.38a gives the indoor air temperatures of three measuring points with the same height of about 1.8 m above the floor. It shows that in Dec. 19, the air temperature for different points was about 22.5 °C, while in Dec. 20 the air temperature was about 22 °C, a bit lower due to a lower outdoor air temperature as indicated in Fig. 8.37. As the height of the check-in hall is very high (about 26.5 m), to investigate the indoor air temperature distribution further, Fig. 8.38b gives the temperature variances during the test period. The supply water temperature T_{sup} was about 35 °C in Dec. 19, while it was about 40 °C during Dec. 20 and 21. The range of the temperature variance was about 21–24 °C, and the vertical temperature distribution was almost uniform. It shows that the temperature was lower during the night (from 0:00 to 6:00) due to the operating mode of this terminal and that the supply water for the radiant floor was turned off at around 0:00 at midnight and restarted at about 6:00 in the morning according to the schedule. On basis of the testing results of the temperature distributions in the horizontal and vertical directions using radiant floor for heating, the indoor thermal environment was appropriate and comfortable.

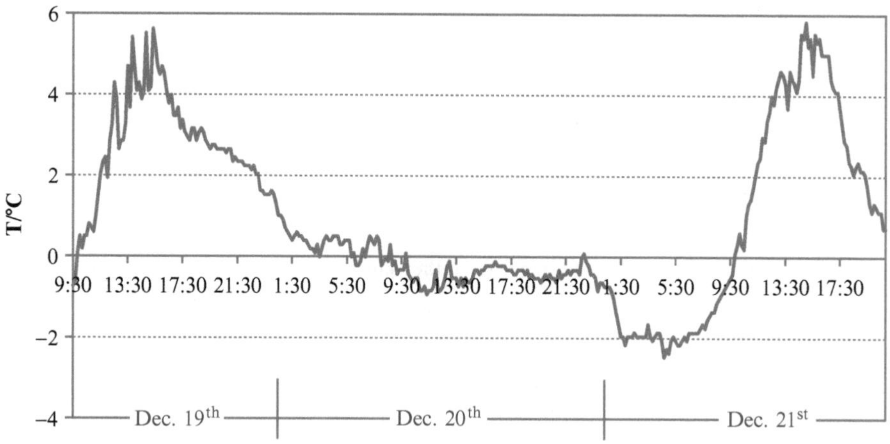

Fig. 8.37 Outdoor climate during the test in winter

8.3.3.2 Operating Performance of Radiant Floors

Figure 8.39 gives the supply and return water temperatures in two branches of the radiant floors, where the supply water temperature was about 35 °C for Dec. 18 and 19 and about 40 °C for Dec. 20 and 21 due to a lower outdoor air temperature. As indicated by this figure, the operating temperature difference was about 3 °C as the supply water temperature was about 35 °C and was about 5 °C as the supply water temperature of the north branch was increased to around 40 °C, mainly due to a lower outdoor air temperature.

Then Fig.8.40a shows the variances of the surface temperatures with different distances from the gate of the check-in hall. As the supply water temperature was about 35 °C, the surface temperatures were among 26–29 °C. When the supply water temperature was increasing to about 40 °C, the surface temperature was then about 27–30 °C, which was a bit higher than those with a lower supply water temperature. As indicated by this figure and Fig.8.38a, the surface temperature of the radiant floor was about 5–8 °C higher than indoor air temperature. Heating capacities of the radiant floors with different distances from the gate are given as Fig. 8.40b: as the supply water temperature was about 35 °C, the heating capacities were within 30–50 W/m^2; as the supply water temperature was increased to around 40 °C, heating capacities were increasing to about 40–70 W/m^2.

8.3.3.3 Performance Comparison Between Different Heating Terminal Devices

As indicated by the tested indoor air temperatures and performances of the radiant floors shown above, the THIC system adopting radiant floor for heating in Terminal 3 could provide a comfortable indoor thermal environment. Figure 8.41a, b

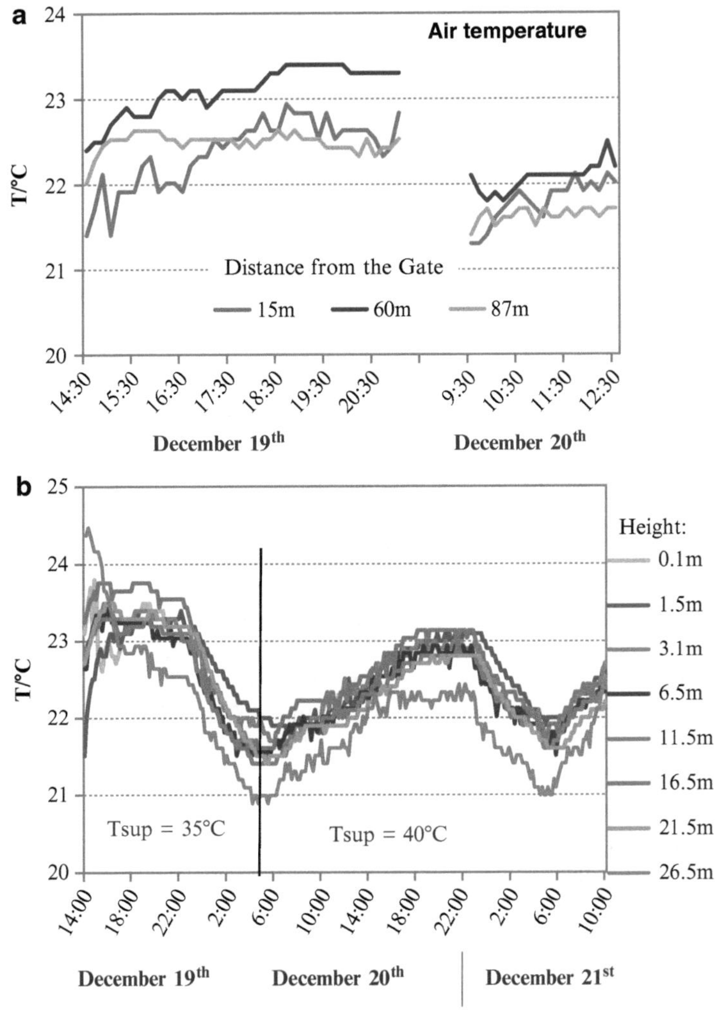

Fig. 8.38 Temperature distributions of indoor environment in the departure hall: (**a**) horizontal direction and (**b**) vertical direction

illustrates the vertical temperature distributions of the arrival hall (using radiant floor as the terminal device for temperature control) and the check-in hall with different measuring points, respectively. It shows that the temperature gradient for both the arrival hall and the check-in hall is relative low and the vertical temperature distributions are almost uniform. However, as the arrival hall is in the first floor just under the check-in hall as shown in Fig. 8.29, the cold air infiltration influences the indoor thermal environment, leading to a lower indoor temperature (about 17 °C as shown in Fig.8.41a compared with that of the check-in hall (about 22 °C as shown in Fig.8.41b).

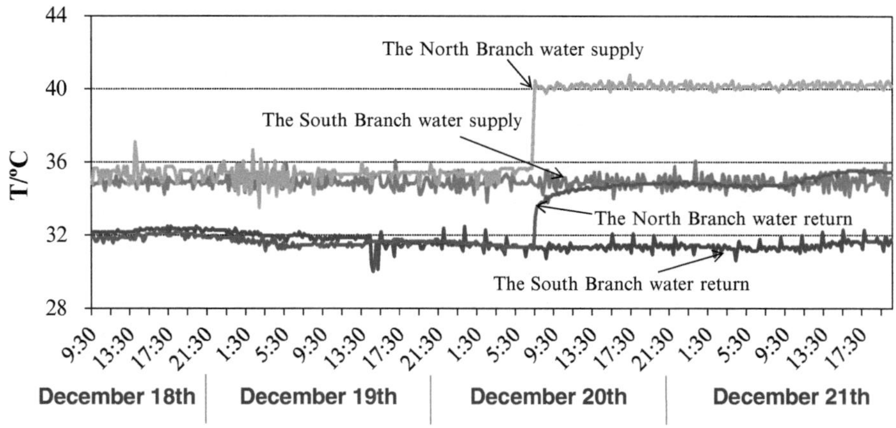

Fig. 8.39 Supply and return water temperatures of the radiant floors

For the arrival hall of Terminal 2 (operated for 10 years), the conventional diffuser from the ceiling is adopted to supply hot air for heating in winter. Figure 8.41c shows the tested temperature distribution in the vertical direction. It indicates that for the area with a height lower than 2 m where occupants always stay, the measured results show that the temperature is lower than 15 °C, indicating a temperature gradient higher than 10 °C. As to the check-in hall of Terminal 2 using the conventional nozzle air supply system, the vertical temperature distributions are given in Fig.8.41d. Compared with Fig.8.41b, there is also a significant temperature gradient for the check-in hall of Terminal 2: temperature for the space with a height of about 6 m is about 22 °C; for the space with a height lower than 2 m, the temperature is about 15–19 °C. Thus, as indicated by the temperature distributions in the vertical direction listed in Fig. 8.41a–d, there would be a more uniform vertical temperature distribution using the radiant floor for heating compared with the conventional diffuser or the nozzle air supply system.

8.3.3.4 Energy Performance of the Total Air-Conditioning System

This new terminal (Terminal 3) has been put into use since May 2012, and the THIC system was adopted for regulating the indoor thermal environment of the large space. For terminal 2 (operated for 10 years), conventional jet ventilation system is adopted. Figure 8.42 gives the monthly HVAC energy consumption of the cooling plant and the terminal devices for Terminals 2 and 3 from the May to the December of 2012, including the cooling season, the transition season, and the heating season. However, as the hot water was supplied by the district heating network in winter,

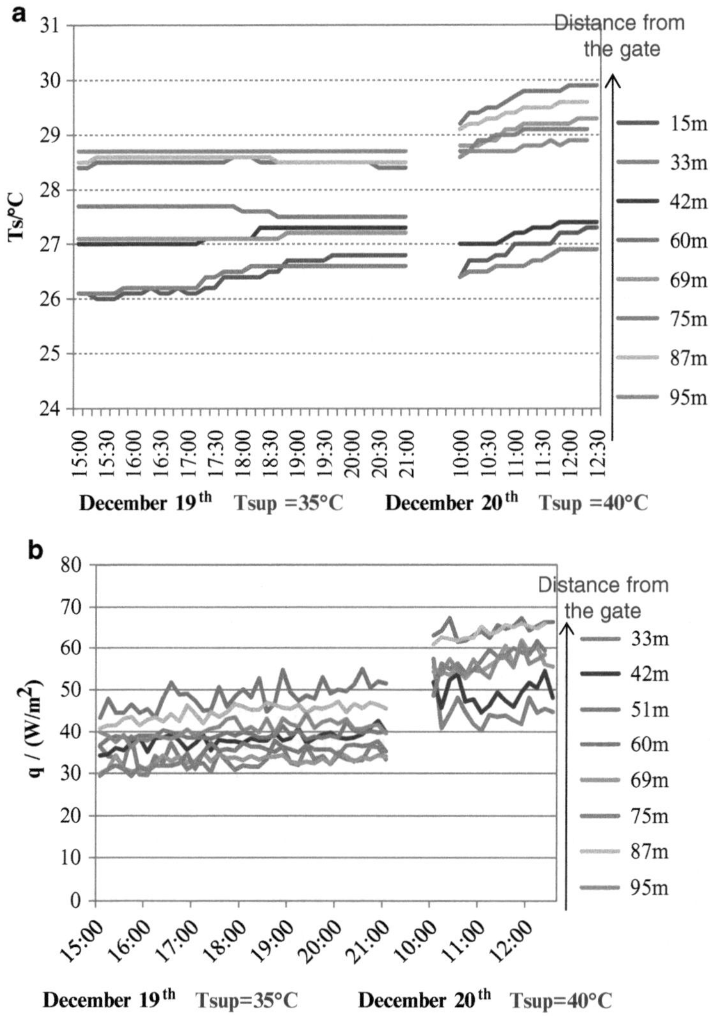

Fig. 8.40 Operating performance of the radiant floor for heating: (**a**) surface temperature and (**b**) heating capacity

the energy consumption in winter only took the power consumptions of the terminal devices, i.e., the fans and water pumps into account.

As indicated by this figure, power consumption for unit area of Terminal 3 is significantly lower than that of Terminal 2. Thus, the new system adopting the THIC air-conditioning idea in Terminal 3 shows a better energy performance than the conventional system in Terminal 2. The total HVAC energy consumption of Terminal 3 from May to December is 49.9 kWh/m^2, which is 36.4 % lower than that of Terminal 2 (78.5 kWh/m^2): power consumption of the terminal devices in Terminal 3 is about 47.2 % lower than that in Terminal 2; power consumption of the cooling plant in Terminal 3 is about 30.6 % lower than that in Terminal 2.

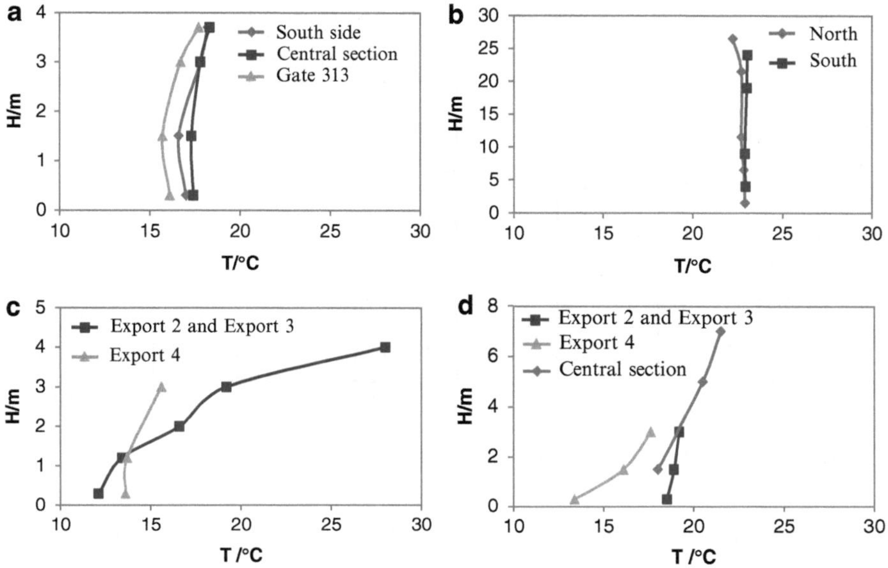

Fig. 8.41 Temperature distributions in the vertical direction: (**a**) arrival hall of Terminal 3, (**b**) check-in hall of Terminal 3, (**c**) arrival hall of Terminal 2, and (**d**) check-in hall of Terminal 2

8.3.4 Conclusion

The air-conditioning system in Terminal 3 in Xi'an Xianyang International Airport has been in use since May 2012, and an advanced THIC system utilizing radiant floors has been adopted there. In this section, the operating principle of the system is introduced, and its performances both in summer and in winter are investigated. The main conclusions can be summarized as follows:

1. In the large spaces of the airport terminal, a THIC air-conditioning system was adopted instead of a system based on the common nozzle air supply method, and the indoor environment in these areas was shown to be comfortable for occupants. The performances of the terminal devices in this THIC system were measured: the liquid desiccant processor was able to dry the outdoor air sufficiently for humidity control, and the two kinds of radiant floors used for temperature control had measured cooling capacities of about 30–40 W/m^2 in a cloudy day.
2. Large Δt of the chilled water is designed, due to the cooling plant is relatively far away from the terminal. The measured operating Δt of the chilled water was about 10 °C, which was higher than the 5 °C in the conventional system. The large Δt helped to lower the flow rate and reduce the energy consumption of the water distribution system. For the typical melting ice condition for cooling, the cooling capacity from the melting ice accounted for about 40 % of the total cooling load, and the EER of the cooling plant was measured to be 2.62.

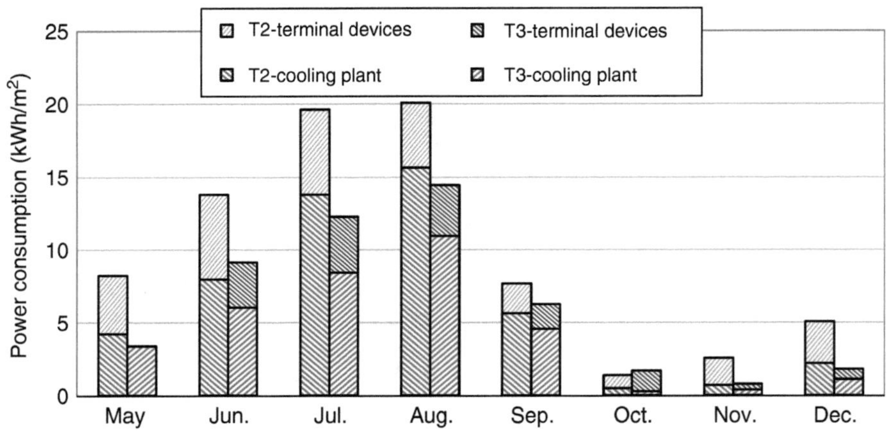

Fig. 8.42 Comparison of HVAC energy consumptions between Terminals 2 and 3

3. In winter, hot water with a temperature of about 35–40 °C is supplied to the radiant floor to achieve a "low-temperature heating" process. The surface temperature of the radiant floor is about 25–30 °C, while the heating capacity varies from about 30 W/m^2 to about 70 W/m^2, sufficient to satisfy the indoor heating requirement. In contrast to the conventional diffuser or nozzle air supply system, vertical temperature distributions of the area adopting radiant floor for heating are more uniform, with a much lower temperature gradient for different heights; the indoor thermal environment adopting radiant floor is more comfortable.
4. Compared with the conventional air-conditioning system adopted in Terminal 2, power consumption of the new system in Terminal 3 is significantly lower. Based on the statistical data of power consumptions in these two terminals from May to December, the energy saving ratio of the new HVAC system for Terminal 3 is as high as 36.4 %.

8.4 Application in an Industrial Factory

8.4.1 Description of the THIC System in an Industrial Factory

8.4.1.1 Basic Information

A workshop for banknote printing located in Beijing is selected as an example to investigate the performance of the THIC system, and its air-conditioning area is 10,000 m^2. The data in this subsection is from the PhD thesis by Liu SQ (tutor: Jiang Y). There are strict demands for indoor temperature and humidity control, i.e.,

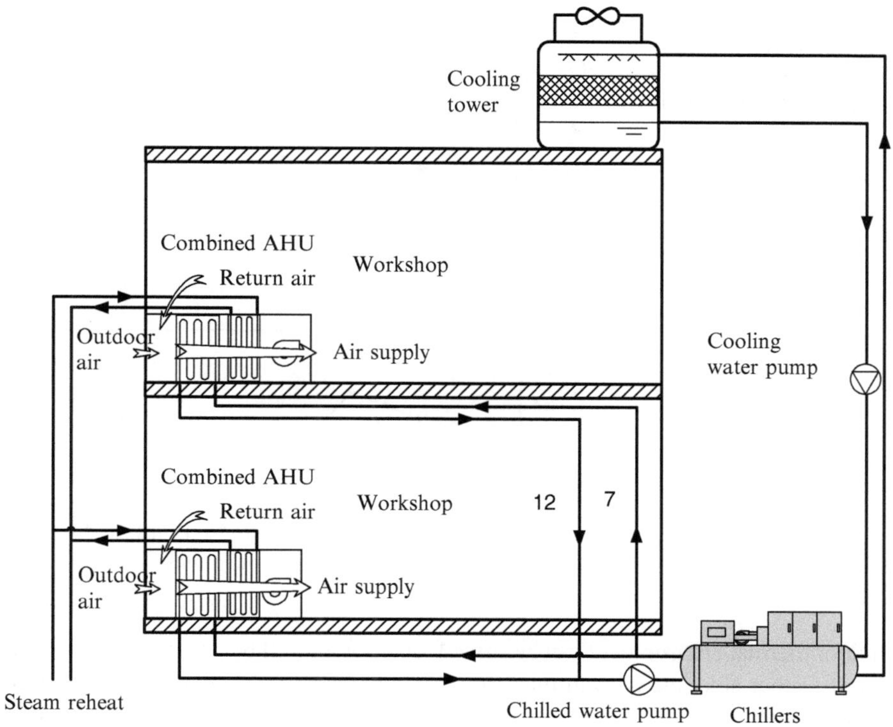

Fig. 8.43 Schematic of primary air-conditioning system

the indoor design parameters are 24 ± 2 °C for temperature and 55 ± 5 % for relative humidity. The typical all-air system with primary return air was adopted in the original air-conditioning system of this workshop (constructed in 1985), as shown in Fig. 8.43. As indicated by this figure, after the mixing process between the outdoor air and the return air, the air is then dehumidified by the condensation dehumidification method using 7 °C/12 °C chilled water to a state satisfying the requirement of humidity ratio. To meet the temperature requirement of the supply air, the dehumidified air is then reheated by steam (using a gas-fired boiler as the heat source).

There exist many shortages for the original air-conditioning system as illustrated in Fig. 8.43. According to the tested indoor parameters in this workshop before retrofitting shown in Fig. 8.44, the indoor temperature was about 21.5–25.5 °C, and the relative humidity is about 46–58 %, out of the range of the required indoor parameters. Besides, in operation of the original system, the valve of the outdoor air is closed due to the requirement of reducing the outdoor air flow rate to maintain an appropriate indoor thermal environment. However, this means the system operates in the all-return air condition without outdoor air. Printing industry uses printing ink widely, in which there may be volatile organic solvent, so lack of outdoor air could be bad for workers' health. And if only the problem of maintaining an appropriate

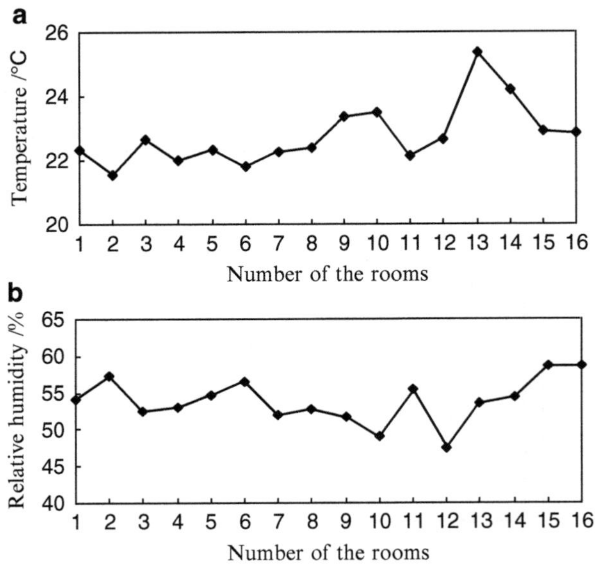

Fig. 8.44 Measured indoor parameters before transformation: (**a**) indoor temperature and (**b**) relative humidity

indoor thermal environment can be solved, outdoor air will be provided adequately. On the other side, the conventional solution adopting the condensation dehumidification method and the reheating method to control indoor temperature and humidity led to significant energy offset and dissipation. It increased cooling load of air-conditioning system and also consumes steam for reheating, increasing the operating cost.

8.4.1.2 THIC Air-Conditioning System

In essence, the shortages of the original air-conditioning system were due to the condensation dehumidification method. Using low-temperature chilled water to cool and dehumidify air simultaneously, the coupled two processes resulted in failure to control indoor temperature and humidity effectively, and reheat led to significant energy offset. Therefore, applying the THIC method is an effective way to solve problems of the original system, with separate methods to control indoor humidity and temperature.

Based on the THIC method, the heat pump-driven liquid desiccant air handling processor is adopted in retrofitting the air-conditioning system. Schematic of the THIC air-conditioning system for this workshop is shown in Fig. 8.45. The dehumidified outdoor air undertakes the tasks of removing indoor moisture and controlling indoor humidity; indoor temperature control is realized by using high-temperature chilled water to cool the return air. After precooled by high-temperature chilled water (14/19 °C), outdoor air flows into the dehumidifier unit

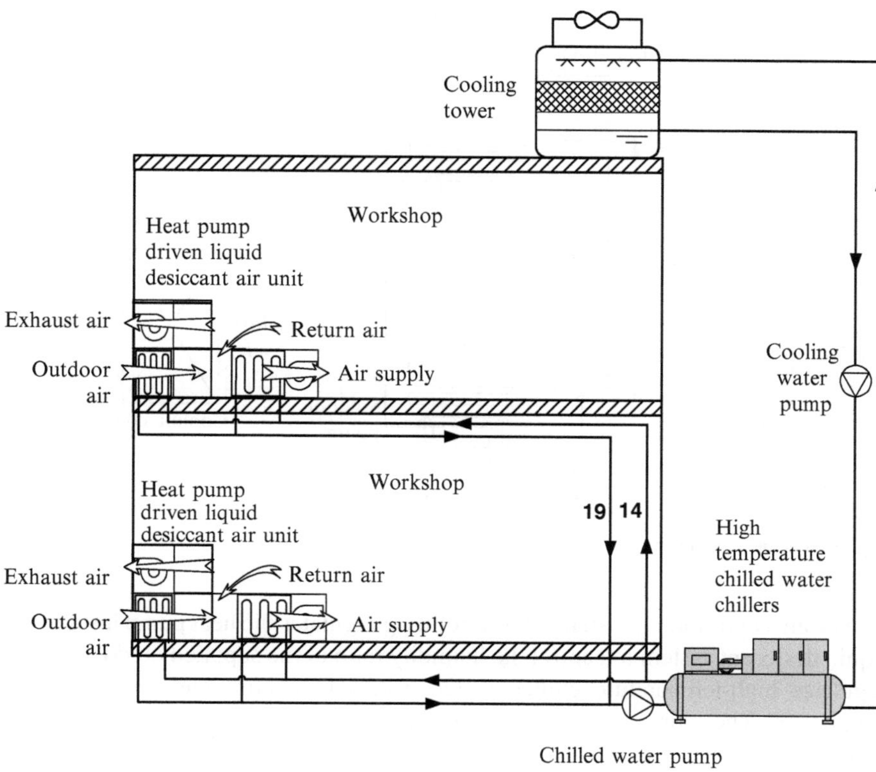

Fig. 8.45 Schematic of THIC air-conditioning system for this workshop

for further dehumidification. Then the dehumidified air mixes with the return air to be cooled to the supply air state using high-temperature chilled water.

Construction schematic diagram of liquid desiccant air handling unit is shown in Fig. 8.46. In air handling units, outdoor air is precooled and dehumidified by 14/19 °C chilled water in cooling coils 1, and then dehumidified independently in liquid desiccant module, and then mixed with return air and cooled to supply air state by 14/19 °C chilled water in cooling coils 2. In dehumidification module, the solution becomes diluted and absorption ability decreases after absorbing moisture from air. Then the diluted solution is pumped into regeneration module, regenerated, and concentrated by regeneration air after releasing water to air. The concentrated solution is pumped to dehumidification module again for next circulation. The evaporator of the heat pump installed in the liquid desiccant system is used to cool the solution sprayed in dehumidification module and enhance its absorption ability. The heat pump's condensation heat is used to heat the solution sprayed in regeneration module for solution concentration.

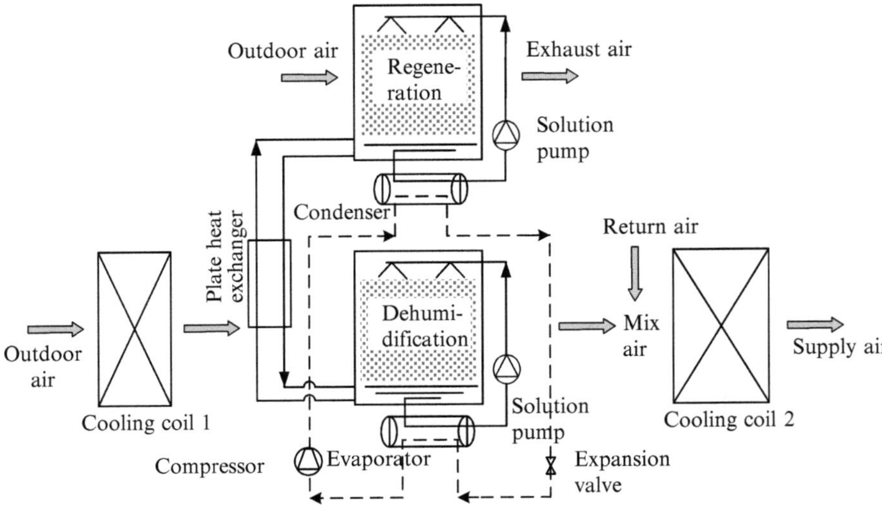

Fig. 8.46 Construction diagram of liquid desiccant air handling units

The air-conditioning system after retrofitting regulates the heat pump installed in liquid desiccant system to control the humidity ratio of the supplied outdoor air and regulates high-temperature chilled water system to control the temperature of supply air. Therefore, both requirements for humidity and outdoor air of the conditioned space are satisfied. As demonstrated, the air-conditioning system after retrofitting can avoid hot and cold offset caused by condensation dehumidification and reheating with steam, and the performance improves significantly due to the increase of the supply chilled water temperature.

8.4.2 Performance of the THIC System

8.4.2.1 Indoor Thermal Environment

During the performance testing period of the THIC air-conditioning system after retrofitting, outdoor air temperature and humidity ratio were within the range of 28–34 °C and 18–24 g/kg, respectively. Figure 8.47a, b shows the supply air temperature and humidity of the air handling processor using liquid desiccant. It indicates that the supply air temperature varies in the range of 16–20 °C and the supply air humidity ratio is about 7–10 g/kg.

Figure 8.48a, b gives the indoor air temperature and relative humidity in this workshop after adopting the THIC system. The indoor temperature and relative

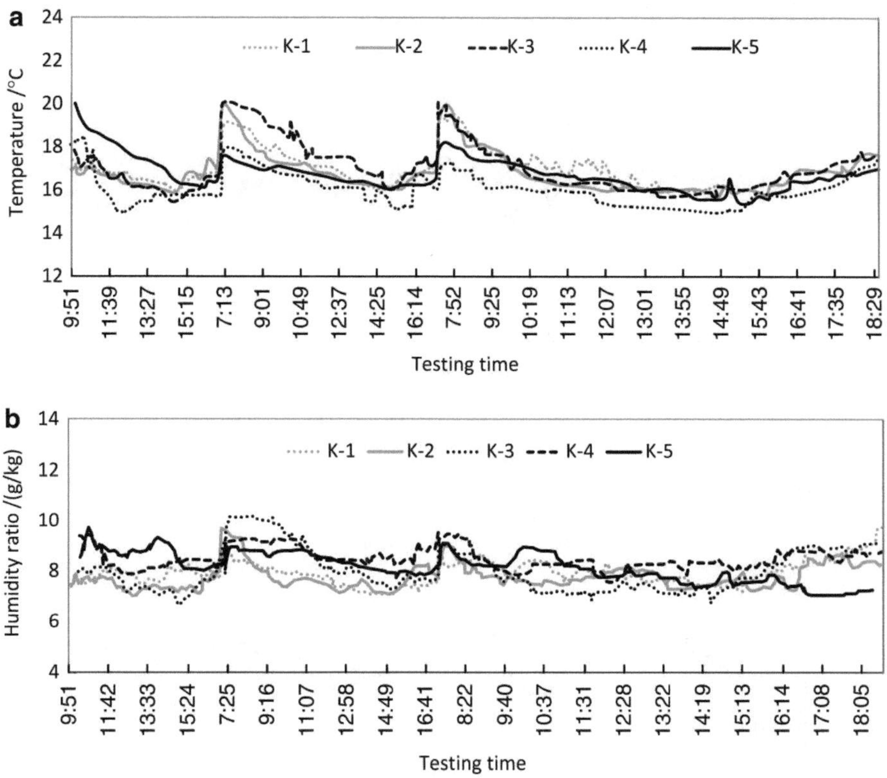

Fig. 8.47 Supply air parameters of the air handling processor: (**a**) temperature and (**b**) humidity ratio

humidity in workshop are in the range of 23–25 °C and 52–58 %, respectively, satisfying the requirements for indoor temperature and relative humidity. Thus adopting the THIC system can meet the demand for temperature and humidity control.

8.4.2.2 Energy Consumption

The operating cost of the air-conditioning system for this workshop before and after the retrofitting is listed in Table 8.14. As indicated by this table, reheat is avoided by adopting the THIC system, and the energy efficiency of the water chiller is also increased due to a higher evaporating temperature of the chiller in the new THIC system. By adopting the THIC system, operating cost could be saved to a great content, and the operating cost saving ratio in summer is as high as 68 %, compared with the original air-conditioning system.

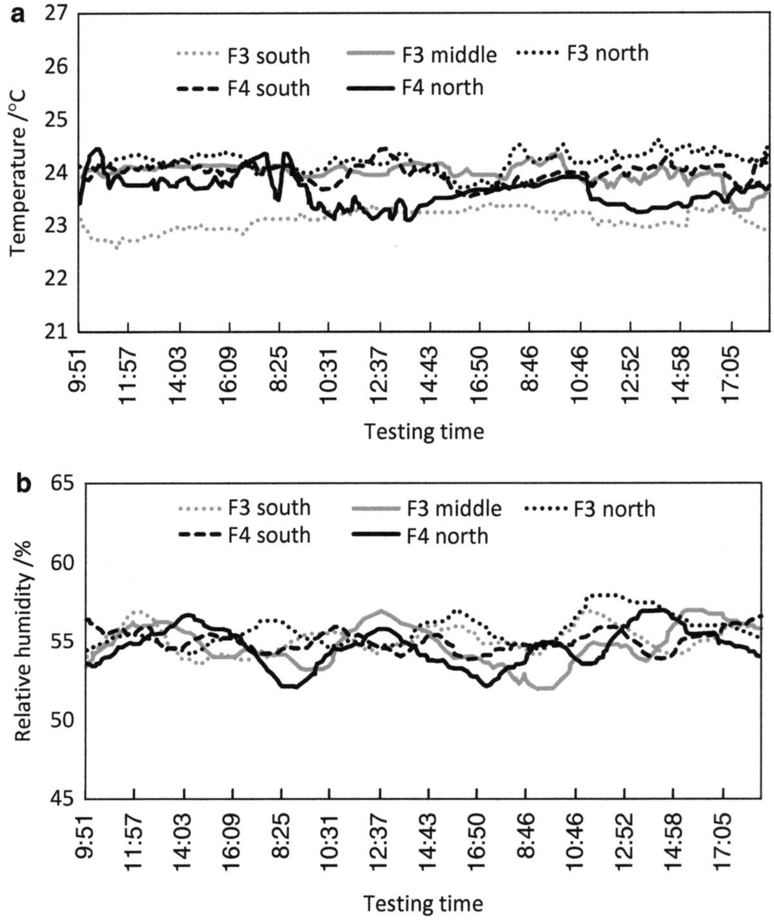

Fig. 8.48 Indoor parameters after adopting the THIC system: (**a**) air temperature and (**b**) relative humidity ratio

Table 8.14 Operating cost of the air-conditioning system before and after retrofitting

Items	Unit	Original system		THIC system	
Devices	–	Chiller	Reheat	Liquid desiccant	Chiller
Rated cooling capacity	kW	1,574	–	206	837
Rated power consumption	kW	463	–	56	164
Rated reheat consumption	kW	–	532	–	–
Power consumption (summer)	MWh	1,167		554	
Gas consumption (summer)	m^3	230,003		0	
Operating cost (summer)	10^4 CNY	137		44	
Cost saved	10^4 CNY	–		93	

References

Chen P, Liu SQ (2009) Temperature and humidity independent control air-conditioning system design for inpatient building of central hospital of Sanmenxia. Refrig Air Cond 9:66–71 (in Chinese)

Fan YY, Lv SY, Li X (2011) Temperature and humidity independent control system design of an art exhibition center. Chin HV&AC 41:39–43 (in Chinese)

Fang Y (2010) Design of temperature and humidity independent control air-conditioning system in the culture relic storeroom. In: Chinese annual HVAC symposium (in Chinese)

Hu JL (2010) Application of temperature and humidity independent control technology to the HVAC design of Expo 2010 Swedish pavilion. Chin HV&AC 40(8):96–98 (in Chinese)

Jiang Y, Zhou M, Zhang Y, Liu Y, Yan D, Gao ZH, Zhang XL (2010) A terminal used for displacement ventilation combined with radiant floor cooling system. CN201020180016.2 (Chinese Patent)

Jin Y (2007) Design for the Sino-Italian ecological and energy efficient building of Tsinghua University. Chinese HV&AC 37:73–75 (in Chinese)

Li Y, Hu R, Feng TT (2011) Application of temperature and humidity independent control air conditioning system to the second phase of Qingdao Xiangxi Villa. Chin HV&AC 41(1):42–47 (in Chinese)

Lin KP, Xu HQ, Zhou XR (2009) Design of air conditioning system for major venues of the 7th China flower exposition. Chin HV&AC 39(5):2–6 (in Chinese)

Liu XH, Li Z, Jiang Y, Lin BR (2006) Annual performance of liquid desiccant based independent humidity control HVAC system. Appl Therm Eng 26:1198–1207

Lu B, Chen M, Zhang YN (2011) Energy-saving design for air-conditioning system in ECO building. Refrig Air-Cond 11(1):77–81 (in Chinese)

Xie XY (2009) Research on indirect evaporative cooling air conditioning system. Doctoral dissertation, Tsinghua University, Beijing (in Chinese)

Xie XY, Jiang Y, Liu SQ, Qu KY, Yu XY (2007) Design and development of an indirect evaporative water chiller. Chin HV&AC 37(7):66–71 (in Chinese)

Zhang T, Liu XH, Zhang L, Jiang JJ, Zhou M, Jiang Y (2013) Performance analysis of the air-conditioning system in Xi'an Xianyang International Airport. Energy Build 59:11–20

Zhao K, Liu XH, Zhang T, Jiang Y (2011) Performance of temperature and humidity independent control air-conditioning system applied in an office building. Energy Build 43:1895–1903

Zhou M (2011) Application of independent temperature and humidity control of terminal T3A to Xi'an Xianyang International Airport. Chin HV&AC 41(11):27–30 (in Chinese)

Zhu WF, Li ZJ, Liu S, Liu SQ, Jiang Y (2010) In situ performance of independent humidity control air-conditioning system driven by heat pumps. Energy Build 42:1747–1752

Zi XQ (2007) Diagnosis of energy efficiency and study on solutions for the air conditioning of the high-rise comprehensive office building in Shenzhen. Master thesis, Chongqing University, Chongqing (in Chinese)

Chapter 9
Development Tendencies and Perspectives of the THIC Systems

Abstract As a burgeoning air-conditioning system for commercial buildings, THIC air-conditioning system can better satisfy the regulating requirements of indoor temperature and humidity and embodies a significant potential for energy conservation compared with the conventional system. The design methodology of the THIC system has been incorporated into the national standards and authoritative design handbooks of China. The key devices for the new THIC system have been developed and realized industrialization. Since 2006, there have been more than 20 million m^2 commercial buildings that adopted the THIC system in China. Although rapid progresses have been made in developing THIC system, unremitting efforts are still needed for relevant researches and applications in the long term.

9.1 Development of the THIC Systems in China

As a burgeoning air-conditioning system for commercial buildings, THIC air-conditioning system can satisfy the regulating requirements of indoor temperature and humidity better and embodies a significant potential for energy conservation compared with the conventional system. The design methodology of the THIC system has been incorporated in the national standards and authoritative design handbooks of China, such as the Design code for heating ventilation and air-conditioning of civil buildings (GB 50736), Design Standard for Energy Efficiency of Public Buildings (GB 50189, MOHURD, AQSIQ 2013), and Practical design manual for heating and air-conditioning (2nd ed.). As the first technical solution for air-conditioning system, the THIC system has already been selected in the national energy conservation technology promotion directory (the third batch, announced by the National Development and Reform Commission of China).

Key devices for the new THIC system have been developed and realized industrialization, including outdoor air handling processors, high-temperature water chillers, and radiant terminals. And the testing standards for these key devices have been formulated to ensure the marketing activities. Since 2006, there have

X. Liu et al., *Temperature and Humidity Independent Control (THIC) of Air-conditioning System*, DOI 10.1007/978-3-642-42222-5_9,

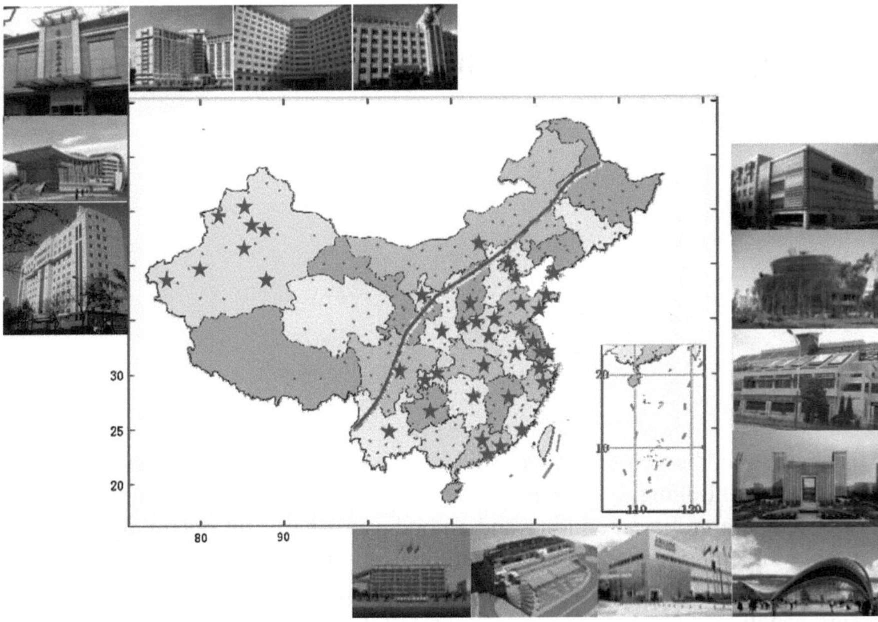

Fig. 9.1 Part of the applications adopting THIC systems in China

been more than 20 million m^2 of commercial buildings adopted the THIC systems in China. Figure 9.1 illustrates part of the buildings adopting the THIC systems, where building types include office building, hotel building, museum building, airport terminal, hospital building, and industry building.

As one aspect indicating the rapid development of the THIC system in China, relative research has been increasingly on-growing. Figure 9.2 shows the literature search results in CNKI (China National Knowledge Infrastructure, the major retrieval platform for searching Chinese literature), using 'THIC' as the key words. It indicates that the quantity of relevant literatures is increasing rapidly from 2008 to 2012. On basis of the THIC idea for air-conditioning system, IEA EBC (Energy in Buildings and Communities, formerly ECBCS) Annex 59 (IEA EBC-Annex 59 http://www.annex59.com/) has been in the running stage since 2011, with a title of "High temperature cooling and low temperature heating in buildings" and the participants from China, Denmark, Belgium, Italy, Japan and so on.

As the rapid developments of the THIC idea for air-conditioning system, there has been an industry alliance constructed to promote the THIC air-conditioning system (http://www.athic.org.cn/), including the manufacturers, real estate developers, design institutes, research institutions, and universities. Thanks to this kind of popularization mode, the basic theory, the key device development, and the applications are expanding with a rapid speed.

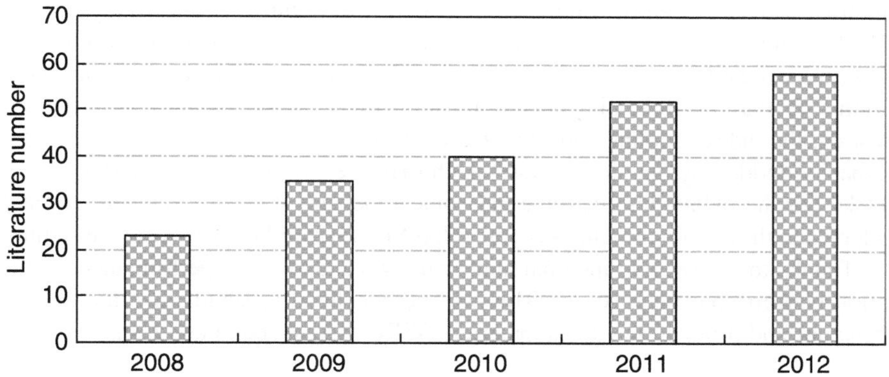

Fig. 9.2 Literature search results of THIC in CNKI

9.2 Standards of Key Components for THIC Systems

Newly developed handling devices of THIC systems have been continuously promoted in recent years. In the early stages, THIC systems consisted of devices commonly used for conventional systems. Now, there are equipment components that are made especially for THIC systems, including high-temperature water chillers, dry fan coil units (FCUs), and outdoor air dehumidifiers using condensation dehumidification method or desiccant dehumidification method. The main types and performances of newly developed key devices for the THIC system have been introduced in Chaps. 3, 4, 5, and 6.

Various kinds of devices can be combined together to form a THIC system, and different natural resources can be utilized for different climate regions. In THIC systems, the requirements for outdoor air handling processes, high-temperature cooling sources, and indoor terminals are different from those of conventional systems. For example, the outdoor air handling processors have to handle the air to a state drier than that in conventional system to satisfy the requirement for extracting indoor moisture load. Meanwhile the change of evaporating condition for the high-temperature water chillers requires the relative changes for the key components of the vapor compression refrigeration cycle. And to regulate the market behavior, the national or industry standards have been set and published to support the rapid development of key devices.

For the outdoor air handling processors using liquid desiccant, the national standard "Heat pump driven liquid desiccant outdoor air processor with heat recovery" (GB/T 27943-2011, AQSIQ, SAC 2011) has been set and published. In this standard, the application environment and performance index have been clearly defined; the testing conditions for cooling and dehumidification, for heating and humidification are also given. The mechanical industry standard "Heat pump driven outdoor air handling processor with heat recovery module" (plan No. 2011-1899 T-JB) is now in setting. For the high-temperature water chillers, the mechanical industry standard "Centrifugal high temperature water chiller" (plan No. 2011-1898 T-JB) is now in the setting stage. For the indoor sensible terminal, the mechanical industry standard "Dry fan coil unit (FCU)" (JB/T 11524-2013) has been published. In this standard, the operating conditions of the dry FCU are given for performance testing; the index "cooling capacity and air flow rate ratio" (defined as the cooling capacity divided by the air flow rate) is chosen to evaluate the performance of dry FCU. For the radiant terminal, the mechanical industry standard "Radiant terminal adopted in the air-conditioning system" (plan No. 2012-1783 T-JB) is now in setting.

Thanks to the relative national and industry standards, the performance evaluation platform has been set up. This will be beneficial to standardize the market behavior and promote the development of THIC system sound and fast.

9.3 Perspectives of the THIC System

Although rapid processes have been made in THIC system, unremitting efforts are still needed for relative research and applications in the long term. Based on current developments, we can try to understand the research priorities and tendencies of the THIC system as follows.

The first is to summarize and refine the design methodology for THIC system. As the idea for air-conditioning and handling devices of THIC system are quite different from those of conventional system, a scientific and reasonable design is the basis to ensure the normal operation and reduce operating energy consumption for applications. As various devices can be utilized to construct THIC systems for buildings with different climate conditions or functions, how to choose an appropriate system scheme becomes a desiderating problem to solve, and the design methodology still needs to be investigated. At the same time, some concerning questions in designing a THIC system are urgently to answer, such as the applicable supplied temperature and temperature difference of the high-temperature chilled water. The research on design methodology of THIC system will also help to ascertain the requirements for key handling devices.

The second is further research and developments of relative equipments. THIC provides a new thinking for the developments of air-conditioning devices. Series of key components, such as the high-temperature water chillers and dry FCUs, have been developed and adopted. However, there is still leeway for performance improvement of existing devices. On a long view, sensible terminals with high energy efficiency and new types of outdoor air handling processors are to be proposed. Besides, nowadays applications of THIC systems concentrate upon the larger-scale commercial buildings in China, and there is still a lack in devices suitable for smaller-scale buildings. For example, to develop a VRF unit especially for temperature control with high efficiency will be a good option to apply THIC systems further.

The last is the feedback and excavation coming from operations and applications. The basic task of an air-conditioning system is to regulate indoor temperature and humidity effectively, and the problems can only be identified through practical applications. Some details neglected or not concerned in system design will be found out and to solve these helps to ameliorate the whole THIC system. In practical operation, the annual and daily operating control strategies can be improved; the operating energy performances of key components can be tested, and the system efficiency can also be investigated; the feedback between system design and practical application will be built, and discrepancies as same as relations will be reflected helping to improve performance.

References

AQSIQ, SAC (2011) GB/T 27943-2011 Heat pump driven liquid desiccant outdoor air processor with heat recovery. China Standards Press, Beijing (in Chinese)

IEA EBC-Annex 59. High temperature cooling and low temperature heating in buildings. http://www.annex59.com

MOHURD, AQSIQ (2013) GB 50189 Design standard for energy efficiency of public buildings. Enquiry stage (in Chinese). http://www.cabr-ibp.com/

References

Appendices

Appendix A: Moisture Load Calculation

Indoor moisture is mainly from moisture generated by occupants and moisture evaporated from open water surface, plant transpiration, and infiltration moisture through building envelope. The total amount of indoor moisture is the sum of the above-mentioned moisture sources.

A.1 Moisture Generated by Occupants

Moisture generated by occupants is related with a variety of factors such as gender, age, clothing, labor intensity, and environment condition (temperature, humidity), and moisture generated by adult men is shown in Table A.1. For the gender, it can be considered that moisture generated by adult women is 84 % of by men and by children is 75 % of by men. Due to different proportions of adult men and women and number of children, for convenient calculation, take adult men as the base, timed by a coefficient considering proportions of various members, which is called clustering coefficient and listed in Table A.2.

Moisture generated by occupants (g/h) can be calculated by Eq. (A.1):

$$W_1 = g \cdot n \cdot \beta \tag{A.1}$$

where g is the moisture generated by adult men, g/h; n is the total number of occupants; and β is the clustering coefficient. For typical offices, the moisture amount generated by an adult man is 102 g/h as indoor temperature is 25 °C. If building area for per person is 5 m^2, moisture generated by occupants for per unit building area is about 20 g/h/m^2.

X. Liu et al., *Temperature and Humidity Independent Control (THIC) of Air-conditioning System*, DOI 10.1007/978-3-642-42222-5,

Table A.1 Moisture generated by adult men under different temperatures

	Temperature/°C														
Labor intensity	16	17	18	19	20	21	22	23	24	25	26	27	28	29	30
Sit	26	30	33	35	38	40	45	50	56	61	68	75	82	90	97
Extremely light labor	50	54	59	64	69	76	83	89	96	102	109	115	123	132	139
Light labor	105	110	118	126	134	140	150	158	167	175	184	194	203	212	220
Medium labor	128	141	153	165	175	184	196	207	219	227	240	250	260	273	283
Heavy labor	321	330	339	347	356	365	373	382	391	400	408	417	425	434	443

Source: Design handbook for air conditioning (2nd edition), 1995

Table A.2 Clustering coefficient

Workplace	Theater	Department store	Hotel	Stadium	Reading room	Factory (light labor)	Bank	Factory (heavy labor)
Clustering coefficient	0.89	0.89	0.93	0.92	0.96	0.90	1.0	1.0

Source: Design handbook for air conditioning (2nd edition), 1995

A.2 Moisture from Open Water Surface

In some buildings, there exist some open water surfaces such as water tanks, pools, and water in sanitary equipment, and the open water surface scatters vapor into air constantly. The calculation of scattered moisture amount (kg/s) is given as Eq. (A.2):

$$W_2 = \gamma \cdot \left(p_{\mathrm{qb}} - p_{\mathrm{q}}\right) \cdot A_{\mathrm{w}} \cdot \frac{B}{B'} \tag{A.2}$$

where p_{qb} is the vapor partial pressure of saturated moist air at water surface temperature, Pa; p_{q} is the vapor partial pressure in air, Pa; A_{w} is the surface area of evaporating water tanks, m^2; γ is the evaporation coefficient, kg/(N·s); B is the standard atmospheric pressure, Pa; and B' is the local atmospheric pressure, Pa.

Computational formula of evaporation coefficient is given as following, in which a and v are diffusion coefficient [kg/(N·s)] under different water temperature and air flow velocity (m/s) around the water surface, respectively. And diffusion coefficients at different water temperature are given in Table A.3.

$$\gamma = (\mathrm{a} + 0.00363\mathrm{v}) \times 10^{-5} \tag{A.3}$$

As indoor temperature is 25 °C, relative humidity is 55 %, and temperature of water tank is the same with air temperature, p_{qb} and p_{q} are 3,156 Pa and 1,729 Pa, respectively. For typical offices, open water surface area is usually less than 1 % of

Table A.3 Diffusion coefficients at different water temperature

Water temperature [°C]	<30	40	50	60	70	80	90	100
Diffusion coefficient [kg/(N·s)]	0.0043	0.0058	0.0069	0.0077	0.0088	0.0096	0.0106	0.0125

Source: Air conditioning (3rd edition) by Zhao et al. (2000)

room building area, so corresponding scattered moisture for unit building area is much less than 2.5 g/h. Compared with moisture generated by occupants, the scattered moisture from open water surface can be ignored.

A.3 Moisture from Plant Transpiration

If there are many plants in buildings, moisture from plant transpiration should be taken into consideration. Table A.4 gives measurement results for evaporation rates of some plants. If the blade area of a pot of large plant reaches 1 m^2, its moisture production is equal with moisture production amount of 2–3 persons as indicated by this table. Therefore, for area with relative more virescence indoor, this part of scattered moisture must be considered.

A.4 Infiltration Moisture Through Building Envelope

The envelopes of most buildings are porous structure, so water can be absorbed, diffused, and transferred with indoor air through internal surface of walls. Moisture transfer through walls and roofs is often neglected in comfort air-conditioning because the actual rate is quite small and the corresponding latent heat gain is insignificant (ASHRAE handbook-Fundamentals 2009). However, in some special buildings such as underground buildings, due to walls in contact with rocks or soil, groundwater in surrounding rocks or soil can infiltrate into indoor environment through porous structure of walls. And because factors influencing walls scattered moisture are very complicated, nowadays there isn't mature calculation formula of walls scattered moisture amount. Under condition without measurement data, it can be estimated on the basis of 0.5 g/(m^2·h) for separate lining or 1–2 g/(m^2·h) for wall lining (Zhang 2005). So moisture scattered from walls (g/h) can be mathematically described by following:

$$W_3 = A_b \cdot g_b \tag{A.4}$$

where A_b is the internal surface area of lining, m^2, and g_b is the scattered moisture amount of unit internal surface area, g/(m^2·h).

Table A.4 Evaporation rates of plants

Plant name	Osmanthus fragrans	Flowering plum	Prunus cerasifera	Fructus forsythiae	Magnolia	Pawpaw
Evaporation rate [g/(cm^2 · h)]	0.0396	0.0441	0.0648	0.0431	0.0511	0.0620
Plant name	Malus spectabilis	Tree falsespiraea bark	Wisteria	Firethorn	Crape myrtle	Chinese redbud
Evaporation rate [g/(cm^2 · h)]	0.0359	0.0954	0.0435	0.0378	0.0677	0.0364

Source: "The evaluation of need power of forty woodplants water loss and the classification of flammability" by Zhang et al. (1999)

Appendix B: Global Climate Analysis and Standards for Water Chillers

B.1 Global Outdoor Humidity Ratios in Summer

Figure B.1 shows the climate conditions all over the world. There are significant discrepancies between different countries. Southeastern China is in the humid subtropical zone, and northwestern of China is in the arid zone. Figure 2.25 in Chap. 2 gives the average outdoor humidity ratios in the wettest month for major cities in China. According to outdoor humidity ratio in summer, whether outdoor air needs to be dehumidified in a certain area can be determined. Figures B.2, B.3, B.4 and B.5 give the design summer outdoor humidity ratios in the United States, Europe, Australia, New Zealand, Japan, etc.; the data is from Chapter 27 Climatic Design Information of ASHRAE Handbook.

Temperate and subtropical climate can be found in most parts of the United States. Due to the vast and varied range of local terrain, climate variation is complex. Figure B.2 shows the design outdoor humidity ratios in main cities of the United States, and the values are with no assurance of 0.4 % (without guaranteeing for 35 h) and 1.0 % (without guaranteeing for 88 h). Outdoor humidity ratio is lower in western area and higher in eastern. The design summer outdoor humidity ratio in eastern part of the United States is around 17–21 g/kg, while in western area like Los Angeles, San Francisco, and Seattle, it is just 11–12 g/kg. Thus, there will be only requirement for cooling but not dehumidification in summer for buildings in western coast.

Most part of Europe is located in the range of 35–60° of north latitude, belonging to Mediterranean climate. Figure B.3 gives outdoor humidity ratios in cities of Europe within low latitudes. It can be seen that outdoor humidity ratio is lower in northern area but higher in southern. In many northern cities, design outdoor humidity ratios are about 11–13 g/kg in summer. Due to the dry outdoor air, there is only requirement for cooling but almost no dehumidification in many buildings. While in the southern area, such as Italy, outdoor design humidity ratio is about 20 g/kg, and thus both cooling and dehumidification are needed in summer.

Figure B.4 shows the design outdoor humidity ratios of Australia and New Zealand in summer. There is more rainfall for windward slopes of mountainous region in eastern of Australia, and the northern is influenced by northwest monsoon in summer which brings more rainfall, while there is less rainfall for central mainland and west bank with a dry climate. There is a mild climate in New Zealand, mild and wet in winter and warm and dry in summer. There is a small difference between outdoor temperatures in summer and winter. Even in the coldest months of July and August, the temperature is no less than 10 °C, and in the hottest months of January and February, temperature is about 25 °C.

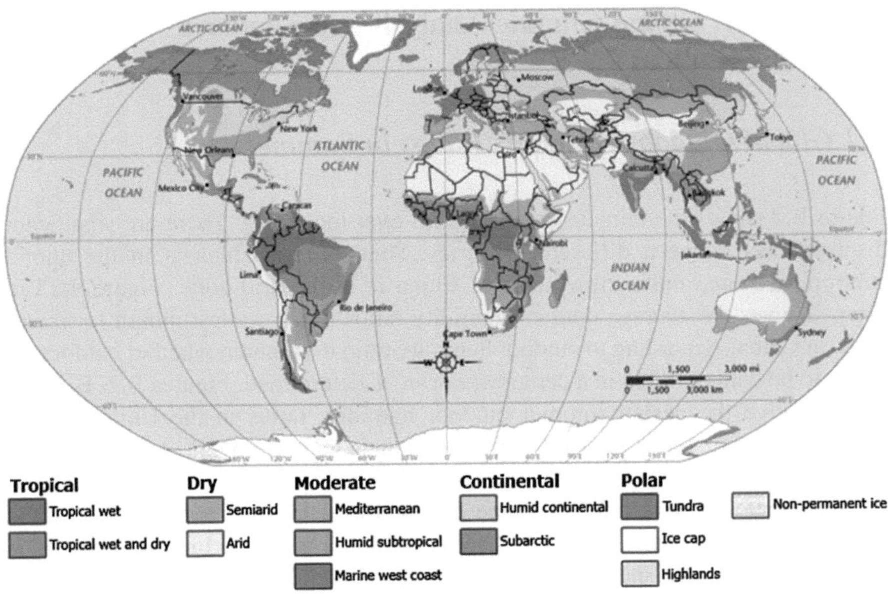

Fig. B.1 Global climate analysis

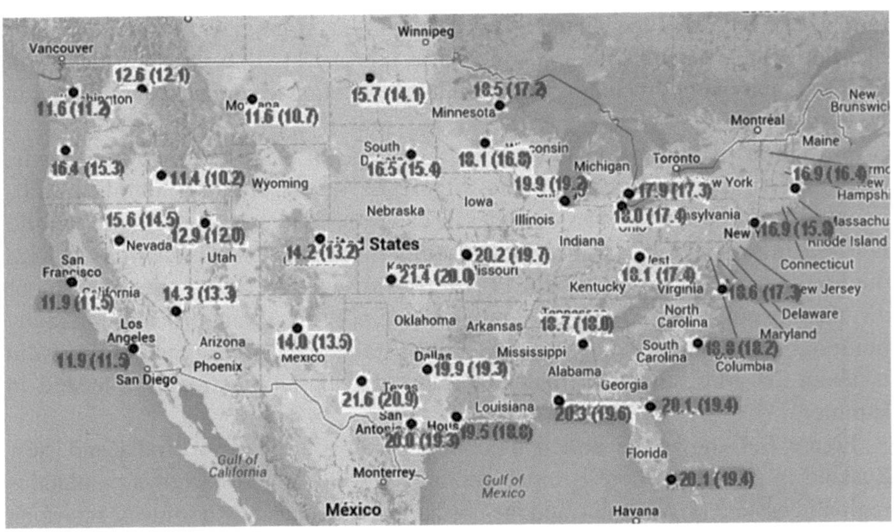

Fig. B.2 Design outdoor humidity ratios in main cities of the United States (apart from Hawaii and Alaska, unit: g/kg) (Note: Data out of brackets is parameter for no assurance of 0.4 % and in brackets is 1.0 %)

Japan covers within a range of 25° in latitude from north to south, where a main feature is the four distinctive seasons. Figure B.5 shows summer design outdoor humidity ratios for main cities in Japan. It can be seen that there is a damp climate in

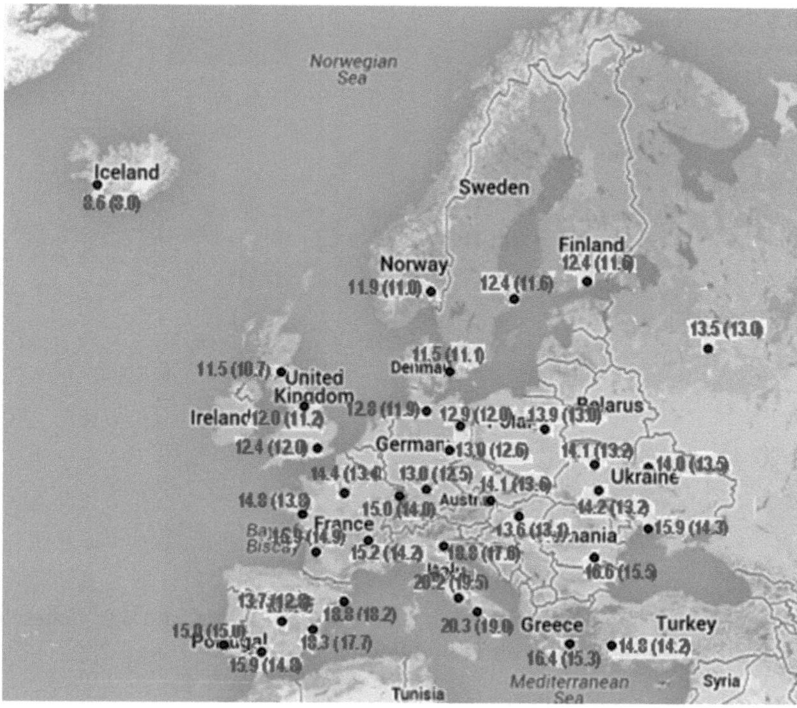

Fig. B.3 Design outdoor humidity ratios in main cities of Europe (unit: g/kg) (Note: Data out of brackets is parameter for no assurance of 0.4 % and in brackets is 1.0 %)

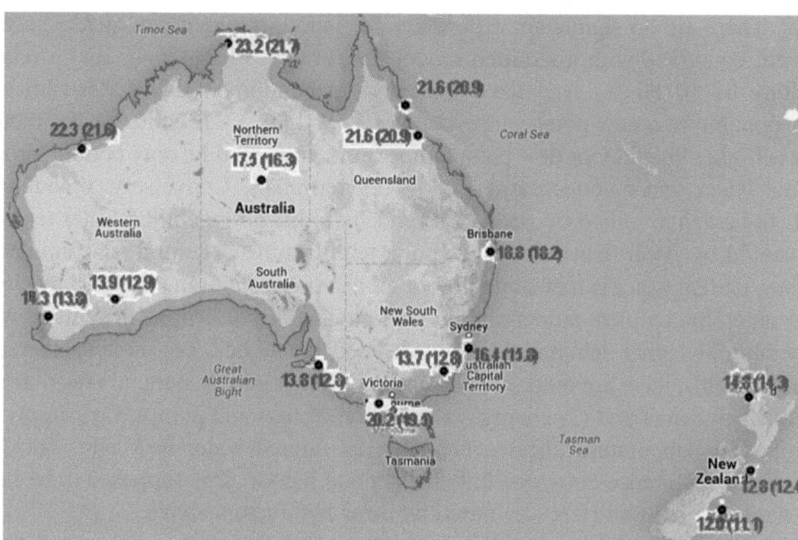

Fig. B.4 Design outdoor humidity ratios in main cities of Australia and New Zealand (unit: g/kg) (Note: Data out of brackets is parameter for no assurance of 0.4 % and in brackets is 1.0 %)

Fig. B.5 Design outdoor humidity ratios in main cities of Japan (unit: g/kg) (Note: Data out of brackets is parameter for no assurance of 0.4 % and in brackets is 1.0 %)

summer with a design outdoor humidity ratio of 17–22 g/kg, and thus dehumidification is required for outdoor air in buildings.

B.2 Standards for Water Chillers in Different Countries

Chilled water temperature reflects the process of air-conditioning system to some extent. There are no significant differences for indoor required parameters between different countries, with a required indoor temperature of 24–26 °C and a relative humidity of 50–60 %. And the corresponding humidity ratio is 9.3–12.6 g/kg, corresponding to a dew point temperature of 12.9–17.6 °C. If chilled water temperature is higher than indoor dew point temperature, there will be only cooling process without water vapor condensation for indoor terminals. Conversely, if indoor air needs to be dehumidified by condensation dehumidification, chilled water temperature should be lower than indoor dew point temperature. In central air-conditioning systems, chilled water is transported to the air handling unit or indoor FCU by chilled water pump to complete indoor air handling process. To determine the chilled water temperature, whether dehumidification is needed should be taken into consideration. Apart from this, there are other two aspects to be taken into account: (1) restriction of heat exchange area and (2) energy consumption of fans and pumps, i.e., supply and return water temperature difference caused by limited water flow rate. Thus, the temperature difference between chilled water and indoor air is supposed to be larger than the temperature difference caused by these two factors above.

Table B.1 gives a list of chilled water temperatures at the inlet and outlet of chillers under rated conditions (or standard conditions) set in some countries or regions, and there are mainly three types of chillers:

Table B.1 Summary of standards for chillers

Source	Number	Name	Rated condition: Outlet temp./°C	Inlet temp./°C	ΔT/°C	Note
China	GB/T 18430.1-2001	Water chilling (heat pump) packages using the vapor compression cycle Water chilling (heat pump) packages for industrial and commercial and similar application	7	12	5	Substitute for JB/T 4329–1997 and JB/T 3355-1998
	GB/T 18430.1-2007	Water chilling (heat pump) packages using the vapor compression cycle – Part 1: Water chilling (heat pump) packages for industrial and commercial and similar application	7	12	5	Substitute for GB/T 18430.1-2001
	GB/T 18430.2-2001	Water chilling (heat pump) packages using the vapor compression cycle Household and similar water chilling (heat pump) packages	7	12	5	Based on JB/T 4329-1997
	GB/T 18430.2-2008	Water chilling (heat pump) packages using the vapor compression cycle – Part 2: Household and similar water chilling (heat pump) packages	7	12	5	Substitute for GB/T 18430.2-2001
The United States	AHRI 550 590-2003	Performance rating of water chilling packages using the vapor compression cycle	6.7	12.3	5.6	Substitute for ARI 550 590-1998
	AHRI 550 590-2011	Performance rating of water chilling and heat pump water-heating packages using the vapor compression cycle	6.7	12.3	5.6	Substitute for ARI 550 590-2003
Japan	JIS B 8613:1994	Water chilling unit	7	12	5	Reaffirmed in 2001 and 2007
Europe	EN 14511-2	Air conditioners, liquid chilling packages, and heat pumps with electrically driven compressors for space heating and cooling – Part 2: Test conditions	7	12	5	For conventional air-conditioning
			18	23	5	For floor cooling and similar application
Australia	AS/NZS 4776.1.1: 2008	Liquid-chilling packages using the vapor compression cycle – Part 1.1: Method of rating and testing for performance – Rating	4–9	–	–	Not find the detailed data

- Supply and return chilled water temperatures are 6.7 °C (44 °F)/12.3 °C (54 °F), respectively, with a temperature difference between supply and return chilled water of 5.6 °C (10 °F) (American standard AHRI 550 590).
- Supply and return chilled water temperatures are 7 °C/12 °C, respectively, with a temperature difference between supply and return chilled water of 5 °C (Chinese standards GB/T 18430.1 and GB/T 18430.2, Japanese standard JIS B 8613 and European standard EN 14511).
- Supply and return chilled water temperatures are 18 °C/23 °C, with a temperature difference between supply and return chilled water of 5 °C (European standard EN 14511 for radiant floor cooling or similar purposes).

From these three sets of chilled water temperatures mentioned above, outlet temperature of chilled water of 6.7–7.0 °C can meet the demand of building dehumidification (by condensation dehumidification), while outlet temperature of 18 °C can only be used for sensible cooling but not dehumidification. From the standards investigated, only European standards provide the conditions where the chiller's outlet temperature of chilled water is 18 °C and 7 °C, respectively, as the standard conditions for test and analysis of chiller performance.

In Europe, as shown in Fig. B.3, there is a relatively low outdoor humidity ratio for the northern part in summer. For example, in Britain and Denmark, design value of outdoor humidity ratio in summer is about 11–12 g/kg, and in Poland, Austria, Ukraine, most parts of Germany, and northern area of France, design value of outdoor humidity ratio in summer is 13 g/kg or so. As to most of Europe, outdoor air is dry in the summer, and thus buildings only need cooling but not dehumidification for air-conditioning system. Terminal devices such as radiant floor and ceiling have been in rapid development in Europe. Many scholars have researched the energy storage methods for cold storage during night and discharging it during the daytime to take advantage of the temperature difference between day and night. To meet the demand for outdoor air, radiant cooling with independent outdoor air handling process has achieved an increasing use in Europe. The IEA Energy in Buildings and Communities (EBC, formerly ECBCS) Annex 49 project (2006–2009), in which many European researchers participated, focused on low-temperature heating and high-temperature cooling system. Because of the dry climate in Europe, it is easy to understand the reason why outlet temperature of 18 °C chilled water is taken as the standard condition in European standards.

Then chillers of different brands are analyzed as follows. The major brands including Carrier, Trane, York and Daikin (McQuay) all take 7 °C as the outlet temperature of chilled water for conventional air-conditioning system. In the 2006 Sino-Italian cooperation project (Office Building for Environment Department in Tsinghua University) designed by Italian designers and HVAC engineers, radiant ceiling and independent outdoor air systems were used. The chiller produced by Daikin (McQuay) was used, which can work for producing chilled water with outlet temperatures of 7 °C (for outdoor air dehumidification) and 14 °C (for radiant panel cooling) together. For this chiller with two operating conditions, COP can be 5.22 under low chilled water temperature condition (7 °C/13 °C, with cooling water

Table B.2 Comparison of chillers' performances

Type of chiller	COP of centrifugal chiller			Thermodynamic perfect degree			Note
	7 °C outlet	16 °C outlet	18 °C outlet	7 °C outlet	16 °C outlet	18 °C outlet	
① Conventional centrifugal chiller in China (*GREE*)	5.78	6.80	7.05	66 %	54 %	51 %	Measured value
② Main centrifugal chiller manufacturers in the United States[a]	–	8.30	–	–	66 %	–	Calculated value
③ High temperature centrifugal chiller in China (*GREE*)	–	8.58	9.47	–	69 %	69 %	Measured value

[a]Data is from Mr. Yungang Pan in China Architecture Design & Research Group

temperatures of 30/35 °C) and 5.98 under high water temperature condition (14 °C/18 °C, with cooling water temperatures of 30/35 °C). According to the COP of Carnot cycle calculated with an evaporating temperature 2 °C lower than outlet chilled water temperature and a condensing temperature 2 °C higher than inlet cooling water temperature, thermodynamic perfect degree (actual COP divided by Carnot COP) for this chiller is about 60 and 52 % under the low water temperature condition and under the high water temperature condition, respectively.

The main results listed in Table B.2 include ① measured COP of 7, 16, and 18 °C water chillers; ② theoretical value given by the main manufacturers mentioned above; and ③ measured performance of high-temperature water chiller newly developed in China. It can be seen that if 7 °C water chiller works for producing 16 and 18 °C chilled water, thermodynamic perfect degree will drop from 66 to 54 % and 51 %, respectively. While for the chillers specially developed for high-temperature condition, thermodynamic perfect degree will realize as high as 69 %. As outlet water temperature is 16 °C, COP of chillers specially developed for high-temperature condition is 8.58, much higher than that of 7 °C water chiller producing 16 °C chilled water (only 6.80).

Appendix C: Typical Buildings' Models and Preferences in DeST Software

C.1 Models and Parameter Settings for Different Buildings

C.1.1 Office Building

The total area of this 11-floor office building (shown as Fig. C.1) is 66,440 m^2; each floor mostly consists of offices and meeting rooms and has an area of 6,040 m^2.

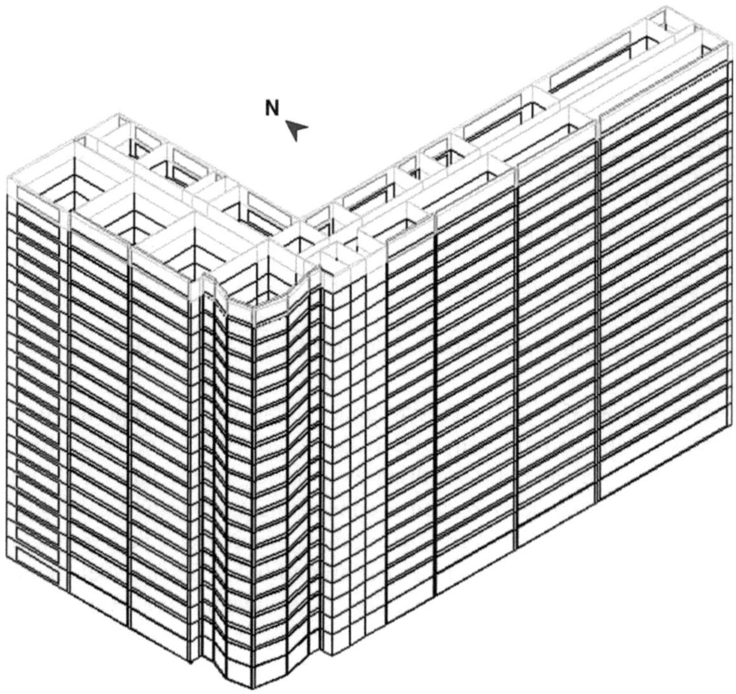

Fig. C.1 Building schematic of the office building in DeST software

Annual load for air-conditioning system could be simulated as this office building could be located in Beijing, Shanghai, Wuhan, or Guangzhou. Design parameters for building envelopes in different regions are listed in Table C.1, with design parameters of indoor thermal environment, time periods for different seasons, and indoor heat sources shown in Tables C.2, C.3 and C.4, respectively. The daily schedule of typical office and meeting rooms on workdays in this building is shown in Fig. C.2.

C.1.2 Hotel Building

The total area of this hotel building (shown in Fig. C.3) is 71,850 m^2 with the main function of guest rooms. There are 22 floors with an area of 4,650 m^2 for the first floor and 3,200 m^2 for the typical floor.

Daily schedule and thermal disturbance of typical guest rooms are shown as Fig. C.4 and Table C.5, respectively.

Table C.1 Design parameters for building envelopes in different cities

Heat transfer coefficient ($W/(m^2 \cdot °C)$)	Beijing	Shanghai	Wuhan	Guangzhou
Roof	0.55	0.7	0.7	0.9
External wall	0.6	1.0	1.0	1.5
External window	3.0	3.5	3.5	4.7

Table C.2 Indoor design parameters in consideration of thermal comfort

Season	Indoor temperature (°C)	Indoor relative humidity (%)
Summer	24–26	40–60
Transition season	22–26	40–60
Winter	20–22	30–60

Table C.3 Time periods for different seasons in different cities

Season	Beijing	Shanghai	Wuhan	Guangzhou
Summer	6.1–9.30	5.1–10.30	5.1–10.30	4.1–11.30
Transition season	3.16–5.30	3.1–4.30	3.1–4.30	12.1–12.31
	10.1–11.14	10.1–11.30	10.1–11.30	3.1–3.31
Winter	11.15–3.15	12.1–2.28	12.1–2.28	1.1–2.28

Table C.4 Design parameters of indoor heat sources

Functions	Light (W/m^2)	Device (W/m^2)	Person (p/m^2)	Outdoor air flow rate ($m^3/(h{\cdot}p)$)
Office room	10	20	0.1	30
Meeting room	15	0	0.3	30
Corridor	5	0	0.05	30

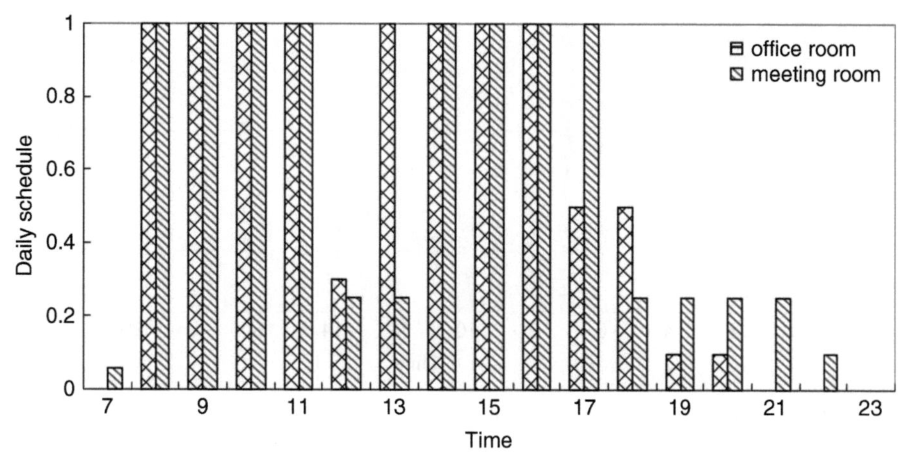

Fig. C.2 Daily schedule of the office building set in the DeST software

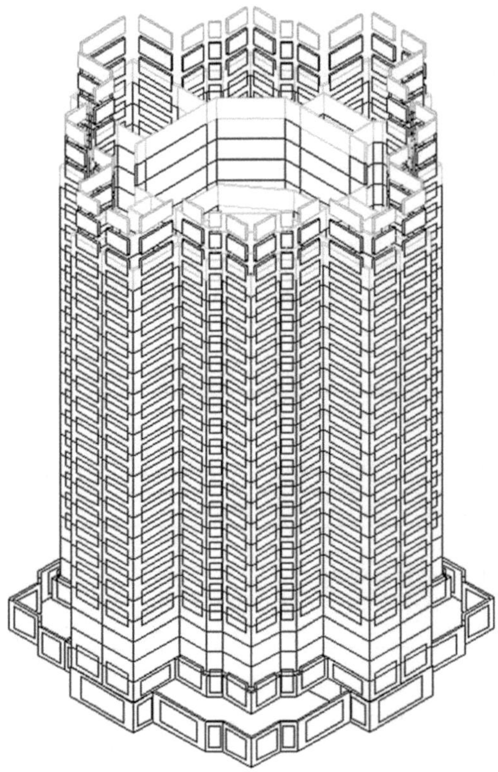

Fig. C.3 Building schematic of hotel building model

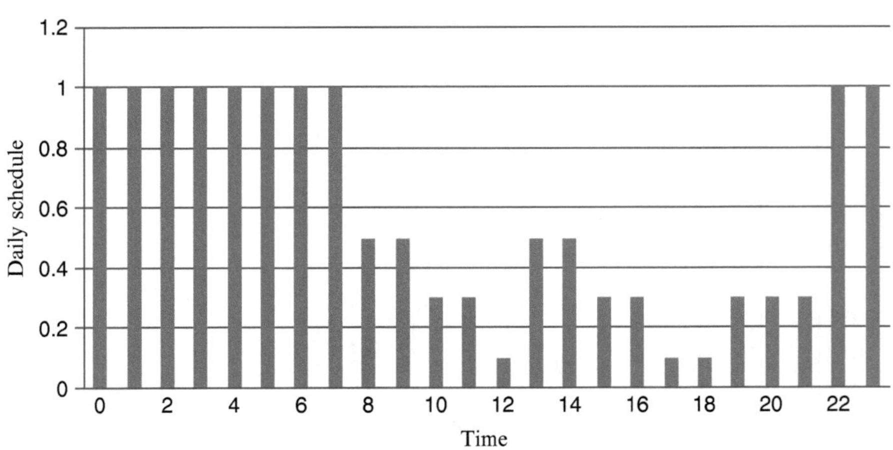

Fig. C.4 Thermal disturbance schedule of hotels

Table C.5 Indoor thermal disturbance parameters of hotels

Functional zone	Light (W/m^2)	Device (W/m^2)	Person (p/m^2)	Outdoor air flow rate (m^3/(h·p))
Guest rooms	15	10	0.04	50

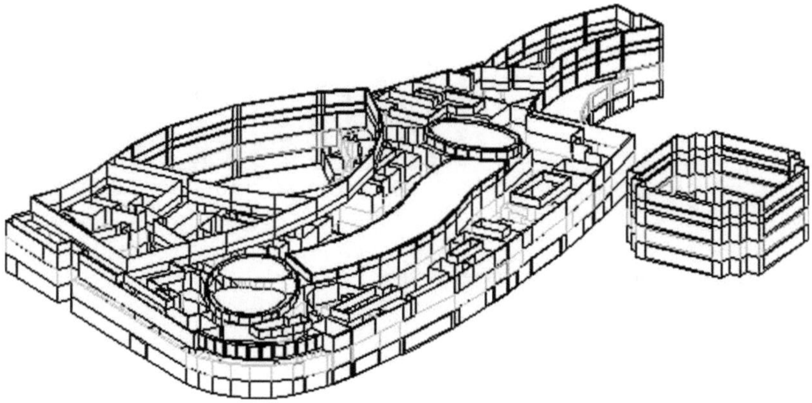

Fig. C.5 Building schematic of the shopping mall

C.1.3 Shopping Mall

The total area of this shopping mall (shown as Fig. C.5) is 119,000 m^2, with six floors including five aboveground level and one underground. There are some commercial facilities such as restaurants, stores, and movie theaters, so the outdoor air flow rate of this building is relative high.

Daily schedules of stores and restaurants in this building for one day are shown as Fig. C.6, and thermal disturbance is listed in Table C.6.

C.1.4 Hospital Building

The total building area of this hospital building (shown as Fig. C.7) is 86,000 m^2, with 22 floors. And the 1st floor composes of waiting rooms, patient rooms, halls, examining rooms, etc., and floors 2–22 include mainly wards and offices.

Daily schedule of wards is shown as Fig. C.8, and thermal disturbance is listed in Table C.7.

C.1.5 Gymnasium

The total building area of this gymnasium building (shown as Fig. C.9) is 63,000 m^2. The air-conditioning systems in competition area and audience area are on when there are competitions. The competition site is in use once every week. Daily schedule for the competition day is shown as Fig. C.10, and thermal disturbance is listed in Table C.8.

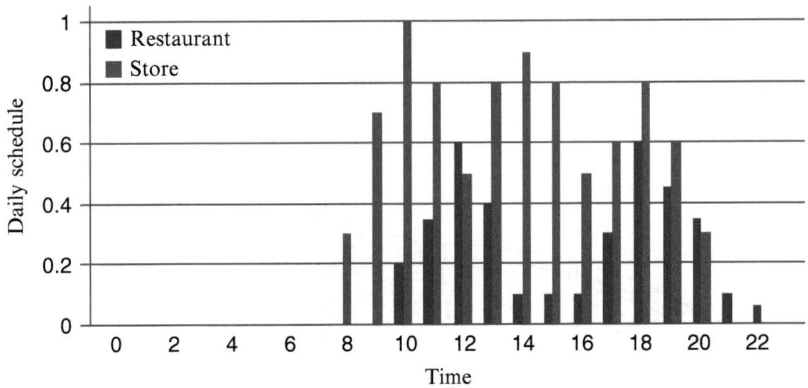

Fig. C.6 Daily schedules of stores and restaurants

Table C.6 Indoor thermal disturbance parameters of the shopping mall

Functional zone	Light (W/m^2)	Device (W/m^2)	Person (p/m^2)	Outdoor air flow rate ($m^3/(h{\cdot}p)$)
Stores	19	13	0.25	20
Restaurants	19	0	0.67	20
Movie theaters	3	0	0.5	20

Fig. C.7 Building schematic of the hospital building

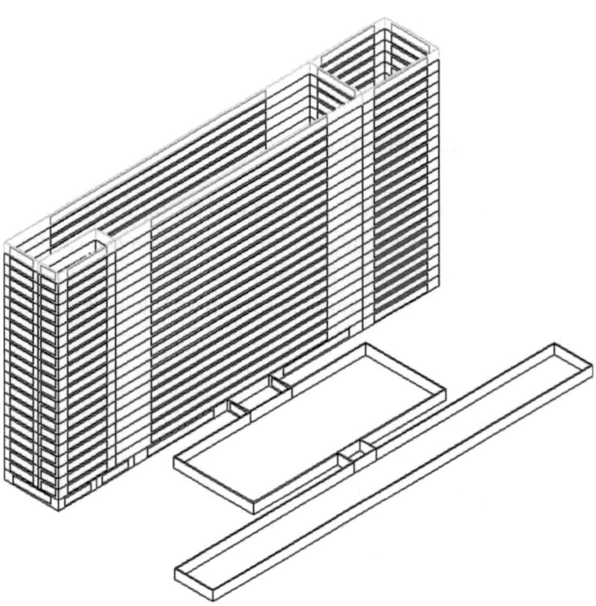

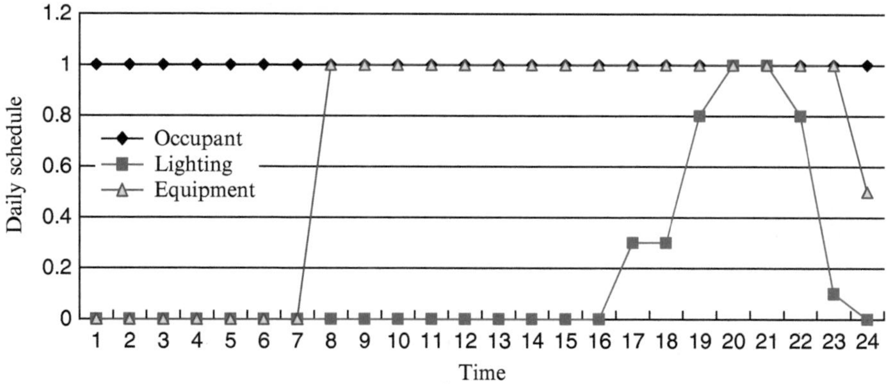

Fig. C.8 Daily schedule of wards

Table C.7 Indoor thermal disturbance parameters of the hospital

Functional zone	Light (W/m^2)	Device (W/m^2)	Person (p/m^2)	Outdoor air flow rate ($m^3/(h{\cdot}p)$)
Wards	5	30	0.1	50

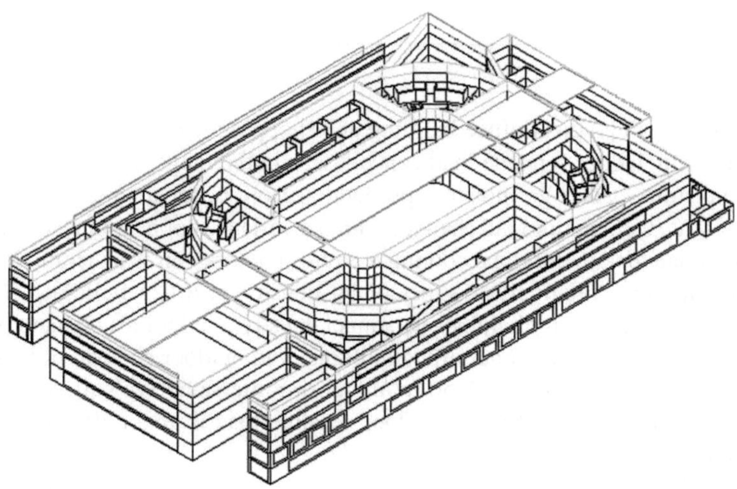

Fig. C.9 Building schematic of the gymnasium building

C.2 Load Calculation Result Analysis

Here takes the typical office building and shopping mall in Beijing as examples to explain the hourly variance and characteristic of indoor sensible load, indoor moisture load, and outdoor air load. Fig. C.11a gives the outdoor hourly ambient parameters in a typical week (July 9–15), and Fig. C.11b, c shows the hourly load

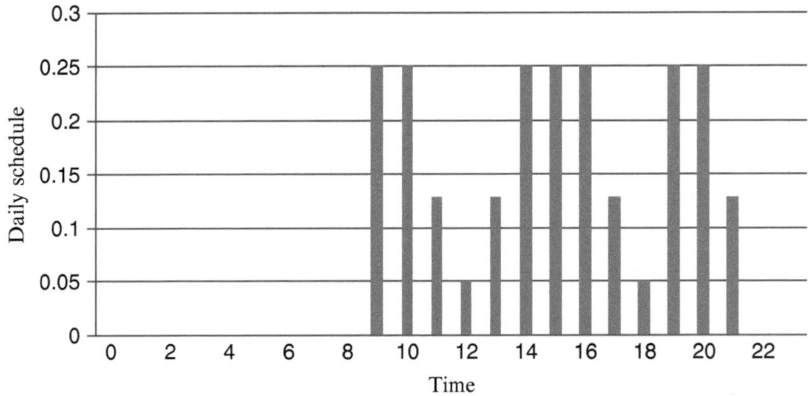

Fig. C.10 Schedule for the competition day of competition site in gymnasium

Table C.8 Indoor thermal disturbance parameters of the gymnasium

Functional zone	Light (W/m^2)	Device (W/m^2)	Person (p/m^2)	Outdoor air flow rate ($m^3/(h{\cdot}p)$)
Audience sites	2.2	0	0.73	30
Competition sites	98	0	0.1	30

per unit area of the typical office building and shopping mall, respectively. For the office building, indoor sensible load concentrates in the range of 30–50 W/m^2, and indoor moisture load (e.g., moisture production from occupants) is generally less than 10 W/m^2. Outdoor air load changes with the variances of personnel density and outdoor climate parameters, and cooling load from outdoor air is generally less than 25 W/m^2 due to a relatively low personnel density in office buildings. For the shopping mall, due to a high indoor personnel density, outdoor air demand per unit area is higher than conventional office building, and outdoor air load per unit area concentrates in the range of 30–50 W/m^2 during the same typical week, with a maximum value beyond 70 W/m^2; indoor sensible load is about 30 W/m^2; indoor moisture load is about 10 W/m^2.

Again taking office building and shopping mall in Guangzhou as examples, Fig. C.12a–c gives outdoor hourly parameters, hourly load per unit area of office building, and hourly cooling load per unit area of shopping mall for the typical week (July 9–15) in Guangzhou, respectively. For this typical office building in Guangzhou, indoor sensible cooling load concentrates in the range of 30–50 W/m^2; indoor moisture load is about 10 W/m^2. For the shopping mall, outdoor air load concentrates in the range of 40–70 W/m^2, with the maximum value beyond 90 W/m^2; indoor sensible heat load is about 30 W/m^2; indoor moisture load is about 10 W/m^2.

Comparing load characteristics of typical buildings in Beijing and Guangzhou, indoor moisture load changes less between different regions; discrepancy of

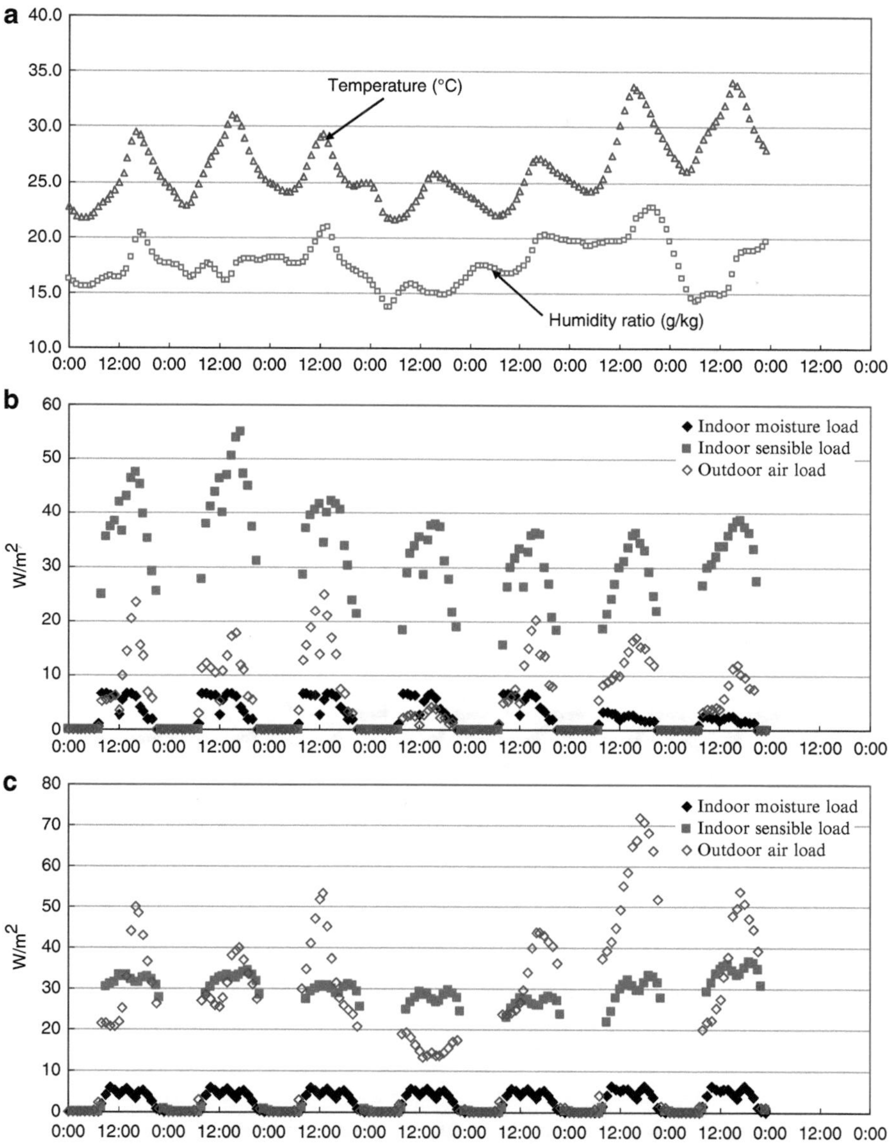

Fig. C.11 Hourly load per unit area of buildings for a typical week (7.9–7.15) in Beijing: (**a**) outdoor climate, (**b**) office building, and (**c**) shopping mall

sensible load for similar buildings in Beijing and Guangzhou is unconspicuous due to similar ambient temperature; discrepancy of outdoor air cooling load is significant because of the significant difference between outdoor air humidity ratios in different regions.

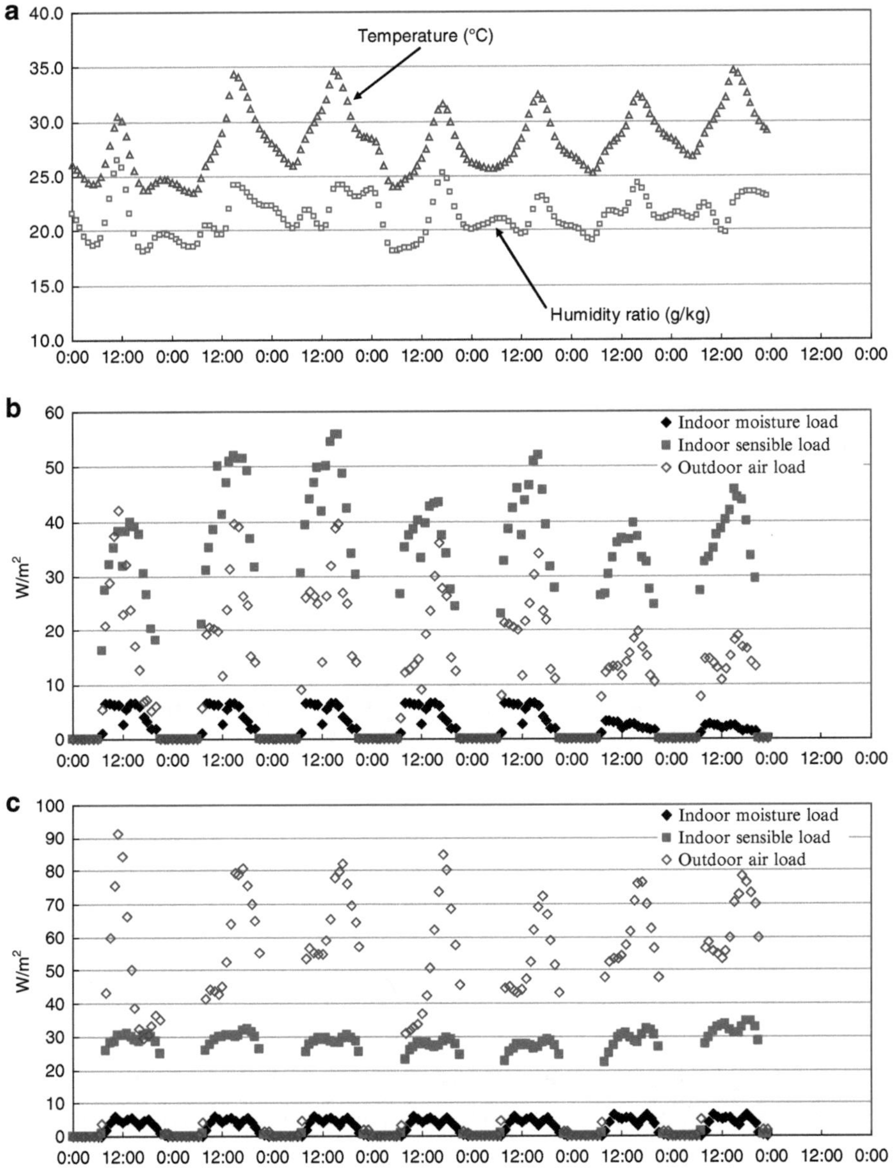

Fig. C.12 Hourly load per unit area of buildings for a typical week (7.9–7.15) in Guangzhou: (**a**) outdoor climate, (**b**) office building, and (**c**) shopping mall

C.3 Load Apportionment Analysis of the THIC System

In the THIC system, the supplied outdoor air is adopted for removing indoor moisture load and regulating indoor humidity. If the supplied air temperature is lower than that of indoor, it means the supplied outdoor air could be adopted for removing part of indoor sensible load, while the rest of the indoor sensible load is removed by indoor sensible terminals. In brief, indoor sensible load is undertaken by the supplied outdoor air and the sensible terminals simultaneously.

The capability of the supplied outdoor air undertaking indoor sensible load is significantly influenced by the temperature and flow rate of the supplied air. Taking the typical buildings shown in this appendix as an example, Fig. C.13 gives the indoor sensible loads undertaken by the supplied outdoor air and the sensible terminals, respectively. The supplied outdoor air temperature is 20 °C and the air flow rate is selected according to the design states shown above. However, if the supplied outdoor air temperature or the flow rate changes, the apportionment of indoor sensible load by the supplied air and the sensible terminals will be correspondingly changed.

Appendix D: Surface Temperature Unevenness of Radiant Panel

Cooling capacity of the radiant panel is closely related to the average surface temperature of the panel. Surface temperature distribution is also an important index for the radiant panel, especially the lowest temperature of the panel surface in the cooling condition. The limitation of the lowest temperature on the surface is the dew point of the air surrounding the panel. The surface temperature distribution is analyzed with the following two conditions:

① Uniform indoor heat sources
② Nonuniform indoor heat sources or shading by furniture

D.1 Uniform Indoor Heat Sources

For the metal radiant panel, its minimum surface temperature is close to the chilled water inlet temperature. Thus, it is convenient to control the supply water temperature to guarantee that the radiant panels run under non-condensing conditions. If the cooling load varies, the temperature distribution of the radiant panel surface will be different, in relevant with the radiant panel spacing L and the thickness δ, as shown in Fig. D.1. As keeping the indoor air temperature $T_a = 26$ °C, indoor wall

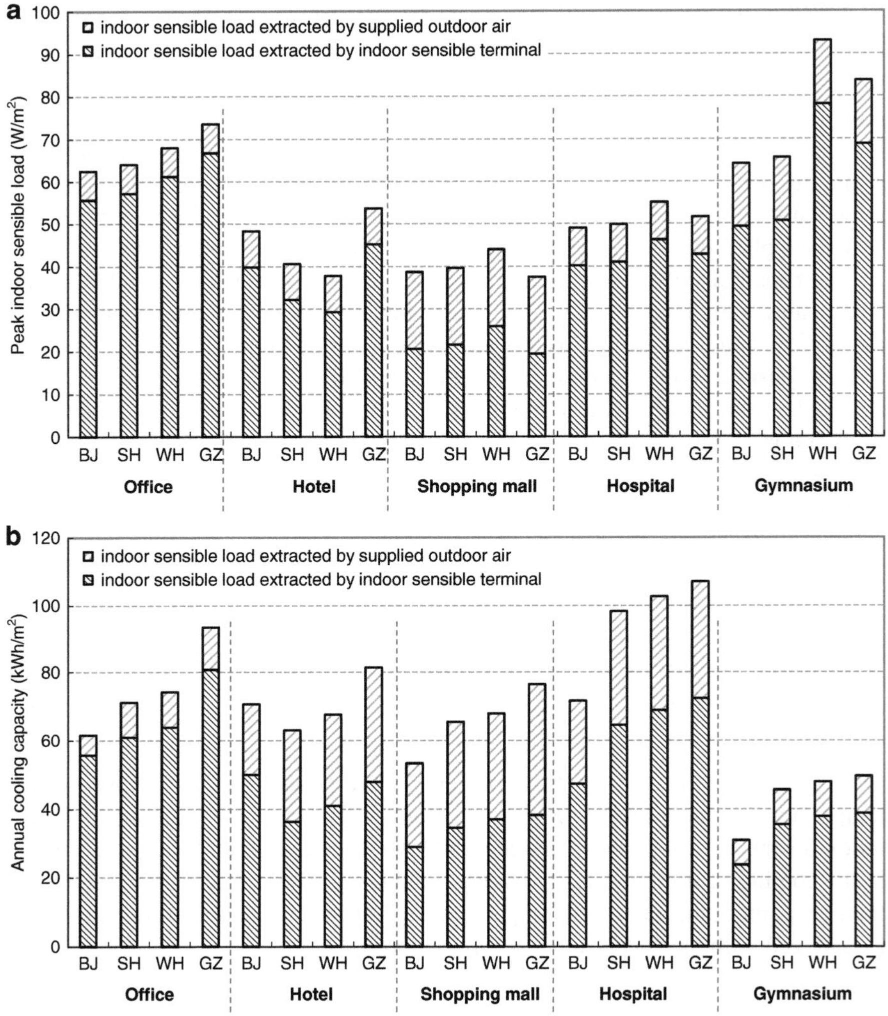

Note: BJ-Beijing; SH-Shanghai; WH-Wuhan; GZ-Guangzhou.

Fig. C.13 Apportionments of indoor sensible load for typical buildings in different regions: (**a**) peak indoor sensible load and (**b**) cooling capacity of the entire cooling season (Note: *BJ* Beijing, *SH* Shanghai, *WH* Wuhan, *GZ* Guangzhou)

temperature AUST increases from 26 to 37 °C, and the corresponding radiant panel cooling capacity increases from 60 to 120 W/m^2:

- $L = 100$ mm, $\delta = 2.0$ mm: Temperature difference between the mean and the lowest points of the radiant panel increases from 0.2 to 0.4 °C.
- $L = 100$ mm, $\delta = 0.5$ mm: Temperature difference increases from 0.8 to 1.6 °C.
- $L = 150$ mm, $\delta = 2.0$ mm: Temperature difference increases from 0.5 to 0.9 °C.
- $L = 150$ mm, $\delta = 0.5$ mm: Temperature difference increases from 1.8 to 3.6 °C.

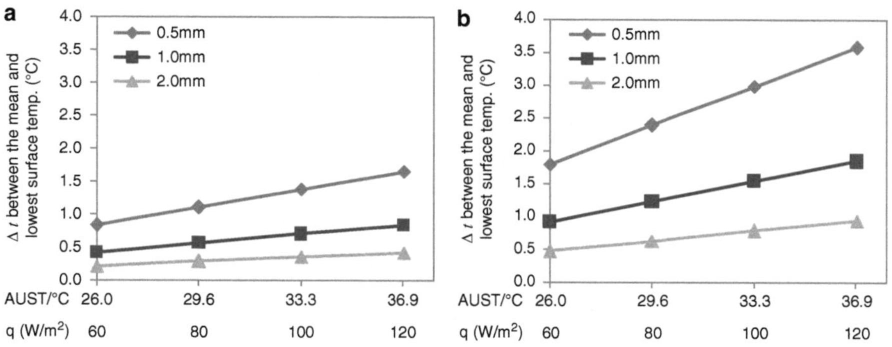

Fig. D.1 Influence of cooling load on the surface temperature uniformity of radiant panel (average surface temperature is 20 °C): (**a**) spacing $L = 100$ mm and (**b**) spacing $L = 150$ mm

Therefore, the surface temperature of radiant panel with smaller spacing L and larger thickness δ is more uniform and less susceptible to variances of the cooling load.

The relationship between the lowest surface temperature and the chilled water temperature requires further analysis, so as to determine a suitable water supply parameter. As for metal radiant panel, its minimum surface temperature is close to the chilled water inlet temperature. However, different characteristic appears in other kinds of radiant panels, such as concrete radiant floor and plastering capillary radiant panel.

Tables D.1 and D.2 list the relationship between the lowest surface temperature and the chilled water temperature in the concrete radiant floor (Structure I in Table 3.4) and plastering capillary radiant panel (Structure III in Table 3.6) in the typical conditions, respectively. It can be seen that the minimum surface temperature is much higher than that of the inlet chilled water. The cooling capacity and average surface temperature of these two kinds radiant panel are almost the same with the same mean water temperature T_w; and the temperature difference between the mean and lowest point of the concrete radiant floor and plastering capillary radiant panel is small, within 0.2 °C under the conditions shown in the Tables D.1 and D.2.

D.2 Nonuniform Indoor Heat Sources or Shading by Furniture

For the radiant panel, the cooling capacity may reach more than 100 W/m^2 when it is exposed to direct sunlight. The effects of solar radiation and wall surface temperature on radiant panels that are shaded by furniture (such as chairs) will be quite different from those on radiant panels without shading. For instance, in an airport terminal or railway station, there are usually many benches in the waiting lounges, as shown in Fig. D.2. Due to the presence of these benches, the radiant

Table D.1 Influence of different supply and return water temperature on concrete radiant floor cooling performance (Structure I)

			Mean water temp. = 16 °C				Mean water temp. = 18 °C				Mean water temp. = 20 °C			
		Supply:	13.5	14	14.5	15	15.5	16	16.5	17	17.5	18	18.5	19
	Main parameters	Return:	18.5	18	17.5	17	20.5	20	19.5	19	22.5	22	21.5	21
Case 1: $T_a = 26$ °C,	Surface lowest temp./°C		13.5	19.3	19.3	19.4	20.6	20.6	20.7	20.7	21.9	21.9	22.0	22.0
AUST = 26 °C,	Surface average temp./°C		19.5	19.5	19.5	19.5	20.8	20.8	20.8	20.8	22.1	22.1	22.1	22.1
$q_{sr} = 0$ W/m^2	Cooling capacity/(W/m^2)		35.4	35.4	35.4	35.4	28.3	28.3	28.3	28.3	21.2	21.2	21.2	21.2
Case 2: $T_a = 26$ °C,	Surface lowest temp./°C		23.1	23.2	23.2	23.3	24.4	24.5	24.5	24.6	25.7	25.8	25.8	25.9
AUST = 28 °C,	Surface average temp./°C		23.3	23.3	23.3	23.3	24.6	24.6	24.6	24.6	25.9	25.9	25.9	25.9
$q_{sr} = 50$ W/m^2	Cooling capacity/(W/m^2)		74.8	74.8	74.8	74.8	67.8	67.8	67.8	67.8	60.7	60.7	60.7	60.7

Table D.2 Influence of different supply and return water temperature on plastering capillary radiant panel cooling performance (Structure III)

		Mean water temp. = 16 °C					Mean water temp. = 18 °C				Mean water temp. = 20 °C			
		Supply:	13.5	14	14.5	15 °C	15.5	16	16.5	17	17.5	18	18.5	19
	Main parameters	Return:	18.5	18	17.5	17 °C	20.5	20	19.5	19	22.5	22	21.5	21
Case 1: $T_a = 26$ °C,	Surface lowest temp./°C		19.0	19.0	19.0	19.1	20.3	20.3	20.4	20.4	21.6	21.7	21.7	21.7
AUST = 26 °C,	Surface average temp./°C		19.2	19.2	19.2	19.2	20.5	20.5	20.5	20.5	21.8	21.8	21.8	21.8
$q_{sr} = 0$ W/m^2	Cooling capacity/(W/m^2)		62.0	62.0	62.0	62.0	48.7	48.7	48.7	48.7	35.6	35.6	35.6	35.6
Case 2: $T_a = 26$ °C,	Surface lowest temp./°C		19.3	19.4	19.4	19.4	20.6	20.7	20.7	20.8	22.0	22.0	22.1	22.1
AUST = 28 °C,	Surface average temp./°C		19.5	19.5	19.5	19.5	20.8	20.8	20.8	20.8	22.2	22.2	22.2	22.2
$q_{sr} = 0$ W/m^2	Cooling capacity/(W/m^2)		68.8	68.8	68.8	68.8	55.6	55.6	55.6	55.6	42.6	42.6	42.6	42.6

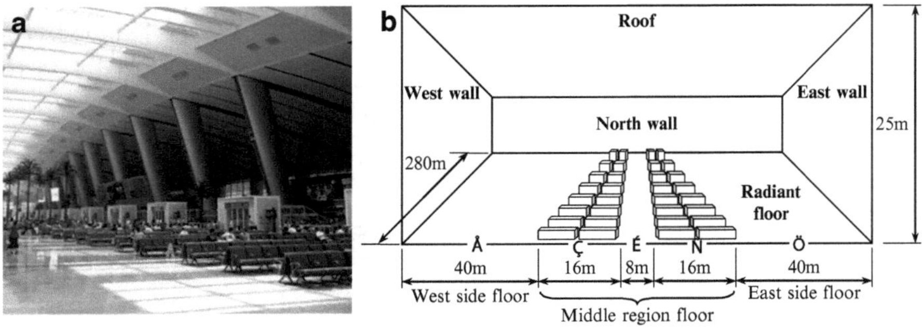

Fig. D.2 Case study of the large space building: (**a**) indoor environment and shaded floor by benches and (**b**) sketch of indoor space

floor under the benches is not affected by solar radiation or high-temperature internal wall surfaces, so its floor surface temperature is lower. Thus, more attention should be paid to the surface temperature of radiant floors shaded by furniture to minimize the risk of moisture condensation.

Taking the large space building in Fig. D.2 as an example, the performance of the radiant floor is analyzed in consideration of surfaces with/without shading. The large space building is 280 m in length (from north to south), 160 m in width (from east to west), and 25 m in height. Double-glass curtain walls with low-E coating are adopted as the walls, with a heat transfer coefficient of 2.3 W/(m^2 · °C) and a shading coefficient of 0.3. On the roof, a daylighting band composed of polycarbonate sheets with a heat transfer coefficient of 2.7 W/(m^2 · °C) and a shading coefficient of 0.4 takes up 30 % of the roof area, while the rest is taken up by 100-mm-thick insulated alloy plates with a heat transfer coefficient of 0.55 W/(m^2 · °C). There are two rows of benches located in the middle of the building, as shown in Fig. D.2b. Figure D.3a shows local solar radiation on a typical summer day if the large space building were located in Beijing. And the solar incident heat flux absorbed by the floor in the same large space building is displayed in Fig. D.3b.

The radiant floor has a thermal resistance of 0.1(m^2 · °C)/W or 0.2(m^2 · °C)/W, and chilled water with a supply temperature of 18 °C and a flow rate of 20 kg/(m^2 · h) is supplied for cooling in the large space. In addition, several fan coil units are set up to provide supplemental cooling to maintain the air temperature in the occupied zone at 26 °C. Indoor humidity control and air quality are performed by the outdoor air processor, in which air is handled and supplied at a temperature and humidity ratio of 26 °C and 10 g/kg with the flow rate of 100,800 m^3/h (ACH = 1 h^{-1} in the height within 3 m), respectively. The overall sketch of the radiant floor combined with outdoor air cooling system is depicted in Fig. D.4. For convenience, the radiant floors from the west side to the east side are numbered ①–⑤.

Three typical outdoor conditions, 8:00 a.m., 12:00 p.m., and 4:00 p.m. with outdoor air temperatures of 29.5, 33.0, and 32.8 °C, are selected to analyze the performance of the radiant floor, respectively. As shown in Tables D.3, D.4, and D.5, the cooling capacity of the radiant floor with a thermal resistance of

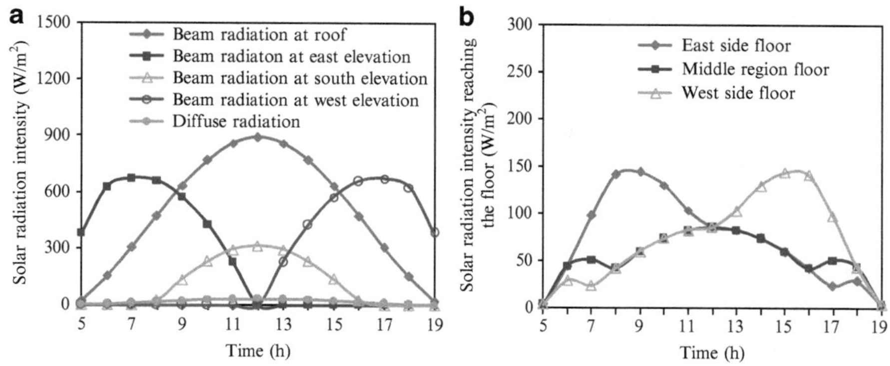

Fig. D.3 Typical solar radiation in Beijing (July 20): (**a**) outdoor solar radiation intensity and (**b**) solar radiation absorbed by radiant floor

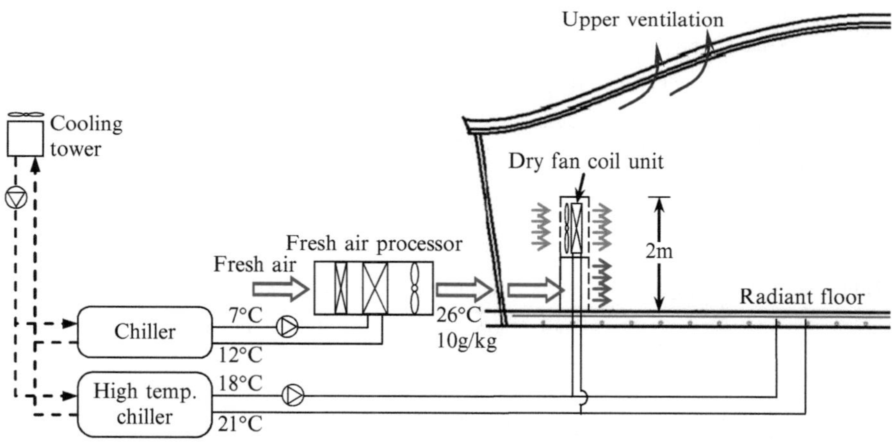

Fig. D.4 Overall sketch of the radiant floor combined with outdoor air cooling system

0.1(m^2 · °C)/W can reach 60–120 W/m^2 when it is exposed to 40–140 W/m^2 solar radiation. Therefore, a higher solar radiation intensity corresponds to a higher cooling capacity of the radiant floor.

The aforementioned cooling capacity is achieved by a radiant floor without shading by benches. The radiant floor under the benches is not exposed to solar radiation or the high-temperature internal wall surfaces. As shown in Fig. D.5, the cooling capacity of the sheltered radiant floor is about 15 W/m^2, which is much lower than that of the unsheltered radiant floor. And the surface temperature of the radiant floor under benches is only 20.3 °C, which is 4.7–7.8 °C lower than that of the unsheltered radiant floor. The temperature difference between the radiant floors with and without shading can be calculated according to Eq. (3.32). For instance, at 12:00 p.m., AUST and solar radiation intensity q_{sr} of the radiant floor under benches and the radiant floor without shading in the middle region are 30.3 °C,

Table D.3 Cooling capacity and surface temperature of radiant floor (8:00 a.m.)

	Floor ①	Floor ②		Floor ③	Floor ④		Floor ⑤
		Under benches	No shading		Under benches	No shading	
Air temp. T_a (°C)	26.0	26.0	26.0	26.0	26.0	26.0	26.0
Average uncooled surface temp. AUST (°C)	28.3	22.0	28.5	28.5	22.0	28.5	28.6
Supply chilled water temp. $T_{w,s}$ (°C)	18.0	18.0	18.0	18.0	18.0	18.0	18.0
Return chilled water temp. $T_{w,r}$ (°C)	20.6	19.6	19.6	20.6	19.7	19.7	23.0
Mean surface temp. of radiant floor T_s (°C)	25.2	20.3	25.0	25.3	20.3	25.0	32.1
Cooling capacity of radiant floor q (W/m^2)	59.7	15.0	61.9	60.1	15.0	62.0	115.7
q_c (W/m^2)	0.8	5.7	1.0	0.7	5.7	1.0	−6.0
q_{lr} (W/m^2)	16.9	9.3	18.9	17.4	9.3	19.1	−19.2
q_{sr} (W/m^2)	42.0	0.0	42.0	42.0	0.0	42	141.0
Cooling supplement by FCU q_{FCU} (W/m^2)	1.8	5.7	5.7	5.7	5.7	5.7	9.9

Notes Internal wall surface temp.: roof = 28.5 °C, west wall = 27.7 °C, east wall = 28.8 °C, south wall = 28.1 °C, and north wall = 28.1 °C

Table D.4 Cooling capacity and surface temperature of radiant floor (12:00 p.m.)

	Floor ①	Floor ②		Floor ③	Floor ④		Floor ⑤
		Under benches	No shading		Under benches	No shading	
Air temp. T_a (°C)	26.0	26.0	26.0	26.0	26.0	26.0	26.0
Average uncooled surface temp. AUST (°C)	30.2	22.2	30.3	30.3	22.2	30.3	30.2
Supply chilled water temp. $T_{w,s}$ (°C)	18.0	18.0	18.0	18.0	18.0	18.0	18.0
Return chilled water temp. $T_{w,r}$ (°C)	21.8	20.3	20.3	21.8	20.3	20.3	21.8
Mean surface temp. of radiant floor T_s (°C)	28.8	20.6	28.4	28.9	20.6	28.4	28.8
Cooling capacity of radiant floor q (W/m^2)	89.3	14.3	92.7	89.6	14.3	92.7	89.3
q_c (W/m^2)	−2.8	5.4	−2.4	−2.9	5.4	−2.4	−2.8
q_{lr} (W/m^2)	7.2	8.9	10.2	7.5	8.9	10.2	7.2
q_{sr} (W/m^2)	85.0	0.0	85.0	85.0	0.0	85	85.0
Cooling supplement by FCU q_{FCU} (W/m^2)	8.1	19.6	19.6	19.6	19.6	19.6	8.1

Notes Internal wall surface temp.: roof = 30.4 °C, west wall = 29.6 °C, east wall = 29.6 °C, south wall = 29.7 °C, and north wall = 29.7 °C

Table D.5 Cooling capacity and surface temperature of radiant floor (4:00 p.m.)

	Floor ①	Floor ②		Floor ③	Floor ④		Floor ⑤
		Under benches	No shading		Under benches	No shading	
Air temp. T_a (°C)	26.0	26.0	26.0	26.0	26.0	26.0	26.0
Average uncooled surface temp. AUST (°C)	29.5	22.0	29.4	29.4	22.0	29.4	29.3
Supply chilled water temp. $T_{w,s}$ (°C)	18.0	18.0	18.0	18.0	18.0	18.0	18.0
Return chilled water temp. $T_{w,r}$ (°C)	23.1	19.7	19.7	20.7	19.7	19.7	20.7
Mean surface temp. of radiant floor T_s (°C)	32.4	20.3	25.4	25.7	20.3	25.3	25.6
Cooling capacity of radiant floor q (W/m^2)	118.6	14.9	65.0	63.0	14.9	64.9	62.6
q_c (W/m^2)	−6.4	5.7	0.6	0.3	5.7	0.7	0.4
q_{lr} (W/m^2)	−16.0	9.3	22.3	20.6	9.3	22.2	20.2
q_{sr} (W/m^2)	141.0	0.0	42.0	42.0	0.0	42	42.0
Cooling supplement by FCU q_{FCU} (W/m^2)	11.7	7.4	7.4	7.4	7.4	7.4	3.6

Notes Internal wall surface temp.: roof = 29.4 °C, west wall = 29.9 °C, east wall = 28.7 °C, south wall = 29.1 °C, and north wall = 29.1 °C

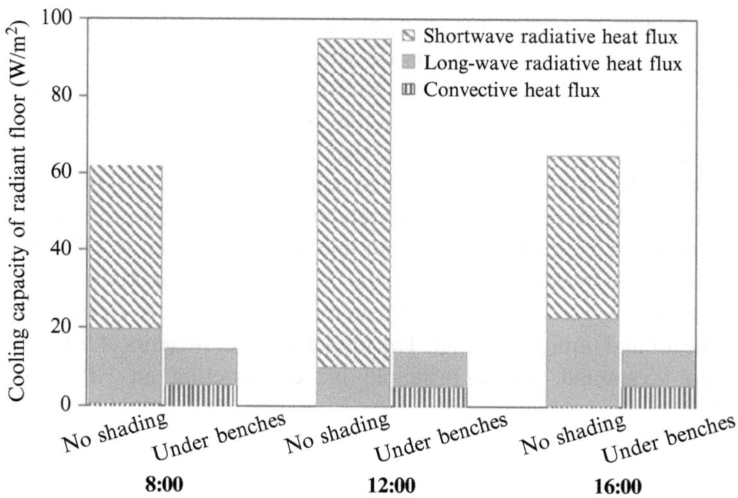

Fig. D.5 Comparison of cooling capacity of radiant floors with/without shading

85.0 W/m^2 and 22.2 °C, 0 W/m^2, respectively. Therefore, surface temperature difference ΔT_s is 7.8 °C, ΔAUST is 8.1 °C, Δq_{sr} is 85 W/m^2, and factors α_2 and α_4 are 0.32 and 0.059, respectively.

Appendix E: Performance Analysis of Heat Pump-Driven Liquid Desiccant Systems

Heat pump-driven liquid desiccant (HPLD) systems have become more and more popular recently due to their compact size and high efficiency in China. There are different ways to utilize the condensing heat for desiccant regeneration for HPLD systems Niu et al. (2012). In the HPLD systems investigated by Yadav (1995), Ma et al. (2006), and Yamaguchi et al. (2011), all the condensing heat from the heat pump was put in the solution for regeneration. In the studies conducted by Kinsara et al. (1996) and Dai et al. (2001), all the condensing heat from the heat pump was removed by the regeneration air rather than by the solution, and the heated air went into the regenerator for desiccant regeneration. The key issues are how to evaluate the performances of different HPLD systems and how to pursue a process with better energy efficiency.

For these HPLD systems, system models can be built with the help of numerical methods, and the operating parameters can then be simulated to investigate performance discrepancies. However, system simulation requires a relatively complex numerical process, and principles or approaches that can help guide the construction of an HPLD system are often limited or absent. In previous research (Zhang et al. 2012), match properties based on entransy analysis in the heat and mass transfer process between the air and water were analyzed, which helped to construct the air-water handling process. The conditions for flow matching and parameter matching were investigated for the air-water process. If the inlet states of two fluids satisfy the conditions for flow matching and parameter matching, entransy dissipation of the heat and mass transfer process will be the lowest with a given heat and mass transfer ability. The heat and mass transfer process close to the saturation line is recommended rather than the process that is close to the isenthalpic line in the air-water process, and it is recommended that the heat source (hot water) be used to heat sprayed water rather than inlet air to achieve a better humidification performance of the air humidifier. Similar approach can be utilized in the coupled heat and mass transfer process between the air and liquid desiccant in HPLD systems.

In the following investigation, we focus on the match properties of key components in the HPLD system, including the coupled heat and mass transfer processes between the air and liquid desiccant. The unmatched coefficients of different HPLD systems are examined, and a principle for constructing an HPLD system is proposed.

E.1 Model for Performance Simulation

There are several key components in the HPLD systems, including the coupled heat and mass transfer components (the dehumidifier and regenerator), the heat pump system model, and the heat exchanger model. The theoretical models of these components can be built and summarized as follows:

- For the dehumidifier/regenerator, the coupled heat and mass transfer process between the air and liquid desiccant can be described with energy conservation, mass conservation, and heat and mass transfer equations, as shown in previous research (Liu et al. 2007).
- For the heat pump system model, the refrigerant cycle using R22 as the refrigerant can be investigated. The performance indexes of the heat pump, including the compressor power, cooling capacity, evaporating temperature, and condensing temperature, are the main output parameters.
- There are different kinds of heat exchangers in the HPLD systems, including air-to-refrigerant heat exchanger, solution-to-refrigerant heat exchanger, and solution-to-solution heat exchanger. The ε-NTU method can be used to describe the performance of heat exchangers.

On the basis of the theoretical models of these key components, a system model for the HPLD process can be established, and numerical simulations can be carried out on the Simulink platform in Matlab software. In previous research (Zhang et al. 2012), a model of this kind of HPLD system was built, and the flow chart of the calculating method was also given; to validate the system models, the tested results and simulated results of a three-stage HPLD system were investigated, including the supply air temperature, humidity ratio, compressor power, and COP_{hp}. It showed that the mean relative biases between the measured and simulated results of the HPLD device were all lower than 10 % and the simulated results showed good agreement with the measured results. Thus, the theoretical models of the various components were validated, and the system models could be used to analyze the performance of different HPLD systems.

The system simulation method helps to calculate the performance parameters of HPLD systems. However, a simple solution for evaluating performance discrepancies is still lacking. To simplify the comparison of different HPLD systems, the current study investigates the match properties in the key components of the HPLD systems. The unmatched coefficient based on entransy dissipation is adopted, including the sensible heat transfer process and the coupled heat and mass transfer processes.

COP_{hp} and COP_{sys} are chosen as indexes to evaluate the performances of various HPLD systems, calculated by Eqs. (E.1) and (E.2), respectively, where Q_{evap} is the cooling capacity of the evaporator, kW, and P_{com} and P_{pump} are power consumptions of the compressor and pumps for solution circulation, kW:

$$COP_{hp} = \frac{Q_{evap}}{P_{com}} \tag{E.1}$$

$$COP_{sys} = \frac{Q_a}{P_{com} + P_{pump}} \tag{E.2}$$

E.2 Performance Analysis of the Two Basic HPLD Systems

There are various kinds of HPLD systems in which condensing heat is usually utilized for desiccant regeneration, and the cooling capacity of the evaporator is adopted for dehumidification. However, to obtain better mass transfer performance, previous studies have recommended that the cooling capacity should be used to cool the inlet solution rather than the air in the dehumidification process (Liu et al. 2009). Accordingly, in the processes analyzed in the present study, the evaporator of the heat pump is utilized to cool the solution for dehumidification.

Operating principles of the basic HPLD systems have been given in Sect. 5.2.2, and system models of Basic Type I and Basic Type II can be established based on the analysis in Sect. E.1. Here, we choose a typical condition for these two basic types to investigate system performance. The parameters of the inlet outdoor air are 32 °C, 18 g/kg, and 1.33 kg/s, and the required humidity ratio of the supplied air is 9.5 g/kg. The circulating solute flow rate of LiBr in the dehumidifier/regenerator is 1.5 kg/s, while the solute flowing between the dehumidifier and the regenerator is 0.2 kg/s. The initial conditions for system simulation are listed in Table E.1, and the input heat transfer capacities of the key components are identical for Basic Types I and II. For example, the *UA* (*U* is the heat transfer coefficient, kW/($m^2 \cdot$ °C); *A* is the heat transfer area, m^2) of the air-cooled condenser in Basic Type I is equal to the *UA* of the solution-cooled condenser in Basic Type II.

The typical operating parameters of key components are shown in Fig. E.1. The unmatched coefficients (described in Sect. 5.2.1) of key components for Basic Type I and Basic Type II in a typical condition are listed in Table E.2. As indicated by the match properties of these components, the unmatched coefficients of the evaporator, the dehumidifier, and the heat exchanger are similar for Basic Types I and II. The unmatched coefficient ξ of the dehumidifiers is 1.07 for these two HPLD systems, indicating that the unmatched flow rates and parameters have a relatively limited effect on the dehumidification process. For the condenser and regenerator, there is significant variation in the unmatched coefficients of Basic Types I and II. In Basic Type I, ξ of the air-cooled condenser is 2.70 in this typical condition, more than two times higher than that of Basic Type II. This means that, for the air-cooled condenser of Basic Type I, the heat transfer resistance caused by the unmatched flow rates increases 170 % on the basis of $1/UA$. Besides, the unmatched coefficient of the regenerator in Basic Type I is 1.34, still significantly higher than that of Basic Type II. In summary, the match properties of Basic Type II are much better than those of Basic Type I.

The significant difference of unmatched coefficients between Basic Types I and II is similar to the performance discrepancy between the simulated HPLD systems. The operating parameters of the HPLD systems in this typical condition are also listed in Table E.2, including the evaporating and condensing temperatures of the heat pump, COP_{hp}, and COP_{sys}. The evaporating temperatures and the supply air temperatures are similar for Basic Types I and II. However, the condensing temperature of Basic Type I (78.8 °C) is much higher than that of Basic Type II

Table E.1 Initial conditions for performance simulation of different HPLD systems

Type	Fresh air parameters	NTU (NTU_M) of key components					
		Condenser	Regenerator	Evaporator	Dehumidifier	Heat exchanger	Evaporative cooler
Basic Type I	32 °C, 18 g/kg, 1.33 kg/s	$NTU_a = 4.8$	2.5	0.8	2.5	3	–
Improved Type I							3.2
Basic Type II		$NTU_s = 0.8$					–
Improved Type II							–

Notes For a single stage of Improved Type II, the circulating solute in the dehumidifier/regenerator is 0.75 kg/s, and that between the dehumidifier and the regenerator is 0.1 kg/s

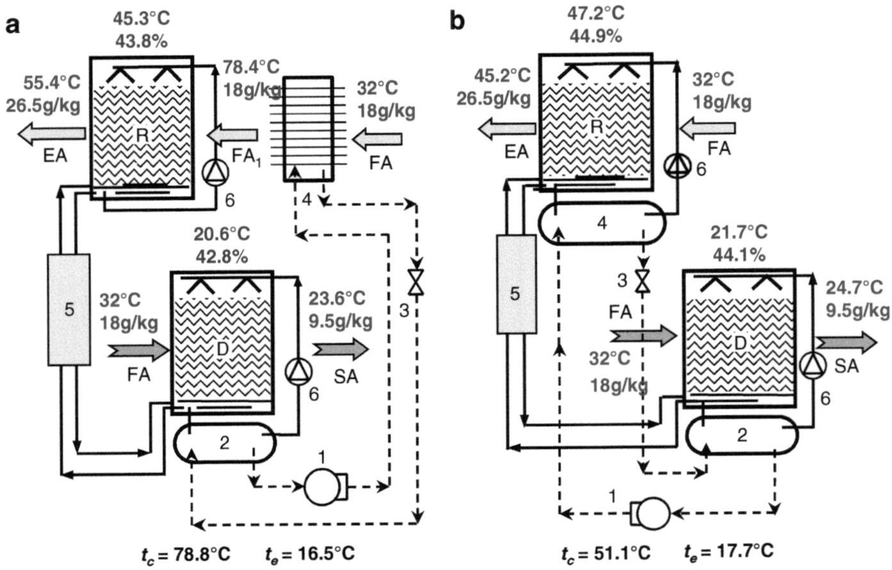

Fig. E.1 Operating parameters of the HPLD systems under a typical condition: (**a**) Basic Type I and (**b**) Basic Type II (*FA* fresh air, *SA* supply air, FA_1 fresh air heated by the condenser, *EA* exhaust air, *D* dehumidifier, *R* regenerator, *1* compressor, *2* evaporator, *3* expansion valve, *4* condenser, *5* solution heat exchanger, *6* pump)

(51.1 °C), indicating that the condensing temperature increases due to the unmatched flow rates of the air-cooled condenser. The increase of the condensing temperature results in a COP_{hp} of 1.89 in Basic Type I, lower than half of the COP_{hp} of Basic Type II (4.53).

E.3 Performance Improvement of Basic Type I

The operating performance and match properties of Basic Type I are significantly inferior to those of Basic Type II. Therefore, a direct evaporative cooler was added to Basic Type I to improve its performance. This proposed system (Improved Type I) is illustrated in Fig. E.2a. The regeneration air first flows through the evaporative cooler and is then heated by the condenser. Taking the same inlet air states and required supply air humidity ratio from the previous analysis, the handling process of Improved Type I is given in Fig. E.2b in a psychrometric chart; the operating parameters of the air and solution are shown in Fig. E.2a.

The unmatched coefficients of key components in this type of HPLD system are listed in Table E.2. Compared to that of Basic Type I, ξ of the condenser for Improved Type I is lower, but ξ of the condenser as well as ξ_M of the regenerator and the dehumidifier are almost identical for these two systems. For the heat pump

Table E.2 Match properties and system performances of HPLD systems under a typical condition (outdoor air: 32 °C, 18 g/kg)

Type	Unmatched coefficients ξ (ξ_M) of key components					Supply air parameters	Performance of the heat pump cycle			
	Condenser	Regenerator	Evaporator	Dehumidifier	Heat exchanger		t_c	t_e	COP_{hp}	COP_{sys}
Basic Type I	2.70	1.34	1.07	1.07	1.00	23.6 °C, 9.5 g/kg	78.8 °C	16.5 °C	1.89	1.83
Improved Type I	2.36	1.34	1.05	1.06	1.00	23.1 °C, 9.5 g/kg	73.2 °C	16.1 °C	2.21	2.12
Basic Type II	1.07	1.05	1.05	1.07	1.00	24.7 °C, 9.5 g/kg	51.1 °C	17.7 °C	4.53	4.17
Improved Type II	1.07 (1.07)	1.03 (1.07)	1.05 (1.05)	1.06 (1.08)	1.00 (1.00)	25.2 °C, 9.5 g/kg	49.6 °C (52.0 °C)	18.6 °C (18.7 °C)	4.70	4.32

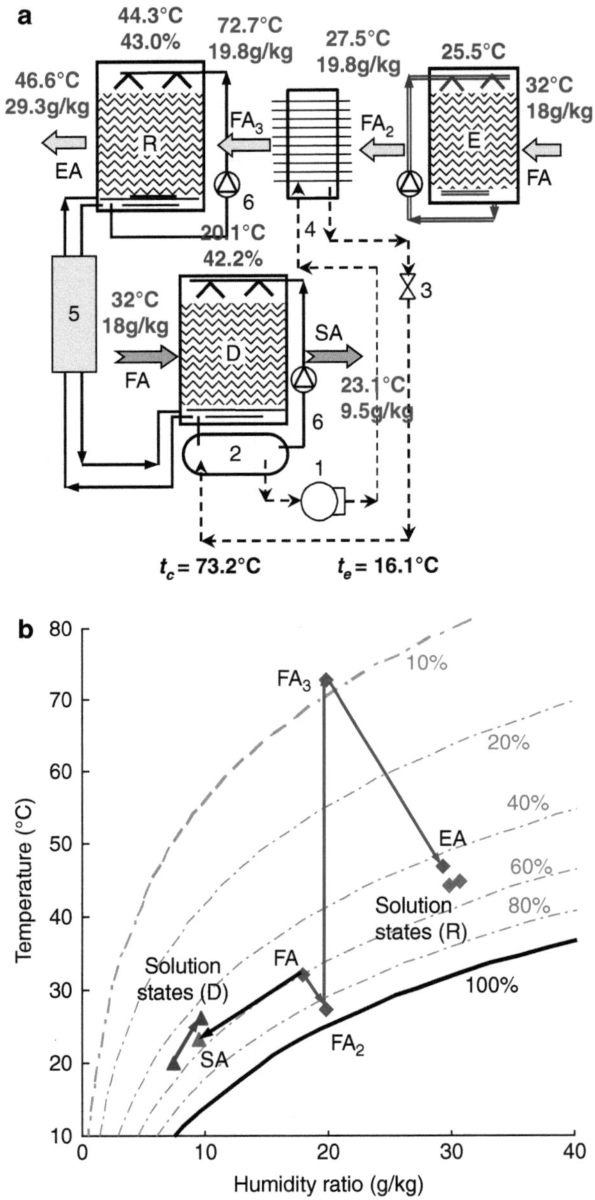

Fig. E.2 Operating principle of Improved Type I (E-direct evaporative cooler): (**a**) system schematic diagram and (**b**) air handling process in the psychrometric chart

cycle, condensing temperature t_c is 73.2 °C, which is 5.6 °C lower than that of Basic Type I, leading to an increase of about 17 % in COP_{hp}. Thus, adding an evaporative cooler helps to improve the match properties to a certain extent, but the performance improvement is limited. Compared to Basic Type II, ξ of the condenser and ξ_M of the regenerator are still much higher for Improved Type I, with COP_{hp} about 50 % lower than that of Basic Type II.

As indicated by the analysis of these three HPLD systems, the operating performance of Basic Type II is much better than the other two types under this typical condition. Therefore, for the HPLD systems, heating the regeneration air leads to a significant increase of heat or mass transfer resistance (or unmatched coefficient) caused by unmatched flow rates and parameters. However, using the condensing heat to heat the solution for regeneration helps to improve the match properties of key components. The energy performance of a system with better match properties is superior to a system with less adequate match properties. Thus, heating the solution rather than the air for regeneration is recommended in HPLD systems.

E.4 Performance Improvement of Basic Type II

As indicated by the analysis on match properties and simulated performances of HPLD systems, the handling process along the iso-concentration line is better than that along the isenthalpic line. And the process of heating the solution for regeneration is recommended rather than heating the regeneration air. The unmatched coefficients of key components in the HPLD systems help to evaluate the performance of the process, providing a relatively simple and effective method to guide the construction of an HPLD system. Thus, a principle for system construction is indicated based on the match property analysis: lower unmatched coefficients mean the heat or mass transfer resistance caused by the unmatched flow rates or parameters is lower, and the system will demonstrate better energy performance as a result. The unmatched coefficient based on entransy dissipation analysis can help to determine the key issues restricting the operating performance, providing the direction for constructing a better HPLD system. Compared with the system simulation method, the match property method is simpler and more feasible.

To improve the performance of the HPLD system further, grading is proposed as an approach. Figure E.3a gives the operating schematic of a two-stage HPLD system (Improved Type II), and the typical air handling process is shown in Fig. E.3b. In contrast to Basic Type II, there are two-stage dehumidifier/regenerator and heat pump cycles in Improved Type II. The humid outdoor air flows through dehumidifiers D2 and D1 successively and is then dehumidified to the required state, while the regeneration air flows through regenerators R1 and R2 step by step for desiccant regeneration. The operating parameters of this two-stage HPLD system are also given in Fig. E.3a, with the same outdoor air parameters and required humidity ratio of the supply air analyzed previously. The unmatched coefficients of the key components and system operating performance of this improved system are listed in Table E.2. The parameters in the brackets of Table E.2 are the unmatched coefficients and performance parameters of the second stage.

As indicated by the match properties of this two-stage HPLD system, ξ_M of the regenerator or the dehumidifier for the first stage is lower than in Basic Type II, but for the second stage, it is a bit higher. There is a little difference between the

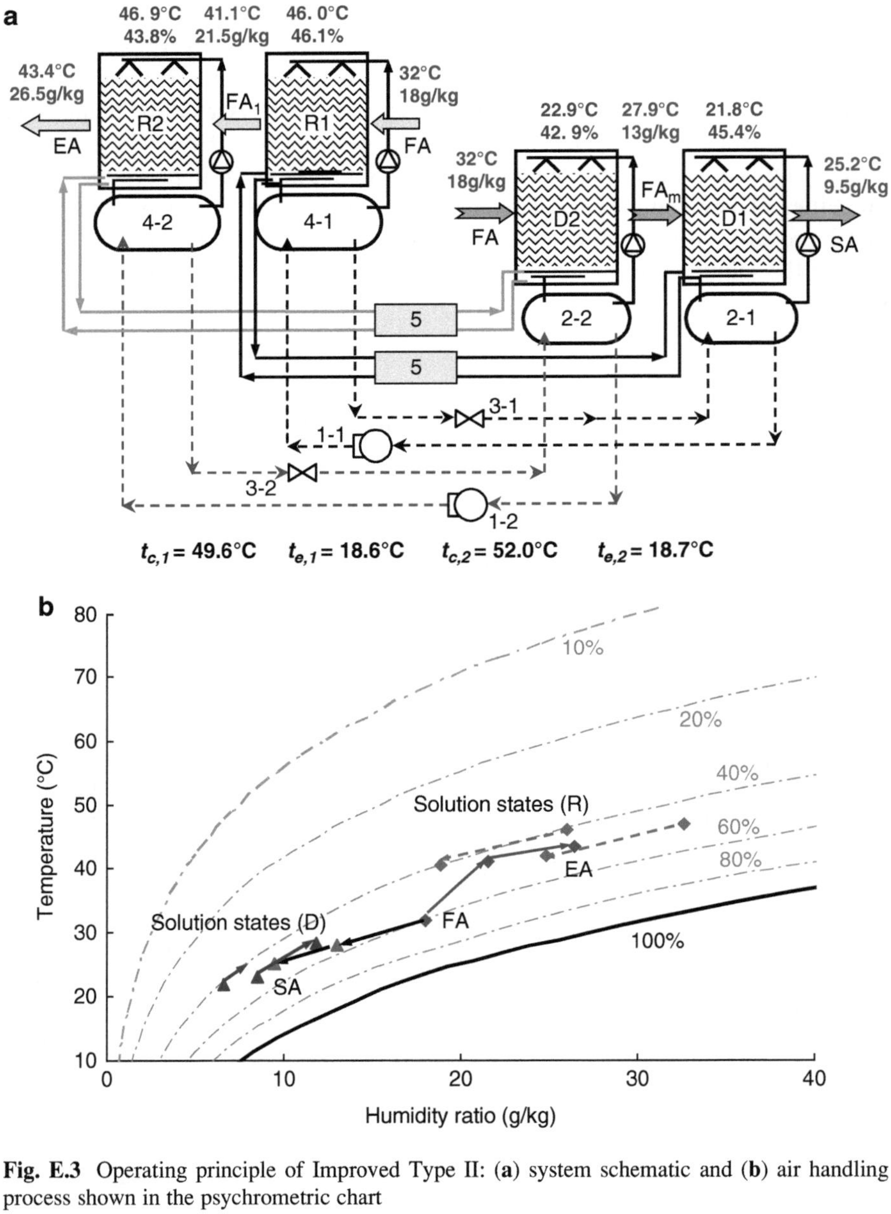

Fig. E.3 Operating principle of Improved Type II: (**a**) system schematic and (**b**) air handling process shown in the psychrometric chart

unmatched coefficients of Basic Type II and Improved Type II. The simulated operating performances of this system confirm the results from the analysis of match properties. The condensing temperature of the first stage in Improved Type II is a bit lower than in Basic Type II, but that of the second stage is a bit higher. COP_{hp} of Improved Type II in this typical condition is 4.70, a bit higher than that of

Basic Type II. Thus, compared with Basic Type II, the grading method (Improved Type II) could help to improve the operating performance of the HPLD system.

Air dehumidification methods that utilize liquid desiccant are becoming more and more popular, and HPLD systems are developing very quickly. Based on the performances of different regeneration modes for HPLD systems simulated and compared above, the main conclusions are summarized for designing a HPLD system as follows:

(1) The unmatched coefficient based on entransy dissipation analysis could be adopted to examine the performances of the key components in HPLD systems. A higher unmatched coefficient means a higher heat or mass transfer resistance caused by the unmatched flow rates or parameters.
(2) The HPLD system in which the solution is heated for regeneration performs much better than the system in which the regeneration air is heated. Moreover, COP_{sys} is significantly improved, rising from 1.83 in Basic Type I to 4.17 in Basic Type II.

For different inlet air and solution states, the unmatched coefficient of the process along the iso-concentration line is significantly lower than that along the isenthalpic line. The unmatched coefficient offers a principle for constructing an HPLD system: lowering the unmatched coefficients helps to improve system performance.

References

AHRI (2003) ARI 550 590-2003 Performance rating of water chilling packages using the vapor compression cycle. Air-conditioning, Heating, & Refrigeration Institute, Arlington

AHRI (2003) ARI 550 590-2011 Performance rating of water-chilling and heat pump water-heating packages using the vapor compression cycle. Air-conditioning, Heating, & Refrigeration Institute, Arlington

AQSIQ, SAC (2004) GB/T 18430.2-2001 Water chilling (heat pump) packages using the vapor compression cycle Household and similar water chilling (heat pump) packages. China Standards Press, Beijing (in Chinese)

AQSIQ, SAC (2005) GB/T 18430.1-2001 Water chilling (heat pump) packages using the vapor compression cycle Water chilling (heat pump) packages for industrial & commercial and similar application. China Standards Press, Beijing (in Chinese)

AQSIQ, SAC (2008) GB/T 18430.1-2007 Water chilling (heat pump) packages using the vapor compression cycle — Part 1: Water chilling (heat pump) packages for industrial & commercial and similar application. China Standards Press, Beijing (in Chinese)

AQSIQ, SAC (2009) GB/T 18430.2-2008 Water chilling (heat pump) packages using the vapor compression cycle — Part 2: Household and similar water chilling (heat pump) packages. China Standards Press, Beijing (in Chinese)

ASHRAE (2009) ASHRAE handbook fundamentals. American Society of Heating Refrigerating and Airconditioning Engineer, Inc., Atlanta

CEN (2007) EN 14511-2 2007 Air conditioners, liquid chilling packages and heat pumps with electrically driven compressors for space heating and cooling – Part 2: Test conditions. European Committee for Standardization, Brussels

Dai YJ, Wang RZ, Zhang HF, Yu JD (2001) Use of liquid desiccant cooling to improve the performance of vapor compression air conditioning. Appl Therm Eng 21:1185–1202

JIS B 8613:1994. Water chilling unit

Joint Technical Committee ME-086 (2008) AS/NZS 4776.1.1: 2008 Liquid-chilling packages using the vapour compression cycle – Part 1.1: Method of rating and testing for performance—rating. The Council of Standards Australia, Sydney

Kinsara AA, Elsayed MM, AlRabghi OM (1996) Proposed energy-efficient air-conditioning system using liquid desiccant. Appl Therm Eng 16:791–806

Liu XH, Jiang Y, Qu KY (2007) Heat and mass transfer model of cross-flow liquid desiccant air dehumidifier/regenerator. Energy Convers Manage 48:546–554

Liu XH, Li Z, Jiang Y (2009) Similarity of coupled heat and mass transfer process between air-water and air-liquid desiccant contact system. Build Environ 44:2501–2509

Ma Q, Wang RZ, Dai YJ, Zhai XQ (2006) Performance analysis on a hybrid air-conditioning system of a green building. Energy Build 38:447–453

Niu XF, Xiao F, Ma ZJ (2012) Investigation on capacity matching in liquid desiccant and heat pump hybrid air-conditioning systems. Int J Refrig 35:160–170

The Japan Refrigeration and Air Conditioning Industry Association (1994) JIS B 8613:1994 Water chilling unit. Japanese Standards Association, Tokyo (in Japanese)

Yadav YK (1995) Vapour-compression and liquid-desiccant hybrid solar space-conditioning system for energy conservation. Renew Energy 6:719–723

Yamaguchi S, Jeong J, Saito K, Miyauchi H, Harada M (2011) Hybrid liquid desiccant air-conditioning system: experiments and simulations. Appl Therm Eng 31:3741–3747

Zhang JQ, Xu Z, Wu KR (1999) The evaluation of need power of forty woodplants water loss and the classification of flammability. J Southwest For Coll 19(3):170–175 (in Chinese)

Zhang LZ (2005) Dehumidification technology. Chemical Industry Press, Beijing (in Chinese)

Zhang T, Liu XH, Jiang Y (2012) Performance optimization of heat pump driven liquid desiccant dehumidification systems. Energy Build 52:132–144

Zhao RY, Fan CY, Xue DH, Qian YM (2000) Air conditioning, 3rd edn. China Architecture & Building Press, Beijing (in Chinese)

MIX
Papier aus verantwortungsvollen Quellen
Paper from responsible sources
FSC® C105338
www.fsc.org

If you have any concerns about our products,
you can contact us on
ProductSafety@springernature.com

In case Publisher is established outside the EU,
the EU authorized representative is:
Springer Nature Customer Service Center GmbH
Europaplatz 3, 69115 Heidelberg, Germany

Printed by Libri Plureos GmbH
in Hamburg, Germany